T0344479

ILLICIT DRUGS IN THE ENVIRONMENT

WILEY-INTERSCIENCE SERIES IN MASS SPECTROMETRY

Series Editors

Dominic M. Desiderio
Departments of Neurology and Biochemistry
University of Tennessee Health Science Center

Nico M. M. Nibbering
Vrije Universiteit Amsterdam, The Netherlands

A complete list of the titles in this series appears at the end of this volume.

ILLICIT DRUGS IN THE ENVIRONMENT
Occurrence, Analysis, and Fate Using Mass Spectrometry

Edited by
SARA CASTIGLIONI
ETTORE ZUCCATO
ROBERTO FANELLI
Department of Environmental Health Sciences
Mario Negri Institute for Pharmacological Research
Milan, Italy

A JOHN WILEY & SONS, INC., PUBLICATION

Published by John Wiley & Sons, Inc., Hoboken, New Jersey.
Published simultaneously in Canada.

For general information on our other products and services or for technical support, please contact our
Customer Care Department within the United States at (800) 762-2974, outside the United States at (317)
572-3993 or fax (317) 572-4002.

Wiley also publishes its books in a variety of electronic formats. Some content that appears in print may
not be available in electronic formats. For more information about Wiley products, visit our web site at
www.wiley.com

Library of Congress Cataloging-in-Publication Data:

Illicit drugs in the environment : occurrence, analysis, and fate using mass spectrometry / edited by Sara
Castiglioni, Ettore Zuccato, Roberto Fanelli.
    p. cm.
  Includes index.
  ISBN 978-0-470-52954-6 (cloth)
 1. Drugs of abuse–Analysis. 2. Drugs of abuse–Environmental aspects. 3. Drugs of abuse–Spectra.
4. Water–Analysis. 5. Organic water pollutants. 6. mass spectrometry. I. Castiglioni, Sara, 1976–
II. Zuccato, Ettore, 1952–   III. Fanelli, Roberto, 1944–
  RS190.D77I65 2011
  363.739′4–dc22
                                        2010036825

Printed in Singapore

10 9 8 7 6 5 4 3 2 1

CONTENTS

PREFACE

PRESENTATION OF THE BOOK

Following the preliminary observation that traces of illicit drugs could be found in the aqueous environment, there was an obvious request for a better characterization of these novel contaminants to assess possible risks for the environment and human health. A less obvious consequence of this finding was the discovery that the residues of illicit drugs in wastewater, and sometimes in surface water receiving untreated wastes, could be used to estimate drug consumption in the group of individuals producing the waste itself. In particular, the potential applications linked to this second issue reinforced the need for a specific, sensitive, and accurate measurement of these substances. Environmental scientists, on the one hand, and social scientists and persons involved in the phenomenon of drug addiction, on the other, have sought analytical methods for the detection and quantification of illicit drugs and metabolites in environmental media, particularly, wastewater. Illicit drugs and their metabolites are commonly measured in forensic sciences, but concentrations in urine, blood, and other fluids or in hair, are much higher and interference much lower than in wastewater. Wastewater is a complex milieu of thousands of different substances, dissolved, mixed, or suspended in water. The list of compounds in wastewater, and by extension, in the downstream environment, is long. Chemicals from industrial or agricultural activities are well known contributors to this milieu, but pharmaceuticals are a recent acquisition, and the same is true for the thousands of products we use daily for personal care. Remnants derived from an enormous number of production and household activities end up in wastewater and contribute to an increase in the complexity of its composition and, thus, the difficulties to detect specific target substances.

Fortunately, given the physicochemical similarities, the previous experience with therapeutic drugs has substantially helped in developing appropriate analytical methods for illicit drugs; the first proposed were actually based on the extension to these molecules of multiresidue methods previously established for therapeutic pharmaceuticals, or alternatively, on methods previously established for forensic investigations, which were adapted to the analysis of these substances in environmental matrices. The methods proposed were mainly based on HPLC-MS, and sometimes on GC-MS. This is not surprising as liquid chromatography is considered the most appropriate technique for the analysis of polar substances and mass spectrometry the most powerful technique for multitrace analysis of compounds in complex matrices, such as wastewater.

This present book "Illicit Drugs in the Environment: Occurrence, Analysis and Fate Using Mass Spectrometry" will describe a new application of mass spectrometry in the detection and measurement of a novel class of environmental contaminants (illicit drugs). This novel application assesses risks of these newly identified pollutants for ecosystem and man and explores the potentials of this innovative approach to estimate illicit drug consumption in the population.

STATE-OF-THE-ART

To our knowledge, there was no trace in the scientific literature of any investigation on illicit drugs in the aqueous environment until year 2001, when Christian Daughton, without having any direct evidence of their presence in the environment, hypothesized that remnants of illicit drugs excreted with the urine of the consumers could end up in wastewater and that their levels could be used to back-calculate the intake of drugs by the population. In a field study in 2004, Jones-Lepp first reported the real occurrence of amphetamines in treated wastewater in the United States and, in 2005, our group measured cocaine and metabolites in rivers and in raw wastewater samples, and used the results to back-calculate the consumption of cocaine in the population. Later, in 2006, the method originally established for cocaine by our group was extended to the analysis of other illicit drugs, which were measured in surface and wastewaters. Thus far, occurrence, behavior, and fate of illicit drugs in waste-, surface, ground, and drinking water has been investigated in several European countries, and traces of various illicit drugs have also been detected in airborne particulates in many places around the world. Later, the approach to estimate cocaine abuse by wastewater analysis was also used to test community-wide consumption of cannabis, heroin, and amphetamines in several countries in Europe and the United States.

The rationale of this approach is known: traces of almost everything we eat, smoke, drink, ingest, or absorb, are excreted with our urine or stool and end up in the sewage system. Therefore, monitoring wastewater has the potential to extract useful epidemiological information from qualitative and quantitative profiling of biological indicators entering the sewage system. This is the basis of what we called "sewage epidemiology;" illicit drugs were the first application of this new branch of environmental epidemiology. Residues of the illicit drugs consumed by a collectivity are excreted in wastewater and their levels, knowing kinetic, metabolism, and behavior

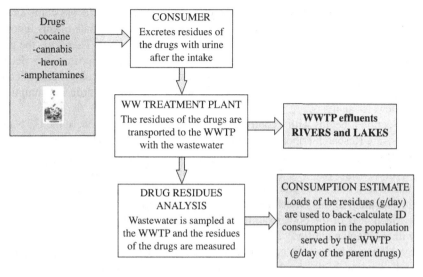

DIAGRAM 1   The pathway of illicit drugs: from the consumer to the wastewater treatment plant (WWTP) and the environment.

in wastewater, and characteristics of the sewage system, such as flow rate and population size, can be used to collectively back-calculate for the type and amount of illicit drugs consumed (Diagram 1).

Thus far, the ecological implications of the presence of illicit drugs in surface water of rivers and lakes have been less explored. Effluents of wastewater treatment plants are probably the major sources (Diagram 1). Concentrations in surface water are generally low but, as previously shown for therapeutic pharmaceuticals, these molecules might also exert potent and specific biological activities on nontargeted organisms. Moreover, interactions with the effects of other licit and illicit drugs are possible and toxicity in the aquatic ecosystem cannot, therefore, be excluded. At the moment, there are several ongoing studies on the effects on the environment of therapeutic pharmaceuticals and is easy to predict the extension in the near future of these investigations to also include illicit drugs.

It is, therefore, expected that the number of scientists interested in these issues will substantially increase in the future. Environmental and social scientists will probably use these new applications of mass spectrometry to measure illicit drugs in waste- and surface water or in air, with the aim of studying their ecological effects or tracking illicit drug consumption in the population. This new branch of science is still in its infancy, but this book will collect all the available knowledge and the new ongoing research on this novel topic. We hope it will become a reference text for future investigations.

ORGANIZATION OF THE BOOK

The focus of this book is on illicit drugs in, mainly, the aqueous environment, and on mass spectrometry, used for their analysis to study occurrence and fate. This twofold novel application of mass spectrometry is : to study risks for ecosystems and man of these newly identified pollutants and to explore the potentials of this innovative approach to estimate illicit drug consumption in communities. Therefore, the goal of the book is to provide information on all the aspects of the mass spectrometry detection of illicit drugs in environmental media, to address the ecotoxicological implications, and to present and discuss this novel approach to estimate drug consumption by wastewater analysis.

Section I, begins with a contribution from Christian Daughton, of the US National Exposure Research Laboratory, entitled "Illicit Drugs and the Environment". It gives an introduction to this issue by providing basic, but detailed, information on what it is an illicit drug, recalls the history of the discovery of illicit drugs in the environment, including ambient air and money supply, and gives some hypotheses on the future development of this branch in the environmental and the social sciences.

Section II of the book examines the physiological properties of the illicit drugs, with a chapter entitled "Metabolism and Excretion of Illicit Drugs in Humans" by Manuela Melis and co-workers. It reviews what is known about metabolism and excretion of these substances in man. As for therapeutic pharmaceuticals, it is supposed that the main source of the spreading of illicit drugs in the environment is the excretion in wastewater of the unmetabolized parent compounds and their break-down products still contained in the urine and the stools of consumers. An extensive knowledge of pharmacokinetic, metabolism, and excretion in human is, therefore, central to study these substances in the environment, to predict which of them will end up in the environment, and to estimate their concentrations, with the aim of identifying the proper target residues for the analysis of illicit drugs.

Section III, entitled "Mass Spectrometry in Illicit Drug Detection and Measurement; Current and Novel Environmental Applications," is the "core" for MS analysis. The first chapter by Bagnati and Davoli reviews the published methods currently used to measure illicit drugs in environmental media and examines novel potential applications of MS to detect these chemical contaminants. This last issue is also discussed by two other chapters in this section, the first by Hernandez et al., of the Spanish Research Institute for Pesticide and Water, which explores the potential of UHPLC-QTOF MS, and the second by de Voogt and co-workers, from the University of Amsterdam, which outlines the chances offered by Orbitrap MS in the analysis of illicit drugs in the environment.

Section IV, "Mass Spectrometric Analysis of Illicit Drugs in the Environment," reports field results of the MS analysis of illicit drugs in waste- and surface waters around the world, with contributions from Spain (Postigo, López de Alda, and Barceló), Italy (Castiglioni and Zuccato), UK (Kasprzyk-Hordern), Nebraska (Bartelt-Hunt and Snow), and the United States (Jones-Lepp and co-workers), in drinking water, exploring the presence and the removal of these substances in conventional drinking water treatment plants in Spain (Huerta-Fontela, Galceran, and Ventura) and the United States (Trenholm and Snyder), and in air and suspended particulate matter around the world (Cecinato and Balducci). Overall, these chapters report on the detection in various environmental media of about 40 different substances, including cocaine and metabolites, cannabinoids, heroin, morphine and their metabolites and conjugates, amphetamine-like molecules, and other related substances, such as, methadone and its metabolite EDDP, codeine, and more.

Section V deals with the applications of MS analysis of illicit drugs in the environment. This twofold analysis was intended to assess risks of these newly identified pollutants for ecosystem and man, and to explore the potentials of an innovative approach to estimate illicit drug consumption in the population. Domingo and Bracale, from Varese University and Schirmer and Pomati from EAWAG, thus explore the implications for ecotoxicology of the presence of illicit drugs in the environment, reviewing what is known on this issue and trying to predict what is still unknown by an original model-based approach.

A second chapter by Norbert Frost, from EMCDDA, introduces the method of estimating illicit drug consumption by wastewater analysis, discussing the potentials offered by this innovative approach from the point of view of a regulatory agency. A third chapter, from Zuccato and Castiglioni, reviews the applications of this approach published in the recent literature and critically discusses its potential and limitations in estimating community-wide illicit drug consumption. A fourth chapter, from van Nuijs and co-workers, shows that this approach, which was established to estimate illicit drug consumption at the community level, can be extended to estimate illicit drug use at nation-wide level, reporting results from a case-study carried out in Belgium. Last, a contribution, from Chiarelli and co-workers from Loyola University, introduces another intriguing potential application of this approach, in studying consumption rates in communities much smaller than those monitored thus far—students of a school. Monitoring the trends of drug use and assessing the overall consumption levels in students are critical to understanding the extent of the drug problem in this age group, with the aim to develop, target, and evaluate preventive interventions.

The book is then concluded by Roberto Fanelli, who sketches a brief history of the MS analysis of environmental contaminants, highlighting potential and limitations of its use in studying illicit drugs in the environment, and discusses the future perspective of the "sewage epidemiology approach" in investigating other substances, such as therapeutic pharmaceuticals, with the aim to monitor patients' compliance to the treatment, or food and air contaminants, and to estimate population exposure.

The most intriguing consequence of the discovery of illicit drugs in wastewater and other environmental media is in the potential to monitor illicit drug use in the population. Diagram 1 shows the mechanism through which the quantification of illicit drug residues in wastewater can be exploited to assess consumption by the population. We think that this application will probably become a major target for the analysis of these substances in environmental media.

ETTORE ZUCCATO
SARA CASTIGLIONI

CONTRIBUTORS

ALVAREZ DAVID, United States Geological Survey, Columbia Environmental Research Center, Columbia, Missouri, USA.

BAGNATI RENZO, Analytical Instrumentation Unit, Mario Negri Institute for Pharmacological Research, Milan, Italy.

BALDUCCI CATIA, Institute of Atmospheric Pollution Research CNR, Monterotondo Stazione, Rome, Italy

BARCELÓ DAMIÀ, Institute of Environmental Assessment and Water Research, Spanish Council for Scientific Research, Barcelona, Spain.

BARTELT-HUNT SHANNON, University of Nebraska-Lincoln, Lincoln, Nebraska, USA.

BERVOETS LIEVEN, Laboratory for Ecophysiology, Biochemistry and Toxicology, Department of Biology, University of Antwerp, Antwerp, Belgium.

BIJLSMA LUBERTUS, Research Institute for Pesticides and Water, University Jaume I Castellón, Castellón, Spain.

BLUST RONNY, Laboratory for Ecophysiology, Biochemistry and Toxicology, University of Antwerp, Antwerp, Belgium.

BRACALE MARCELLA, University of Insubria, Varese, Italy.

CASTIGLIONI SARA, Mario Negri Institute for Pharmacological Research, Milan, Italy.

CECINATO ANGELO, Institute of Atmospheric Pollution Research CNR, Monterotondo Stazione, Rome, Italy.

CHIARELLI PAUL, Loyola University, Chicago, Illinois, USA.

COVACI ADRIAN, Toxicological Centre, University of Antwerp, Antwerp, Belgium and Laboratory for Ecophysiology, Biochemistry and Toxicology, University of Antwerp, Antwerp, Belgium.

DAUGHTON CHRISTIAN G., Environmental Chemistry Branch, National Exposure Research Laboratory, U.S. Environmental Protection Agency, Las Vegas, Nevada, USA.

DAVOLI ENRICO, Mass Spectrometry Laboratory, Mario Negri Institute for Pharmacological Research, Milan, Italy.

DE VOOGT PIM, Earth Surface Science, Institute for Biodiversity and Ecosystem Dynamics, University of Amsterdam, Amsterdam, The Netherlands and KWR Watercycle Research Institute, Chemical Water Quality and Health, Nieuwegein, The Netherlands.

DOMINGO GUIDO, University of Insubria, Varese, Italy.

EMKE ERIK, KWR Watercycle Research Institute, Chemical Water Quality and Health, Nieuwegein, The Netherlands.

FANELLI ROBERTO, Mario Negri Institute for Pharmacological Research, Milan, Italy.

FROST NORBERT, Addiction Medicine, European Monitoring Centre for Drugs and Drug Addiction (EMCDDA), Addiction Medicine-Lisbon, Lisbon, Portugal.

GALCERAN MARIA TERESA, University of Barcelona, Barcelona, Spain.

HELMUS RICK, KWR Watercycle Research Institute, Chemical Water Quality and Health, Nieuwegein, The Netherlands.

HERNÁNDEZ FÉLIX, Research Institute for Pesticides and Water, University Jaume I Castellón, Castellón, Spain.

HUERTA-FONTELA MARIA, AGBAR-Aigües de Barcelona, Barcelona, Spain, University of Barcelona, Barcelona, Spain.

JONES-LEPP TAMMY, Environmental Sciences Division, National Exposure Research Laboratory, Office of Research and Development, United States Environmental Protection Agency, Las Vegas, Nevada, USA.

JORENS PHILIPPE G., University of Antwerp, Antwerp University Hospital, Antwerp, Belgium.

KASPRZYK-HORDERN BARBARA, Department of Chemistry, University of Bath, Bath, UK.

LOGANATHAN BOMMANNA, Center for Reservoir Research, Murray State University, Murray, Kentucky, USA.

LÓPEZ DE ALDA MIREN, Institute of Environmental Assessment and Water Research, Spanish Council for Scientific Research, Barcelona, Spain.

MELIS MANUELA, Mario Negri Institute for Pharmacological Research, Milan, Italy.

NEELS HUGO, Toxicological Centre, University of Antwerp, Antwerp, Belgium.

PANAWENNAGE DEEPIKA, Loyola University, Chicago, Illinois, USA.

PANTELIADIS PAVLOS, Earth Surface Science, Institute for Biodiversity and Ecosystem Dynamics, University of Amsterdam, Amsterdam, The Netherlands.

POMATI FRANCESCO, Eawag, Swiss Federal Institute of Aquatic Science and Technology, Kastanienbaum, Switzerland.

POSTIGO CRISTINA, Institute of Environmental Assessment and Water Research, Spanish Council for Scientific Research, Barcelona, Spain.

SANCHO JUAN V., Research Institute for Pesticides and Water, University Jaume I Castellón, Castellón, Spain.

SCHIRMER KRISTIN, Eawag, Swiss Federal Institute of Aquatic Science and Technology, Dübendorf, Switzerland.

SNOW DANIEL D., Water Sciences Laboratory, University of Nebraska-Lincoln, Lincoln, Nebraska, USA.

SNYDER SHANE A., Southern Nevada Water Authority, Applied Water Quality Research and Development Center, River Mountain Warer Treatment Facility, Las Vegas, Nevada, USA.

TRENHOLM REBECCA A., Southern Nevada Water Authority, Applied Water Quality Research and Development Center, River Mountain Warer Treatment Facility, Las Vegas, Nevada, USA.

VAN LEERDAM JAN A., KWR Watercycle Research Institute, Chemical Water Quality and Health, Nieuwegein, The Netherlands.

VAN NUIJS ALEXANDER L. N., Toxicological Centre, University of Antwerp, Antwerp, Belgium.

VENTURA FRANCESC, AGBAR-Aigües de Barcelona, Barcelona, Spain.

ZUCCATO ETTORE, Mario Negri Institute for Pharmacological Research, Milan, Italy.

SECTION I

INTRODUCTION

CHAPTER 1

ILLICIT DRUGS AND THE ENVIRONMENT

CHRISTIAN G. DAUGHTON

1.1 INTRODUCTION

The spectrum of chemicals recognized as contributing to widespread contamination of the environment began to be extended to pharmaceutical ingredients as early as the 1970s. However, the topic did not begin to attract broader scientific attention until the mid-1990s (Daughton, 2009a). Occurring generally at levels below 1 μg/liter in ambient waters, the near ubiquitous presence of pharmaceuticals in a wide variety of environmental compartments serves as a stunning measure of advancements in analytical chemistry in expanding our understanding of the scope of environmental pollution.

The extent of progress and effectiveness of pollution regulation, mitigation, control, and prevention over the last 40 years is now reflected by a focus on trace-level chemical contaminants—a phenomenon only hypothesized as a possibility in the early 1970s. This focus is particularly embodied with the so-called "emerging" contaminants (Daughton, 2009b) and the myriads of others not yet noticed or identified and which could be referred to as the "quiet contaminants."

Up through the 1990s, the emerging study of pharmaceuticals in the environment (PiE) inexplicably excluded from consideration the contributions by the so-called "illicit" drugs. Involving a structurally diverse group of chemical agents possessing extremely high potential for biological effects in humans and nontarget organisms, the magnitude of worldwide illicit drug trafficking is presumably enormous, but can only be very roughly estimated. The potential for illicit drugs to enter the environment should not differ markedly from that of medical pharmaceuticals—with contributions from excretion, bathing, disposal, and discharge of manufacturing waste. While known for many decades that illicit drugs and metabolites (just as with medicinal pharmaceuticals) are excreted in urine, feces, hair, and sweat, not until 1999

Illicit Drugs in the Environment: Occurrence, Analysis, and Fate Using Mass Spectrometry, Edited by Sara Castiglioni, Ettore Zuccato, and Roberto Fanelli
Copyright © 2011 John Wiley & Sons, Inc.

(Daughton and Ternes, 1999) and 2001 (Daughton, 2001a, 2001c) was the scope of concerns surrounding PiE expanded to include illicit drugs. In characterizing and assessing risks incurred from PiE, both licit and illicit drugs need to be seamlessly considered.

Perhaps the first published indication that illicit drugs might be pervasive contaminants of our immediate surroundings and the larger environment was a 1987 FBI study in response to a newspaper report 2 years earlier that cocaine was present on money in general circulation (Aaron and Lewis, 1987). Over the intervening 20 years, analogous seminal surveys of illicit drugs as ambient contaminants have been published for sewage wastewaters (Khan, 2002), surface waters (Zuccato et al., 2005), air (Cecinato and Balducci, 2007), sewage sludge (Kaleta et al., 2006) and biosolids (Jones-Lepp and Stevens, 2007), and, most recently, drinking water (Huerta-Fontela et al., 2008b). An examination of the US EPA's bibliographic database on PiE (USEPA, 2009a), shows that the core journal references having a major focus on illicit drugs in wastewaters, ambient waters, drinking water, or air total around 60 (this excludes those published on the topic of drugs on money). References (in any type of technical publication) dealing with illicit drugs in the environment total fewer than 200—composing only 2% of the documents (approaching 10,000) surrounding the broader topic of PiE, in general.

Presented here is a broad overview of illicit drugs as environmental contaminants. Perspectives are provided on their occurrence in various environmental compartments, what their occurrence might mean with regard to risk, and how their occurrence can be used as an analytical measurement tool to assess society-wide usage of illicit drugs.

A chronology of seminal publications on significant aspects of illicit drugs and the environment is presented in Table 1.1. The topic is transdisciplinary, involving a variety of disparate, but intersecting, fields, including healthcare, pharmacology, criminology, forensic sciences, epidemiology, toxicology, environmental and analytical chemistry, and sanitary engineering, among others.

1.2   WHAT IS AN "ILLICIT" DRUG?

Discussions regarding illicit drugs can become confused by the ambiguity in what exactly defines an "illicit" drug. Confusion stems from the fact that illicit drugs are not necessarily illegal. Many are licit medical pharmaceuticals having valuable therapeutic uses—two common examples being morphine and oxycodone. Instead, whether a drug is illicit is defined by international convention or national law, not necessarily by any inherent property of the drug. Some discussion is essential to better understand the scope of drug substances that can be considered illicit.

1.2.1   Terminology

There is no single, widely used term that accurately captures the myriad substances that become abused by habitual or addictive use. Although widely used, the term

TABLE 1.1   Chronology of Some Selected Seminal Publications Regarding Illicit Drugs in the Environment

Year	Aspect[a]	Unique Features of Study	Reference
1987	M	First report in a journal confirming the presence of an illicit drug (cocaine) on banknotes in general circulation) (objective to distinguish "drug" money from "innocent" money)	Aaron and Lewis (1987)
1998	A	Perhaps first data on an illicit drug in the ambient environment; non-target analysis revealed cocaine associated with fractions of particulate matter in outdoor air (Los Angeles)	Hannigan et al. (1998)
2000	M	First comprehensive overview of drugs on banknotes	Sleeman et al. (2000)
2001	F	Use of residues in sewage to reconstruct community-wide drug usage first proposed (later to be termed "sewage epidemiology" or "sewage forensics", or sometimes "community drug testing" or "community urinalysis"); first discussion to broaden the topic of drugs as environmental contaminants to include illicit drugs	Daughton (2001a)
2004	WW, Mon	Methamphetamine and MDMA in WWTP effluent; first report by US EPA of an illicit drug in the environment; first use of integrative time-weighted sampling for illicit drugs in wastewaters	Jones-Lepp et al. (2004)
2005	WW	First report of widespread occurrence of an illicit drug in surface water and wastewater (cocaine and BZE in WWTP influent and river)	Zuccato et al. (2005)
2005	F	First in-field implementation of "sewage epidemiology" to reconstruct community-wide drug usage	Zuccato et al. (2005)
2006	WW	First study to target a spectrum of illicit drugs and metabolites (in WWTP influents and effluents); those not identified in prior studies: norbenzoylecgonine, norcocaine, cocaethylene, 6-acetylmorphine, morphine-3-D-glucuronide, amphetamine, MDA, MDEA, EDDP, 11-nor-9-carboxy-9-THC	Castiglioni et al. (2006)
2006	SS	First report in peer-reviewed literature of an illicit drug in sewage sludge (amphetamine in sewage sludge)	Kaleta et al. 2006

(Continued)

TABLE 1.1 *(Continued)*

Year	Aspect[a]	Unique Features of Study	Reference
2006	F, mon	First nationwide monitoring in the US of illicit drugs in sewage; study by ONDCP targeted about 100 WWTPs across two dozen regions in the US (results never published)	See Bohannon (2007)
2006	F	First multi-country monitoring of cocaine in wastewaters to estimate usage	See UNODC (2007)
2007	A	First targeted analysis of ambient air for an illicit drug; cocaine quantified in particulates from all air sampled around Rome and several other Mediterranean locations (also in air samples archived several years prior)	Cecinato and Balducci (2007)
2007	SS	First report of an illicit drug in biosolids (methamphetamine in sewage biosolids)	Jones-Lepp and Stevens (2007)
2007	R	First conference devoted to topic of illicit drugs in the environment; led to first published overview of many of the aspects of the topic (including scientific, technical, social, privacy, ethical, and legal concerns)	EMCDDA (2007); Frost and Griffiths (2008)
2008	DW	First data on the occurrence and stepwise removal of illicit drugs at a municipal drinking water treatment plant	Fontela et al. (2008b)
2008	F	First use of the term "sewage epidemiology" in peer-reviewed literature; perhaps first mentioned in a 2007 interview, by Fanelli (Bohannon 2007)	Zuccato et al. (2008b)
2008	F	Creatinine in urine first assessed as means of normalizing drug concentrations across WWTPs (and therefore to facilitate drug usage comparisons across communities); creatinine first analyzed in sewage. Creatinine first proposed as a means for normalizing data by Daughton (2001a)	Chiaia et al. (2008)
2008	WW, mon	First systematic survey of illicit drugs in surface waters	Zuccato et al. (2008a)
2008	M, R	First overview of an illicit drug (cocaine) from banknotes from multiple countries	Armenta and de la Guardia (2008)
2008–2009	R	First major overviews of illicit drugs in the environment	Kasprzyk-Hordern et al. (2009b); Postigo et al. (2008); Zuccato et al. (2008b); Zuccato and Castiglioni (2009)

TABLE 1.1 (*Continued*)

Year	Aspect[a]	Unique Features of Study	Reference
2008-2009	R	First major overviews of the analytical approaches used for illicit drugs in the environment	Castiglioni et al. (2008); Postigo et al. (2008); Zuccato and Castiglioni (2009)
2008-2009	R, M	First major overview of the analytical approaches used for illicit drugs on money	Armenta and de la Guardia (2008)
2008-2009	EF	First studies regarding the sorption of illicit drugs to sediments, soils, and sewage sludge	Barron et al. (2009); Stein et al. (2008); Wick et al. (2009)
2009	DW	First data on the occurrence and stepwise removal of cannabinoids at a municipal drinking water treatment plant	Boleda et al. (2009)
2009	R	First major overview of illicit drugs in airborne particulates	Postigo et al. (2009)
2009	WW	First time that illicit drugs (cocaine, BZE, and morphine) monitored monthly in the sewage from an entire city over the course of a year	Mari et al. (2009)
2009	Sw	Sweat first proposed as a means of general transfer of drugs not just to sewage (via bathing and laundry) but also to any object in the surrounding environment contacted by skin (dermal transfer)	Daughton and Ruhoy (2009)
2009	Mon	First geographic spatial surveys; 24-h composite WWTP influent samples representing 65% of population of State of Oregon analyzed for BZE, methamphetamine, and MDMA, and Belgium-wide survey of cocaine, BZE, and ecgonine methyl ester	Banta-Green et al. (2009); van Nuijs et al. (2009a; 2009b)
2009	A	First qualitative report of cannabinols in ambient air aerosols (in Rome)	Cecinato et al. (2009b)
2009	A, mon	First quantitative study of cocaine in ambient air across several continents	Cecinato et al. (2009a)

[a]A, air; DW, drinking water; EF, environmental fate; F, forensics; M, money (banknotes); mon, monitoring; R, review; SS, sewage sludge (and biosolids); sw, sweat; WW, wastewater.

"illicit drug" is not accurate in the sense that most of the widely known abused drugs have bona fide medical uses as licit pharmaceuticals; the few that do not are incorporated in various listings or schedules of controlled substances maintained by various countries.

A variety of terms are used, often interchangeably, including: street drugs, designer drugs, club drugs, drugs of abuse, recreational drugs, clandestinely produced drugs, and hard and soft drugs. The term "designer" drug gained popularity in the 1980s when 3,4-methylenedioxymethamphetamine (MDMA, ecstasy) was introduced to the black market. Perhaps the most notable first designer drugs were introduced in the 1920s—dibenzoylmorphine and acetylpropionylmorphine.

Regardless of the terminology, much overlap exists with licit pharmaceuticals (those with approved medical uses). This can lead to much confusion or ambiguity as to exactly what the scope of the topic is. Discussion of the confusion surrounding illicit drug terminology is provided by Sussman and Ames (2008). In the overview provided here, the guiding definition used is that of the United Nations Office on Drugs and Crime (UNODC), which focuses not on the chemical identity of the drug itself, but rather on the life-cycle pathway traveled by a drug. The UNODC does not recognize any distinction between the chemical identity of licit and illicit drugs—only the way in which they are used (UNODC, 2009a). In this sense, the term "illicit" refers to the way in which these drugs are manufactured, distributed, acquired, and used, and by the fact that they are being used for nonmedical purposes.

This definition allows the inclusion of legal pharmaceuticals—that is, when they are manufactured, distributed, trafficked, or used illegally, or diverted from legal sources. The wide spectrum of sources and routes by which legal drugs become diverted for illicit use range from the relatively large-scale diversion from pharmaceutical distributors, pharmacies, and healthcare facilities, to the smaller scale (e.g., "theft" from home storage locations, such as for teen "pharming") and reuse of used medical devices, especially dermal medical patches, which present lethal hazards for both intentional and accidental exposures (Daughton and Ruhoy, 2009).

Whether a drug is classified as illicit is a complicated function of mores and evidence-based health studies, which are sometimes at odds with one another and under increasing scrutiny and debate [e.g., see Nutt (2009)]. Illicit substances (drugs and the precursors used for their manufacture) are captured on various government lists (controlled-substance schedules) that specify their allowable use. The primary criteria evaluated for listings are health risks, potential for abuse/addiction, therapeutic value, and utility as precursors for illicit manufacturing. The unifying worldwide scheme, used by the EU, for regulation comprises the schedules of the three UN Conventions of: 1961 (United Nations Single Convention on Narcotic Drugs, New York, amended 1972), 1971 (Convention on Psychotropic Substances, Vienna), and 1988 (Convention against Illicit Traffic in Narcotic Drugs and Psychotropic Substances, introducing control on precursors, Vienna). Combined, these schedules currently comprise about 250 explicitly named controlled substances (EMCDDA, 2009a).

Showing how the lines of demarcation become blurred, prescription analgesic opioids have overcome heroin and cocaine in the United States in leading to fatal drug overdoses (Leonard and Yongli, 2008). Indeed, the use of certain licit drugs, including over-the-counter (OTC) medications, for nonmedical purposes has recently surpassed the use of illicit drugs (NIDA, 2008). For example, of the top 10 drugs misused by high-school seniors in the United States, seven were legal prescription or OTC medications.

Numerous other illicit substances (such as structural analogs) exist but can only be captured implicitly by generalized chemical criteria that preemptively ban their synthesis; not all countries, however, have control acts that implicitly capture chemical analogs. Unknown numbers of additional substances exist but their chemical identities are elucidated only after they have experienced sufficient illegal use. A resource that provides the chemical structures for many of these substances (those listed by the Canadian Controlled Drugs and Substances Act) is maintained by Chapman (2009).

Adding further confusion regarding the distinctions between illicit drugs and medical pharmaceuticals, the laws dealing with illicit drugs dramatically vary from country to country. Long-standing drug policies in certain countries are also in a state of flux, because various changes are underway and adjustments are under consideration. These range from "reducing harm" (e.g., via decriminalization of possession and use) to acknowledgment from the American Medical Association regarding the medical benefits of a Schedule I drug (namely, cannabis) and calling for its clinical research (AMA, 2009). Beginning with Portugal in 2001 with the decriminalization of drug use, possession, and acquisition by drug end-users (Law no. 30/2000, which focuses on harm reduction) [see Greenwald [2009]], the array of laws dealing with illicit drugs has become quite diverse, but growing, illegal manufacturing, and trafficking remain criminal offenses. Among the EU States, the spectrum of law is captured by EMCDDA (2009b).

1.3   DIFFERENCES BETWEEN ILLICIT AND LICIT DRUGS AS ENVIRONMENTAL CONTAMINANTS

With respect to understanding their overall significance in the environment, seven aspects of illicit drug use contrast sharply with legitimate pharmaceutical use:

1. For most illicit drugs, there are no accurate quantitative data available concerning production or usage. For regulated pharmaceuticals, sales figures and regional real-time prescribing data can be used in models to calculate predicted environmental concentrations (PECs); these values can then be compared with measured environmental concentrations (MECs).

2. Although the chemical identities for the core group of illicit drugs are known, an ever-increasing number of new drugs (such as structural analogs with minor modifications of regulated pharmaceuticals and of previously known illicit drugs) can elude detection by forensics laboratories for years before they are noticed and identified. The myriad numbers of designer drugs and constant synthesis of new ones will pose challenges for mass spectrometrists for years to come and also introduce great uncertainty regarding the true scope of synthetic chemicals that contaminate the environment. Even though many of these unique chemicals are probably produced in relatively small quantities, the fact that they belong to relatively few chemical classes possibly means that they share only a few mechanisms of biological action. This makes additive action very likely, especially with substantial numbers of licit and illicit drugs often sharing the same mechanism of action. Since some have extremely low

effective doses (e.g., in the range of 1 µg per human use), this has relevance especially for aquatic exposure. As examples, *cis*-3-methylfentanyl and β-hydroxy-3-methylfentanyl (as with carfentanyl, a large-animal tranquilizer) are extraordinarily potent designer drugs, being three to five orders of magnitude more potent than morphine.

3. Drugs manufactured via illicit routes are commonly contaminated with unintended impurities and purposeful adulterants. These are often present at extremely high levels (e.g., sometimes more than one-half of the total mass, as opposed to milligram per kilogram [parts per million] levels for impurities in registered medicines) and are often more toxic than the sought-after drug.

4. The manufacture of illicit drugs (particularly methamphetamine) can cause extensive ecological damage as well as irreversible damage to infrastructures, such as buildings (USEPA, 2009b).

5. To date, the primary interest in residues of illicit drugs in the environment has been their occurrence in sewage (mainly untreated raw sewage) for use as a tracking tool to calculate community-wide consumption. This relatively new tool has been termed sewage (or sewer) forensics or epidemiology. In contrast to the licit use of pharmaceuticals, interest in their potential as biological stressors in the environment has been secondary, and very little is known.

6. Much less is known regarding the toxicology (including pharmacokinetics) of most illicit drugs.

7. With respect to environmental impact, numerous measures can be implemented to reduce the entry of licit pharmaceuticals into the environment. Routes of entry span an enormous spectrum of possibilities (Daughton and Ruhoy, 2008). With illicit drugs, pollution prevention measures—that is, to discourage their manufacture, distribution (e.g., via unapproved Internet pharmacies), and end use—are straightforward, but more difficult to implement.

Note that the frequent changes in the introduction of new pharmaceuticals with potential for abuse, as well as new illicit substances, preclude any comprehensive definitive worldwide compilation of chemicals. The INCB (International Narcotics Control Board) maintains three major listings (INCB, 2009): Yellow List (Narcotic Drugs Under International Control), Green List (Psychotropic Substances under International Control), and Red List (Precursors and Chemicals frequently used in the Illicit Manufacture of Narcotic Drugs and Psychotropic Substances under International Control). A convenient listing of many of the corresponding chemical structures is provided by Chapman (2009).

1.4 THE CORE ILLICIT DRUGS AND THE ENVIRONMENT

The types of drugs commonly abused are described in various ways, depending on their origin and biological effect. They can either be naturally occurring, semisynthetic (chemical manipulations, such as analogs, of substances extracted from natural materials), or synthetic (created entirely by laboratory synthesis and manipulation). The primary categories are opiates, other CNS depressants (sedative-hypnotics), CNS stimulants, hallucinogens, and cannabinoids.

The scope of chemicals that could be considered illicit can be viewed in terms of the following categories of medical efficacy:

1. No known medical use (and are illegal in all circumstances according to various conventions) (e.g., benzylpiperazine; heroin in the United States),
2. Limited established medical use, but which are also manufactured illegally and used primarily for nonmedical purposes (e.g., methamphetamine),
3. Firmly established wide medical use, but are diverted illegally (e.g., theft; illegal prescribing such as via unapproved Internet "pharmacies"),
4. Firmly established wide medical use and are obtained "legally," but for non-medical use (e.g., doctor/hospital shopping or by other con schemes),
5. Similar biological action to prescription drugs but are synthesized as analogs (which are not individually and explicitly categorized as illegal; examples include the numerous analogs of phosphodiesterase type-5 inhibitors).

All of these categories comprise drugs with high potential for abuse or that enjoy recreational use. Methadone is usually included in these discussions even though most of its use is legal; it serves to track opiate addiction but is also used and abused as an analgesic.

Some drug residues in the environment have substantial multiple origins (both legal and illegal), making it difficult to ascribe monitored levels to illicit use. Morphine is one example. It can originate from medical use of morphine itself or from codeine (via O-demethylation). It can also originate from diverted morphine or codeine, as well as from heroin. By collecting data on other (and more unique) metabolites, these pathways can be teased apart. Using morphine as the example, by monitoring for the heroin metabolite 6-AM (6-acetylmorphine), a more representative picture can be obtained for that portion of morphine originating from heroin.

While drug usage patterns and prevalence vary among countries as well as with time, those drugs in frequent use in the United States can serve as an organizing framework for further discussion. The annual reports of the US DEA's National Forensic Laboratory Information System, NFLIS (USDEA, 2008), provide the best insights regarding which known drugs are most used in nonmedical circumstances (see Table 1.2).

Of all the samples analyzed in 2008 by US local and state forensic labs for the presence of nonmedically used drugs, 25 controlled substances (Table 1.2) composed 90% of all samples. The most frequent four were tetrahydrocannabinol (THC), cocaine (benzoylmethylecgonine), methamphetamine, and heroin. Of these 25, only 15 have been targeted in environmental studies of illicit drugs: amphetamine, cocaine, codeine, heroin, hydrocodone, MDA, MDMA, methadone, methamphetamine, methylphenidate, morphine, oxycodone, PCP (phencyclidine), pseudoephedrine, and THC (Δ^9-tetrahydrocannabinol).

TABLE 1.2 Drugs of Abuse Frequently Detected by United States Forensics Labs[a]

25 Abused Drugs	Other Abused Drugs
Most frequent	**Narcotic analgesics**
Tetrahydrocannabinol (THC)	Butorphanol
Cocaine (benzoylmethylecgonine)	Dihydrocodeine
Methamphetamine	Fentanyl
Heroin (diacetylmorphine; diamorphine)	Meperidine
	Nalbuphine
Narcotic analgesics	Opium
Buprenorphine	Oxymorphone
Codeine	Pentazocine
Hydrocodone	Propoxyphene
Hydromorphone	Tramadol
Methadone	
Morphine	**Benzodiazepines**
Oxycodone	Chlordiazepoxide
	Flunitrazepam
Benzodiazepines	Midazolam
Alprazolam	Temazepam
Clonazepam	Triazolam
Diazepam	
Lorazepam	**"Club" drugs**
	1-(3-Trifluoromethylphenyl)piperazine
Others	(TFMPP)
1-Benzylpiperazine (BZP)	3,4-Methylenedioxy-N-ethylamphetamine
3,4-Methylenedioxyamphetamine (MDA)	(MDEA)
3,4-Methylenedioxymethamphetamine (MDMA)	5-Methoxy-N,N-diisopropyltryptamine
Amphetamine	(5-MeO-DIPT)
Carisoprodol	γ-Hydroxybutyrate/γ-butyrolactone
Methylphenidate	(GHB/GBL)
Phencyclidine (PCP)	Ketamine
Pseudoephedrine	
Psilocin	**Stimulants**
	Cathinone
	Ephedrine
	Phentermine
	Anabolic steroids
	Methandrostenolone
	Nandrolone
	Stanozolol

[a]USDEA's National Forensic Laboratory Information System (USDEA, 2008).

Note that the top 25 detected by NFLIS are all among the most commonly abused drugs in the United States. The major ones missing (but which are captured in the remaining 10% of samples analyzed by NFLIS) are barbiturates (e.g., seconal and phenobarbital, but whose rate of abuse has been declining), certain benzodiazepines (except flunitrazepam, such as alprazolam, diazepam, and chlordiazepoxide), methaqualone, mescaline (3,4,5-trimethoxyphenethylamine), and dextromethorphan (NIDA, 2009). Extensive statistics on rates of drug use worldwide (including those maintained by the UNODC) can be found on the ONDCP web page (ONDCP, 2009). The UNODC World Drug Report (UNODC, 2009b) provides comprehensive statistics on world illicit drug supply and demand.

From a comprehensive examination of the published literature on illicit drugs and their metabolites in a variety of environmental compartments (wastewaters, surface waters, drinking water, sewage sludge, sewage biosolids, air, and banknotes), positive occurrence data as well as indications of negative occurrence (data of absence) were compiled (data not shown here). From these data, those analytes with absence of data (i.e., those that have yet to be targeted in monitoring studies) can be deduced. For example, Postigo et al. (2008) noted that norcocaethylene and ecgonine ethyl ester have not been targeted in any monitoring study. Major reviews of illicit drugs in the environment are provided by Huerta-Fontela et al. (2010) and Zuccato and Castiglioni (2009).

The published data reveal that the drugs with the most positive occurrence data across all environmental compartments are among the top 25 detected by NFLIS —notably codeine, morphine, methadone, amphetamine, methamphetamine, cocaine, and THC, and the primary metabolites of methadone (i.e., EDDP), cocaine (i.e., BZE, benzoylecgonine), and THC (i.e., 11-nor-9-carboxy-9-THC [THC-COOH]). Although widely detected in drug screens, the occurrence of heroin (diacetylmorphine) in an environmental compartment is limited primarily to banknotes—because of its propensity to hydrolyze in water. Likewise, the cannabinoids are detected most frequently in air. Not surprisingly, no illicit drug (or metabolite) frequently reported with environmental occurrence data is missing from the 25 most frequently identified by forensic labs.

Nine of the remaining 25 drugs most frequently identified by the forensic testing labs have not yet been targeted in environmental studies focused on illicit drugs: alprazolam, buprenorphine, BZP (1-benzylpiperazine), carisoprodol, clonazepam, diazepam, hydromorphone, lorazepam, and psilocin (4-hydroxydimethyltryptamine, 4-HO-DMT). Of these nine drugs, environmental occurrence data have been published in studies targeted at medical pharmaceuticals for: alprazolam, carisoprodol, diazepam, and lorazepam. Data do not exist for buprenorphine, BZP, clonazepam, hydromorphone, and psilocin. Depending on their pharmacokinetics and extent of excretion unchanged, these latter five drugs could be considered for targeting in future environmental monitoring.

Some illicit drug analytes, when targeted, are infrequently reported possibly as a result of their considerably higher detection limits. Normorphine and THC-COOH are examples, sometimes having limits of detection one to two orders of magnitude higher than other analytes. Other targeted analytes are not detected because they are extensively metabolized or excreted as conjugates. Conjugation undoubtedly plays a critical role in determining whether a free parent drug will be found in waters. Many drugs are extensively conjugated, and without a hydrolysis step in analysis, these will be missed (Pichini et al., 2008; Daughton and Ruhoy, 2009).

Important to note is that some illicit drugs are metabolic/transformation daughter products of others, explaining why their concentrations in sewage or receiving waters are routinely higher than their parents. One example is heroin, which is quickly deacetylated to 6-AM followed by hydrolysis to morphine. This means that the probability is higher that these parent drugs, when detected in waters (especially waters distanced from impact by sewage), are present because they were directly flushed down the toilet (or excreted via sweat), rather than being excreted via urine; an alternative source is run-off into streams, such as during clandestine manufacturing. Another example is fentanyl, which is extensively excreted as norfentanyl.

Environmental occurrence data from most of the major studies on illicit drugs have been reviewed by Huerta-Fontela et al. (2010) and Zuccato and Castiglioni (2009).

1.4.1 Adulterants and Impurities

In contrast to pharmaceuticals produced under Good Manufacturing Practices, drugs made illegally contain myriad other chemical substances in addition to (or sometimes even in place of) the sought-after drug. Adulterants are often used to enhance desired biological effects. Included are diluents, which are added to mimic the physical appearance of the sought-after drug when the objective is economic gain (to extend the doses per mass). Impurities are sometimes integral to the natural chemistry of the native plant from which a drug is isolated and, other times, a function of the synthetic route to the desired drug (as dictated by the skill of the operator/ chemist).

Many dozens of impurities and adulterants are possible for any given drug synthesis. Impurities, in turn, can each yield numerous metabolites, most of which are not yet known. Adulterants can range from common substances such as caffeine (albeit in very high concentrations) to more insidious chemicals, such as the cytotoxic veterinary dewormer drug levamisole, which has led to a number of deaths; in this way, illicit drugs can serve as a route of entry to the environment for licit drugs that otherwise would never themselves experience nonmedical use. Adulteration of illicit drugs has grown to become a major health risk for drug users.

These substances are often present at very high levels, especially in intentionally mislabeled drugs. They sometimes represent the bulk of the purported drug (e.g., noscapine can be present at levels up to 60% in heroin, or phenacetin, which can be present at levels up to 50% in cocaine). These contaminants include products of synthesis or processing (precursors, intermediates, by-products), natural impurities (e.g., natural product alkaloids), products of degradation (e.g., oxidation during storage), and pharmacologically active adulterants (e.g., many licit drugs and other chemicals, obtained illegally, such as levamisole, xylazine, lidocaine, phenacetin, hydroxyzine, and diltiazem). Some of these impurities or adulterants are more potent than the sought-after drug (cocaethylene being one example, which is a synthesis by-product as well as a metabolite of cocaine when consumed together with ethanol). Some have considerable toxicity. In the course of reviewing the literature, over 90 common adulterants and impurities were noted just for the four illicit drugs cocaine, MDMA, methamphetamine, and heroin. These represent but a very small sampling of the variety of chemicals that can compose illicit drugs.

1.5   LARGE-SCALE EXPOSURE OR SOURCE ASSESSMENTS VIA DOSE RECONSTRUCTION

Interest in illicit drugs in the environment has both prospective and retrospective dimensions. The prospective dimension concerns the questions surrounding the exposure of aquatic organisms and of humans to environmental residues. Of the environmental studies conducted, however, the major objective in collecting data on the presence and scope of illicit drugs in sewage and wastewaters has not been for prospectively assessing their significance as environmental contaminants and their potential for ecological or human health exposure. Rather, the objective has been the use of these data as a retrospective tool for reconstructing society-wide drug usage. This could be considered a large-scale version of exposure assessment called "dose reconstruction" [e.g., see ATSDR (2009)].

Separate but analogous approaches have also been attempted making use of the presence of drug residues on banknote currency and in airborne particulates. These could be more accurately referred to not as dose reconstruction, but rather as source reconstruction (deciphering the source and intensity of the origin of the drugs).

1.5.1   Sewage Epidemiology or Forensics

First proposed in 2001 (Daughton, 2001a), the analysis of sewage for residues of illicit drugs unique to actual consumption (rather than originating from disposal or manufacture) for the purpose of back-calculating estimates of community-wide usage rates has since been discussed under a variety of terms, including "sewage epidemiology" (a term first reported in the literature by Zuccato et al. (2008b)], "sewage forensics," and "community-wide urinalysis" or "community drug testing." None of these terms, however, fully captures the multiple purposes that can be served by the methodology.

Epidemiology can be defined simply as the study of populations sharing similar characteristics of disease (or health status). Among its uses are identifying at-risk subpopulations, monitoring the incidence of exposure/disease, and detecting/controlling epidemics. Elements of illicit drug use fit all of these. In its simplest state, "forensics" involves the extraction of pertinent information to support an argument or investigation (Daughton, 2001b). One of its best known modern renditions is to assist in resolving legal issues—and the worldwide legal system plays an integral role in all aspects of illicit drug use.

Since this still-evolving approach for measuring drugs in sewage to estimate collective drug usage has elements of both forensics and epidemiology, it would be more accurately captured under the newer term "Forensic Epidemiology," which integrates the principles and methods used in public health epidemiology with those used in forensic sciences (Goodman et al., 2003; Loue, 2010).

With this in mind, a more accurate descriptive term should be considered in order to better unify the published literature. One possibility could be "Forensic Epidemiology Using Drugs in Sewage" (FEUDS). Use of a unique term and acronym would have the added benefit of more easily facilitating communication across disciplines and would greatly facilitate literature searches. In the remainder of this discussion, however, the shorthand term "Sewage (or Sewer) Forensic Epidemiology" (SF/E) will be used.

1.5.2 SF/E Used in Community-Wide Dose Reconstruction for Illicit Drugs

After its conceptualization in 2001 (Daughton, 2001a), SF/E was first implemented in a field-monitoring study by Zuccato et al. (2005). SF/E was originally proposed as the first evidence-based approach for measuring drug use, because the long-practiced approaches that use population surveys are fraught with limitations, not the least of which involve numerous sources of potential error that are difficult to define, control, or measure (especially self-reporting bias) (Daughton, 2001a). This has been corroborated in "concordance" studies (comparisons of self-report data with empirical bioanalysis data), which point to gross underreporting by self-reports (often at rates as low as one-half). These conventional approaches to estimating illicit drug usage also suffer from two inherent limitations: (a) extreme delays in times before results can be compiled and reported and (b) costs associated with data collection and interpretation.

Like public surveys, SF/E also suffers from a large number of sources of potential error. However, SF/E is in its infancy, and its error derives from variables still under investigation and which could be better controlled. While conceptually rather straightforward, the back-calculations used in SF/E are a function of numerous variables, including demographics, population flows (transient visitors and commuters) served by a sewage treatment facility, sewage flows, and pharmacokinetics. Combined, these pose a major challenge for modeling to accurately reconstruct dose. The numerous problems facing SF/E are discussed in Frost and Griffiths (2008). Most SF/E investigators couple drug concentrations in sewage with per-capita sewage flows to calculate what is sometimes called "index loads" or "per capita loads," expressed as milligram/person/day. Many of the sources of uncertainty are covered by Banta-Green et al. (2009) and Zuccato et al. (2008b).

Despite the plethora of uncertainties in the many variables involved in SF/E back-calculations, the ability to provide estimates of near real-time community-wide usage is something that is not possible with any other known approach. This also opens the possibility of detecting real-time trends or changes in drug use. Example applications include verifying reductions in drug use as a result of interdictions, or detecting the emergence of newly available drugs or overall changes in drug-use patterns. Data on real-time usage could better inform decisions regarding drug control and mitigation. Correlating policy actions with resulting society-wide impacts cannot be effectively done when collected data are significantly delayed in reporting.

Of great potential significance, there is also no apparent technical obstacle to designing automated continuous monitors for use in sewage collection/distribution systems. Implementing continuous monitoring to support SF/E would serve to better inform decisions regarding control and mitigation of drug use.

Another advantage with SF/E as opposed to population surveys is that not all drug use is necessarily known to the users themselves, who then unintentionally report to surveys incorrect drug identities and usage quantities. Illicit-drug users often do not know the identity or the quantity of the active substances they have consumed because the purity is unknown. Often the active substance or quantity is not what the distributor claims. Adulterants are often substituted, in part or in whole, for the purported drug. One general route of uninformed exposure is the surreptitious incorporation of designer drugs into otherwise legal OTC diet supplements or recreational or life-style products. An example is the relatively new (and probably incompletely characterized) synthetic analogs of the approved phosphodiesterase type-5 (PDE-5) inhibitors (used primarily in treating erectile dysfunction), such as sildenafil, vardenafil, and tadalafil (Poon et al., 2007; Venhuis and de Kaste, 2008). The legal registered versions of PDE-5 inhibitors have only recently been detected in wastewaters (Nieto et al., 2010). The extent of such adulteration in the drug and supplements industry is unknown, largely because the targets for analysis are often not known to forensic analysts.

Hagerman (2008) provides a brief history of SF/E research in the United States. The ONDCP performed the first SF/E monitoring in the United States in 2006, targeting about 100 wastewater treatment plants (WWTPs) across two dozen regions (Bohannon, 2007). The first conference devoted to SF/E was organized by EMCDDA in Lisbon, Portugal in April of 2007 (EMCDDA, 2007). It led to the first published overview of many of the aspects of the topic (including scientific, technical, social, privacy, ethical, and legal concerns), as provided by Frost and Griffiths (2008).

1.5.3 Summary of Published Research in SF/E

Overviews and discussion of the SF/E studies published up until 2008 are provided by Postigo et al. (2008) and Zuccato et al. (2008b). The major published articles regarding the SF/E approach are compiled in the chronology of Table 1.3. As of the beginning of 2010, there had been fewer than two dozen studies. All but a handful have been published after 2007.

Published SF/E studies have been conducted in a number of countries, with assessments at the local, regional, or national levels, primarily in Belgium, Germany, Ireland, Italy, Spain, Switzerland, the United States (i.e., Oregon), and Wales. To date, SF/E assessments have focused on a select few parent drugs (primarily cannabis, cocaine, heroin, and MDMA) using various metabolites. They have been performed using a broad range of sampling methodologies ranging from single-event discrete grab sampling to longer-term (e.g., 12-month) integrative continuous sampling over numerous WWTPs or rivers, servicing regions with populations exceeding millions. Many of these studies have searched for temporal usage patterns—comparing yearly seasons or the day of the week (e.g., higher cocaine use on weekends). Usage rates are reported on various comparative bases, often involving per capita (e.g., g/day/1000 population, usually ranging only up to several grams), total consumption (e.g., tonnes/year/geographic area), or flows (mass/river/day). Discrete monitoring must acknowledge the cyclic or episodic drug use pattern fluctuations in concentrations that can result from diurnal cycles, seasons, or day of the week. This can be

TABLE 1.3 Selected SF/E Studies (Arranged Roughly According to Chronology)

Year	Title (and reference)
2001	Illicit drugs in municipal sewage: Proposed new nonintrusive tool to heighten public awareness of societal use of illicit/abused drugs and their potential for ecological consequence (Daughton, 2001a)
2005	Cocaine in surface waters: New evidence-based tool to monitor community drug abuse (Zuccato et al., 2005)
2006	High cocaine use in Europe and US proven – Stunning data for European countries: First ever comparative multicountry study of cocaine use by a new measurement technique (Sörgel, 2006)
2007	Using environmental analytical data to estimate levels of community consumption of illicit drugs and abused pharmaceuticals (Bones et al., 2007)
2008	Occurrence of psychoactive stimulatory drugs in wastewaters in northeastern Spain (Huerta-Fontela et al., 2008a)
	Estimating community drug abuse by wastewater analysis (Zuccato et al., 2008a)
2009	Cocaine and metabolites in waste and surface water across Belgium (van Nuijs et al., 2009b)
	Cocaine and heroin in waste water plants: A 1-year study in the city of Florence, Italy (Mari et al., 2009)
	Monitoring of opiates, cannabinoids and their metabolites in wastewater, surface water and finished water in Catalonia, Spain (Boleda et al., 2009)
	Can cocaine use be evaluated through analysis of wastewater? A nation-wide approach conducted in Belgium (van Nuijs et al., 2009a)
	Illicit drugs and pharmaceuticals in the environment – Forensic applications of environmental data, Part 1: Estimation of the usage of drugs in local communities (Kasprzyk-Hordern et al., 2009a)
	Assessing illicit drugs in wastewater: Potential and limitations of a new monitoring approach (Frost and Griffiths, 2008)
	Municipal sewage as a source of current information on psychoactive substances used in urban communities (Wiergowski et al., 2009)
2010	The spatial epidemiology of cocaine, methamphetamine and 3,4-methylenedioxymethamphetamine (MDMA) use: a demonstration using a population measure of community drug load derived from municipal wastewater (Banta-Green et al., 2009)
	Drugs of abuse and their metabolites in the Ebro River basin: Occurrence in sewage and surface water, sewage treatment plants removal efficiency, and collective drug usage estimation (Postigo et al., 2010)

particularly pronounced for recreational drugs. Limits of detection (LOD) will dictate the extent to which a monitoring study will produce meaningful data-of-absence (negative data).

An enormous published literature surrounds the forensic chemistry of illicit drugs. The numbers of illicit drugs analyzed in the environment, however, is but a small fraction of those that have been targeted in countless studies published on biological tissues and fluids for the purposes of forensics and patient compliance monitoring and for the study of pharmacokinetics in animals. Accurate mass identification of unknowns (for example, via LC/TOF−MS) plays a central role especially when authentic reference standards are not available. While this conventional forensics literature can serve as a guide for environmental analysis, it is not directly relevant. There are numerous variables involved with (and impacting) the procedural steps used in analysis for SF/E, ranging from sampling design and matrix interferences to analyte determination and need for extremely low limits of detection. Some major overviews and discussion of the analytical approaches for measuring illicit drugs in wastewaters and other waters are available (Castiglioni et al., 2008; Postigo et al., 2008; Zuccato and Castiglioni, 2009).

An issue little addressed in SF/E studies has been the complications posed by chirality. Possibly the majority of illicit drugs have at least one chiral center (Smith, 2009). The alkaloid truxilline, as an example, occurs in coca leaf as 11 stereoisomers. Amphetamines can each have a pair of enantiomers, sometimes distinguishing the licit and illicit forms (as well as dictating toxicology). This may account for a portion of some of the large variance in estimated amphetamine usage across SF/E studies. While chiral isomers can pose difficult challenges for analytical chemists, they can also provide a wealth of forensics information in terms of chemical "fingerprinting," for example, in distinguishing legal from illegal origins.

1.6 ILLICIT DRUGS AND ENVIRONMENTAL IMPACT

With the exception of the immediate and overt (as well as hidden) environmental impacts from clan labs, little is known with respect to the potential actions of illicit drugs in the environment.

Compared with pharmaceuticals, little attention has been devoted to the environmental fate and transport of illicit drugs. Most illicit drugs have never been monitored in sewage biosolids or sediments. Domènech et al. (2009) used fugacity modeling to predict the fate of cocaine and BZE. The microbial degradation of methamphetamine has been reported by Janusz et al. (2003). Wick et al. (2009) examined biological removal in activated sludge and found rapid removal for morphine, codeine, dihydrocodeine, oxycodone, and methadone but not for tramadol. Two studies report on the sorption of illicit drugs to sediments (Stein et al., 2008; Wick et al., 2009). Wick et al. (2009) and Barron et al. (2009) acquired low distribution coefficients (K_d) for amphetamine, cocaine, cocaethylene, BZE, MDMA, morphine, codeine, dihydrocodeine, methadone, and tramadol, showing that removal via sorption to sewage sludge is possibly negligible.

Far more is known regarding the ecotoxicology of licit pharmaceuticals than of illicit drugs, especially with regard to low-level mixed-stressor exposures. Almost nothing is known regarding the potential for biological effects in aquatic systems or the bioconcentration of illicit drugs in biota. Gagne et al. (2006) report some nominal effects data for morphine in mussels. The potential for effects from low-level exposure of fish is further complicated by the complexities in extrapolating across species. The first in-depth study of an ectotherm with any analgesic (i.e., morphine) comports with extreme variability between species (Newby et al., 2006).

1.7 THE FUTURE

Future work addressing the various environmental aspects of illicit drugs in the environment would benefit from a comprehensive assessment of what has been accomplished to date and what new research needs to be conducted. While the knowledge base regarding all aspects of illicit drugs in the environment is extremely small compared with that of pharmaceuticals, the body of published data is perhaps sufficiently large that we risk duplication of efforts while failing to address the more important remaining gaps or needs (Daughton, 2009a). The first step in ensuring better-targeted research could be creation of a centralized, publically accessible database of results from research conducted worldwide. Such data could include environmental occurrence (sewage influent, effluent, and sludge/biosolids; surface and drinking waters; air; and money), ecotoxicity, and especially data generated from SF/E studies; occurrence data should include data of absence (with detection limits).

1.7.1 Advancing the Utility of SF/E

Advancement of SF/E as a topic of research as well as a survey tool could occur on two fronts. First, numerous improvements could be made to better define and control the many variables contributing to uncertainty in SF/E back-calculations for gauging collective drug usage. Needed are standardized methodologies with better understood and controlled sources of error. This is especially important for facilitating more meaningful intercomparison of SF/E data.

For SF/E to succeed in gauging illicit drug usage, one variable in particular needs to be better understood—the pharmacokinetics (PK) of each drug, especially as it pertains to the excretion of parent drug and metabolites (especially conjugates). PK parameters are key to accurate dose reconstruction. While excretion rates for many pharmaceuticals are not well-defined, even less is known regarding the PK of illicit drugs. PK and its poorly defined variance among a population contributes great uncertainty to the back-calculations used with SF/E. A comprehensive sensitivity analysis (which has yet to be performed) would probably reveal that small changes in variables, such as excretion rates (especially for extensively metabolized drugs), can lead to large errors in SF/E calculations. For those drugs/metabolites with highly variable excretion rates, the error range could be substantial. As a case in point, with a study of 12 methamphetamine addicts, the urine ratio of amphetamine/methamphetamine ranged over two orders of magnitude, from 0.03 to 0.56 (Kim et al., 2008). This would also prove problematic for allocating amphetamine loadings in sewage to methamphetamine use versus medical use. A host of factors contributes to PK variability, including route and size of dose, gender, age, body mass, kidney and liver function, chronobiology, diet, polypharmacy interactions, and genetics/epigenetics (namely pharmacogenomics, which dictates the spectrum of PK variability). Similarly, it is important to be able to distinguish bacterial transformations in sewage (and the ambient environment) from those of human metabolism (Boleda et al., 2009).

Other ways to reduce the error boundaries in SF/E calculations could be viewed as analogous to internal correction methods, such as isotope dilution or standard additions. For example, instead of using correction factors based on modeling assumptions for dilution by waste streams and sewage transformations, correction factors could possibly be empirically derived by monitoring particular pharmaceuticals. Pharmaceuticals that would be most useful for "calibrating" a WWTP system would be those that (i) are widely prescribed, (ii) are not abused or used recreationally, (iii) have real-time prescribing/sales data, (iv) are known to have high patient compliance (minimal leftovers, resulting in little disposal into sewers) and are used in short-term courses (not maintenance medications), (v) have a potential similar to that of the target illicit drug with regard to biodegradation and sorption to sewage solids, and (vi) have well-understood pharmacokinetics (preferably poorly metabolized, resulting in extensive excretion unchanged). By comparing the known consumption rates of the pharmaceutical calibrant (from prescribing databases) with the levels actually detected in the sewage stream, more accurate correction factors could possibly be derived and then applied to the illicit drug. By gathering long-term time-course data for the calibrant pharmaceutical, even more uncertainty could possibly be removed from the calibration factor. This approach, however, cannot remove the confounding of dual inputs from excretion and disposal of the targeted illicit drug; the latter, however, probably leads to episodic spikes in underlying baseline levels, which would become clearer with sustained monitoring.

Second, the current scope of SF/E could possibly be expanded to tackle questions other than simply monitoring or gauging illicit drug consumption. Unexplored possibilities range from early detection of emerging trends in abuse of mainstream pharmaceuticals and in their illegal trafficking (e.g., from diversion or Internet purchases) to better gauging medication compliance rates for patients. For example, with access to real-time local prescribing data, those pharmaceutical ingredients in sewage, whose back-calculated usage rates are substantially higher than the prescribed rates, could be targeted for investigating the possibility of illegal trafficking. A possible example can be seen in the data presented by Kasprzyk-Hordern et al. (2009a; see Table 7 therein), where calculated usage rates for over two dozen prescribed and OTC pharmaceuticals are compared with known nationwide (not local) dispensing rates. Of these drugs, the calculated average usage rates exceeded the national average sales by over an order of magnitude for only one drug: tramadol. Indeed, tramadol (an opioid) is recognized for its growing incidence of misuse and abuse. Real-time prescribing data are greatly confounded, however, by the inability of current tracking systems to correlate location of dispensing with place of actual use (e.g., because of transient populations and mail-order prescribing). Another expanding source of data that could potentially be used to verify calculated usage rates is the growing network of collection programs that take back leftover medications [see Glassmeyer et al., [2009)].

An important aspect of SF/E for illicit drug use is that it has set the foundation for the use of SF/E for other purposes—some unrelated to drug use. A fascinating possibility would be the use of sewage monitoring for measuring indicators of community-wide health status via the presence of various biomarkers of health or disease.

ACKNOWLEDGMENT

USEPA Notice: The United States Environmental Protection Agency through its Office of Research and Development funded and managed the research described here. It has been subjected to Agency's administrative review and approved for publication. Review comments by Dr. Don Betowski (USEPA) and Dr. Stevan Gressitt (Department of Health and Human Services, State of Maine) are much appreciated.

REFERENCES

Aaron, R., and P. Lewis. 1987. Technical article: Cocaine residues on money. *Crime Lab. Dig.* **14**(1):18.

Agency for Toxic Substances and Disease Registry (ATSDR). 2009. Exposure-Dose Reconstruction Program (EDRP). ATSDR, DHS. Atlanta, GA; http://www.atsdr.cdc.gov/edrp/.

AMA. 2009. Use of cannabis for medicinal purposes, CSAPH report 3: Council on Science and Public Health (CSAPH), American Medical Association; http://americansforsafeaccess. org/downloads/AMA_Report.pdf.

Armenta, S., and M. de la Guardia. 2008. Analytical methods to determine cocaine contamination of banknotes from around the world. *Trends Anal. Chem.* **27**(4):344–351.

Banta-Green, C. J., J. A. Field, A. C. Chiaia, D. L. Sudakin, L. Power, and L. de Montigny. 2009. The spatial epidemiology of cocaine, methamphetamine and 3,4-methylene-dioxymethamphetamine (MDMA) use: A demonstration using a population measure of community drug load derived from municipal wastewater. *Addiction* **104**(11): 1874–1880.

Barron, L., J. Havel, M. Purcell, M. Szpak, B. Kelleher, and B. Paull. 2009. Predicting sorption of pharmaceuticals and personal care products onto soil and digested sludge using artificial neural networks. *Analyst (Cambridge UK)* **134**: 663–670.

Bohannon, J. 2007. Hard data on hard drugs, grabbed from the environment: Fieldwork in new and fast-growing areas of epidemiology requires wads of cash and a familiarity with sewer lines. *Science* **316**(5821):42–44.

Boleda, M. R., M. T. Galceran, and F. Ventura. 2009. Monitoring of opiates, cannabinoids and their metabolites in wastewater, surface water and finished water in Catalonia, Spain. *Water Res.* **43**(4):1126–1136.

Bones, J., K. V. Thomas, and B. Paull. 2007. Using environmental analytical data to estimate levels of community consumption of illicit drugs and abused pharmaceuticals. *J. Environ. Monit.* **9**(7):701–707.

Castiglioni, S., E. Zuccato, E. Crisci, C. Chiabrando, R. Fanelli, and R. Bagnati. 2006. Identification and measurement of illicit drugs and their metabolites in urban wastewater by liquid chromatography—tandem mass spectrometry. *Anal. Chem.* **78**(24):8421–8429.

Castiglioni, S., E. Zuccato, C. Chiabrando, R. Fanelli, and R. Bagnati. 2008. Mass spectrometric analysis of illicit drugs in wastewater and surface water. *Mass Spectrom. Rev.* **27**(4):378–394.

Cecinato, A., and C. Balducci. 2007. Detection of cocaine in the airborne particles of the Italian cities Rome and Taranto. *J. Sep. Sci.* **30**(12):1930–1935.

Cecinato, A., C. Balducci, and G. Nervegna. 2009a. Occurrence of cocaine in the air of the World's cities: An emerging problem? A new tool to investigate the social incidence of drugs? *Sci. Total Environ.* **407**(5):1683–1690.

Cecinato, A., C. Balducci, G. Nervegna, G. Tagliacozzo, and I. Allegrini. 2009b. Ambient air quality and drug aftermaths of the Notte Bianca (White Night) holidays in Rome. *J. Environ. Monit.* **11**(1):200–204.

Chapman, S. 2009. Consolidated Index of Drugs and Substances. Isomer Design. Toronto, Ontario; http://www.isomerdesign.com/Cdsa/scheduleNDX.php.

Chiaia, A. C., C. Banta-Green, and J. Field. 2008. Eliminating solid phase extraction with large-volume injection LC/MS/MS: Analysis of illicit and legal drugs and human urine indicators in US wastewaters. *Environ. Sci. Technol.* **42**(23):8841–8848.

Daughton, C. G. 2001a. Illicit drugs in municipal sewage: Proposed new non-intrusive tool to heighten public awareness of societal use of illicit/abused drugs and their potential for ecological consequence. In *Pharmaceuticals and Personal Care Products in the Environment: Scientific and Regulatory Issues*, edited by C. G. Daughton and T. Jones-Lepp. Washington, DC: American Chemical Society, Chapter 20; pp. 348–364; http://www.epa.gov/nerlesd1/bios/daughton/book-conclude.htm.

Daughton, C. G. 2001b. Literature forensics? Door to what was known but now forgotten. *Environ. Forensics* **2**(4):277–282.

Daughton, C. G. 2001c. Pharmaceuticals and personal care products in the environment: Overarching issues and overview. In *Pharmaceuticals and Personal Care Products in the Environment: Scientific and Regulatory Issues*, edited by C. G. Daughton and T. L. Jones-Lepp. Washington, DC: American Chemical Society, Chapter 1, pp. 2–38; http://www.epa.gov/nerlesd1/bios/daughton/book-summary.htm.

Daughton, C. G. 2009a. Chemicals from the practice of healthcare: Challenges and unknowns posed by residues in the environment. *Environ. Toxicol. Chem.* **28**(12):2490–2494.

Daughton, C. G. 2009b. Peering into the shadows of chemical space. Emerging contaminants and environmental science: Is either being served by the other? In *2nd International Conference on Occurrence, Fate, Effects, and Analysis of Emerging Contaminants in the Environment (EmCon09)*, opening address Fort Collins, Colorado, 4–7 August 2009; http://www.epa.gov/esd/bios/daughton/Daughton-abstract-EmCon09.pdf.

Daughton, C. G., and I. S. Ruhoy. 2008. The afterlife of drugs and the role of pharmEcovigilance *Drug Saf.* **31**(12):1069–1082.

Daughton, C. G., and I. S. Ruhoy. 2009. Environmental footprint of pharmaceuticals - The significance of factors beyond direct excretion to sewers. *Environ. Toxicol. Chem.* **28**(12):2495–2521.

Daughton, C. G., and T. A. Ternes. 1999. Pharmaceuticals and personal care products in the environment: Agents of subtle change? *Environ. Health Perspect.* **107**(Suppl 6): 907–938.

Domènech, X., J. Peral, and I. Muñoz. 2009. Predicted environmental concentrations of cocaine and benzoylecgonine in a model environmental system. *Water Res.* **43**(20):5236–5242.

European Monitoring Centre for Drugs and Drug Addiction (EMCDDA). 2007. In aquae veritas? First European meeting on drugs and their metabolites in waste water. European Monitoring Centre for Drugs and Drug Addiction, Lisbon, Portugal; http://www.emcdda.europa.eu/html.cfm/index31432EN.html.

European Monitoring Centre for Drugs and Drug Addiction (EMCDDA). 2009a. Classification of controlled drugs. European Monitoring Centre for Drugs and Drug Addiction, Lisbon, Portugal; http://eldd.emcdda.europa.eu/html.cfm/index5622EN.html.

European Monitoring Centre for Drugs and Drug Addiction (EMCDDA). 2009b. Illicit consumption of drugs and the law - Situation in the EU Member States. European Monitoring Centre for Drugs and Drug Addiction, Lisbon, Portugal; http://eldd.emcdda. europa.eu/html.cfm/index5748EN.html.

Frost, N., and P. Griffiths. 2008. Assessing illicit drugs in wastewater: Potential and limitations of a new monitoring approach. Vol. Insights Series No. 9. Lisbon, Portugal: European Monitoring Centre for Drugs and Drug Addiction (EMCDDA); http://www.emcdda. europa.eu/attachements.cfm/att_65636_EN_EMCDDA-insights9-wastewater.pdf.

Gagne, F., C. Blaise, M. Fournier, and P. D. Hansen. 2006. Effects of selected pharmaceutical products on phagocytic activity in Elliptio complanata mussels. *Comp. Biochem. Physiol. C Toxicol. Pharmacol.* **143**(2):179–186.

Glassmeyer, S. T., E. K. Hinchey, S. E. Boehme, C. G. Daughton, I. S. Ruhoy, O. Conerly, R. L. Daniels, L. Lauer, M. McCarthy, T. G. Nettesheim, K. Sykes, and V. G. Thompson. 2009. Disposal practices for unwanted residential medications in the United States. *Environ. Int.* **35**(3):566–572.

Goodman, R. A., J. W. Munson, K. Dammers, Z. Lazzarini, and J. P. Barkley. 2003. Forensic epidemiology: Law at the intersection of public health and criminal investigations. *J. Law Med. Ethics* **31**(4):684–700.

Greenwald, G. 2009. *Drug Decriminalization in Portugal: Lessons for Creating Fair and Successful Drug Policies*. Washington, DC: Cato Institute; http://www.cato.org/pubs/ wtpapers/greenwald_whitepaper.pdf.

Hagerman E. 2008. Your sewer on drugs. *Popular Sci.* Posted 02.21.2008; http://www. popsci.com/scitech/article/2008-02/your-sewer-drugs.

Hannigan, M. P., G. R. Cass, B. W. Penman, C. L. Crespi, A. L. Lafleur, W. F. Busby, W. G. Thilly, and B. R. T. Simoneit. 1998. Bioassay-directed chemical analysis of Los Angeles airborne particulate matter using a human cell mutagenicity assay. *Environ. Sci. Technol.* **32**(22):3502–3514.

Huerta-Fontela, M., M. T. Galceran, J. Martin-Alonso, and F. Ventura. 2008a. Occurrence of psychoactive stimulatory drugs in wastewaters in north-eastern Spain. *Sci. Total Environ.* **397**(1–3):31–40.

Huerta-Fontela, M., M. T. Galceran, and F. Ventura. 2008b. Stimulatory drugs of abuse in surface waters and their removal in a conventional drinking water treatment plant. *Environ. Sci. Technol.* **42**(18):6809–6816.

Huerta-Fontela, M., M. T. Galceran, and F. Ventura. 2010. Illicit drugs in the urban water cycle. In *Xenobiotics in the Urban Water Cycle*. Netherlands: Springer, Chapter 3, pp. 51–71; http://dx.doi.org/10.1007/978-90-481-3509-7_3.

INCB. 2009. International Narcotics Control Board: Narcotic drugs, psychotropic substances, and precursors. International Narcotics Control Board. Vienna, Austria; http://www.incb.org/.

Janusz, A., K. P. Kirkbride, T. L. Scott, R. Naidu, M. V. Perkins, and M. Megharaj. 2003. Microbial degradation of illicit drugs, their precursors, and manufacturing by-products: Implications for clandestine drug laboratory investigation and environmental assessment. *Forensic Sci. Int.* **134**(1):62–71.

Jones-Lepp, T. L., D. A. Alvarez, J. D. Petty, and J. N. Huckins. 2004. Polar organic chemical integrative sampling and liquid chromatography–electrospray/ion-trap mass spectrometry for assessing selected prescription and illicit drugs in treated sewage effluents. *Arch. Environ. Contam. Toxicol.* **47**(4):427–439.

Jones-Lepp T. L., R. Stevens. 2007. Pharmaceuticals and personal care products in biosolids/sewage sludge: The interface between analytical chemistry and regulation. *Anal. Bioanal. Chem.* **387**(4):1173–1183.

Kaleta, A., M. Ferdig, and W. S. Buchberger. 2006. Semiquantitative determination of residues of amphetamine in sewage sludge samples. *J. Sep. Sci.* **29**(11):1662–1666.

Kasprzyk-Hordern, B., R. M. Dinsdale, and A. J. Guwy. 2009a. Illicit drugs and pharmaceuticals in the environment – Forensic applications of environmental data, Part 1: Estimation of the usage of drugs in local communities. *Environ. Pollut.* **157**(6):1773–1777.

Kasprzyk-Hordern, B., R. M. Dinsdale, and A. J. Guwy. 2009b. The removal of pharmaceuticals, personal care products, endocrine disruptors and illicit drugs during wastewater treatment and its impact on the quality of receiving waters. *Water Res.* **43**(2):363–380.

Khan, S. J. 2002. Occurrence, behaviour and fate of pharmaceutical residues in sewage treatment. Doctoral Dissertation, University of New South Wales, New South Wales, Australia, 2002. 383 pp.

Kim, E., J. Lee, H. Choi, E. Han, Y. Park, H. Choi, and H. Chung. 2008. Comparison of methamphetamine concentrations in oral fluid, urine and hair of twelve drug abusers using solid-phase extraction and GC-MS. *Ann. Toxicol. Anal.* **20**(3):145–153.

Leonard, J. P., and X. Yongli. 2008. Recent changes in drug poisoning mortality in the United States by urban-rural status and by drug type. *Pharmacoepidemiol. Drug Saf.* **17**(10):997–1005.

Loue, S. 2010. Forensic Epidemiology: Integrating Public Health and Law Enforcement. Jones and Bartlett, Boston, MA; http://www.jbpub.com/catalog/9780763738495/.

Mari, F., L. Politi, A. Biggeri, G. Accetta, C. Trignano, M. Di Padua, and E. Bertol. 2009. Cocaine and heroin in waste water plants: A 1-year study in the city of Florence, Italy. *Forensic Sci. Int.* **189**(1–3):88–92.

Newby, N. C., P. C. Mendonça, K. Gamperl, and E. D. Stevens. 2006. Pharmacokinetics of morphine in fish: Winter flounder (Pseudopleuronectes americanus) and seawater-acclimated rainbow trout (Oncorhynchus mykiss). *Comp. Biochem. Physiol. C Toxicol. Pharmacol.* **143**(3):275–283.

NIDA. 2008. Monitoring the future survey. National Institute on Drug Abuse (NIDA), National Institutes of Health (NIH); http://www.nida.nih.gov/Drugpages/MTF.html.

NIDA. 2009. Drugs of Abuse Information. National Institute on Drug Abuse (NIDA), National Institutes of Health (NIH); http://www.drugabuse.gov/drugpages/.

Nieto, A., M. Peschka, F. Borrull, E. Pocurull, R. M. Marcé, and T. P. Knepper. 2010. Phosphodiesterase type V inhibitors: Occurrence and fate in wastewater and sewage sludge. *Water Res.* **44**(5):1607–1615.

Nutt, D. J. 2009. Equasy – An overlooked addiction with implications for the current debate on drug harms. *J. Psychopharmacol.* **23**(1):3–5.

Office of National Drug Control Policy (ONDCP). 2009. Federal Drug Data Sources. http://www.whitehousedrugpolicy.gov/DrugFact/sources.html.

Pichini, S., M. Pujadas, E. Marchei, M. Pellegrini, J. Fiz, R. Pacifici, P. Zuccaro, M. Farré, and R. de la Torre. 2008. Liquid chromatography-atmospheric pressure ionization electrospray mass spectrometry determination of "hallucinogenic designer drugs" in urine of consumers. *J. Pharm. Biomed. Anal.* **47**(2):335–342.

Poon, W. T., Y. H. Lam, C. K. Lai, A. Y. Chan, and T. W. Mak. 2007. Analogues of erectile dysfunction drugs: an under-recognised threat. *Hong Kong Med. J.* **13**(5):359–363.

Postigo, C., M. J. Lopez de Alda, and D. Barceló. 2008. Analysis of drugs of abuse and their human metabolites in water by LC-MS2. *Trends Anal. Chem.* **27**(11):1053–1069.

Postigo C., M. J. Lopez de Alda, M. Viana, X. Querol, A. Alastuey, B. Artiñano, D. Barceló. 2009. Determination of drugs of abuse in airborne particles by pressurized liquid extraction and liquid chromatography-electrospray-tandem mass spectrometry. *Anal. Chem.* **81**(11):4382–4388.

Postigo, C., M. J. López de Alda, and D. Barceló. 2010. Drugs of abuse and their metabolites in the Ebro River basin: Occurrence in sewage and surface water, sewage treatment plants removal efficiency, and collective drug usage estimation. *Environ. Int.* **36**(1):75–84.

Sleeman, R., F. Burton, J. Carter, D. Roberts, and P. Hulmston. 2000. Drugs on money. *Anal. Chem.* **72**(11):397A–403A.

Smith, S. W. 2009. Chiral toxicology: It's the same thing…only different. *Toxicol. Sci.* **110**(1):4–30.

Sörgel, F. 2006. High cocaine use in Europe and US proven – Stunning data for European Countries: First ever comparative multi-country study of cocaine use by a new measurement technique. Nürnberg-Heroldsberg, Germany: Institute for Biomedical and Pharmaceutical Research (IBMP); http://www.sharedresponsibility.gov.co/en/download/drug_consumption/IMDB_Cocaine_River_Study_2006.pdf.

Stein, K., M. Ramil, G. Fink, M. Sander, and T. A. Ternes. 2008. Analysis and sorption of psychoactive drugs onto sediment. *Environ. Sci. Technol.* **42**(17):6415–6423.

Sussman, S., and S. L. Ames. 2008. Concepts and classes of drugs. In *Drug Abuse: Concepts, Prevention, and Cessation*, edited by S. Sussman and S. L. Ames. Cambridge University Press, Cambridge; pp. 1, 3–56.

United Nations Office on Drugs and Crime (UNODC). 2007. World Drug Report: Section 4-Methodology. Vienna, Austria; http://www.unodc.org/pdf/research/wdr07/WDR_2007_4.0_methodology.pdf. p. 272–274.

United Nations Office on Drugs and Crime. (UNODC). 2009a. Information about drugs. Vienna, Austria; https://www.unodc.org/unodc/en/illicit-drugs/definitions/index.html.

United Nations Office on Drugs and Crime (UNODC). 2009b. World Drug Report - Global Illicit Drug Trends. Vienna, Austria; http://www.unodc.org/unodc/data-and-analysis/WDR.html.

US Drug Enforcement Administration (USDEA). 2008. National Forensic Laboratory Information System (NFLIS) Year 2008 Annual Report: National Forensic Laboratory System, Office of Diversion Control; https://www.nflis.deadiversion.usdoj.gov/Reports/NFLIS2008AR.pdf. 32 p.

US Environmental Protection Agency (USEPA). 2009a. Pharmaceuticals and Personal Care Products (PPCPs): Relevant literature (a comprehensive database of literature references compiled by Daughton CG, Scuderi MST; first implemented 19 February 2008); Las Vegas, NV; 2009a. http://www.epa.gov/ppcp/lit.html.

US Environmental Protection Agency, (USEPA). 2009b. U.S. EPA voluntary guidelines for methamphetamine laboratory cleanup. Office of Solid Waste and Emergency Response; 2009b.

van Nuijs, A. L. N., B. Pecceu, L. Theunis, N. Dubois, C. Charlier, P. G. Jorens, L. Bervoets, R. Blust, H. Meulemans, H. Neels, and A. Covaci. 2009a. Can cocaine use be evaluated through analysis of wastewater? A nation-wide approach conducted in Belgium. *Addiction* **104**(5):734–741.

van Nuijs, A. L. N., B. Pecceu, L. Theunis, N. Dubois, C. Charlier, P. G. Jorens, L. Bervoets, R. Blust, H. Neels, and A. Covaci. 2009b. Cocaine and metabolites in waste and surface water across Belgium. *Environ. Pollut.* **157**(1):123–129.

Venhuis, B. J., and D. de Kaste. 2008. Sildenafil analogs used for adulterating marihuana. *Forensic Sci. Int.* **182**(1–3):e23–e24.

Wick, A., G. Fink, A. Joss, H. Siegrist, and T. Ternes. 2009. Fate of beta blockers and psychoactive drugs in conventional wastewater treatment. *Water Res.* **43**(4):1060–1074.

Wiergowski, M., B. Szpiech, K. Reguła, and A. Tyburska. 2009. Municipal sewage as a source of current information on psychoactive substances used in urban communities. *Prob. Foren. Sci.* **79**(79):327–337.

Zuccato, E., and S. Castiglioni. 2009. Illicit drugs in the environment. Philos. Trans. *Royal Soc. A Math. Phys. Eng. Sci.* **367**(1904): 3965–3978.

Zuccato, E., C. Chiabrando, S. Castiglioni, D. Calamari, R. Bagnati, S. Schiarea, and R. Fanelli. 2005. Cocaine in surface waters: New evidence-based tool to monitor community drug abuse. *Environ. Health A Global Access Sci. Source* (14):7.

Zuccato, E., S. Castiglioni, R. Bagnati, C. Chiabrando, P. Grassi, and R. Fanelli. 2008a. Illicit drugs, a novel group of environmental contaminants. *Water Res.* **42**(4–5):961–968.

Zuccato, E., C. Chiabrando, S. Castiglioni, R. Bagnati, and R. Fanelli. 2008b. Estimating community drug abuse by wastewater analysis. *Environ. Health Perspect.* **116**(8):1027–1032.

THE PHYSIOLOGY OF ILLICIT DRUGS

CHAPTER 2

METABOLISM AND EXCRETION OF ILLICIT DRUGS IN HUMANS

MANUELA MELIS, SARA CASTIGLIONI, and ETTORE ZUCCATO

2.1 INTRODUCTION

Drug abuse is a worldwide phenomenon. According to official statistics, between 172 and 250 million people have used illicit drugs at least once in 2007 (UNODC, 2009). However, estimating the real number of subjects who use illicit drugs is difficult. The prevalence of drug abuse is currently estimated integrating both direct approaches, mainly based on population surveys and consumer interviews, and indirect approaches based on drug-related indicators (that is, drug-related morbidity and mortality or treatment demand). The former are often unreliable, since they are related to the willingness of subjects to report their habits related to an illegal and usually stigmatized behavior, that is, the use of illicit substances. On the other hand, indirect estimates, do not refer to the general population, but mainly to heavy drug users. Moreover, these methods are time-consuming, complex, and require considerable resources.

A new methodology, which is described in this book (see Chapter 16), was recently developed to estimate drug consumption in a population by measuring the residues of the drugs conveyed in wastewater after consumers' excretion. This method has been applied in several communities, providing evidence-based estimates of drug consumption, which were in line with national profiles of drug use based on annual prevalence data. This novel method was proposed to complement official data giving a more complete picture of this issue (Zuccato et al., 2008) and its potential was later considered and also discussed by the European Monitoring Centre for Drugs and Drug Addiction (EMCDDA, 2008). Estimates of drug consumption obtained by wastewater analysis are based on direct measurement of the residues of drugs, excreted unchanged in urine (parent drug) or as metabolites and conveyed with the wastewater to the treatment plants. For instance, cocaine consumption has been estimated by measuring

Illicit Drugs in the Environment: Occurrence, Analysis, and Fate Using Mass Spectrometry, Edited by Sara Castiglioni, Ettore Zuccato, and Roberto Fanelli
Copyright © 2011 John Wiley & Sons, Inc.

its major urinary metabolite, benzoylecgonine, and the same was true for heroin and cannabis, which were estimated by analyzing their metabolites; amphetamines were back-calculated using levels of the parent compounds. Extensive knowledge of human metabolism of illicit drugs is, therefore, essential to apply this new approach. In this chapter, we will review the available knowledge on human metabolism of the major illicit drugs. It will focus on identifying the target metabolites, which can be used to estimate drug consumption in the population using wastewater analysis.

2.2 COCAINE

Cocaine is an alkaloid extracted from the *Erythroxylon coca* bush, which grows in the Andes Mountains, Mexico, Indonesia, and the West Indies. In particular, the bush's leaves show the highest concentrations of cocaine, approximately 2% by weight. In the 1800s, cocaine was first used as an anesthetic and analgesic drug. However, it also caused behavioral responses, such as euphoria, excitation, agitation, and addictive behavior. During the 1980s, cocaine became one of the most abused illicit drugs. The UNODC estimated that in 2007 the annual prevalence of cocaine use worldwide ranged from 0.4% to 0.5%, equivalent to 16 to 21 million people aged 15−64 years (UNODC, 2009). Recent estimates suggest that the largest market is North America, followed by western and central Europe and South America.

Cocaine is used in two chemical forms: cocaine hydrochloride, usually snorted (the dose varies between 20 and 100 mg, but can be much higher in heavy users) or injected, and cocaine as free base (crack), usually smoked (Goldstein et al., 2009). An average dose of cocaine (Table 2.2) was reported as 100 mg (UNODC, 2004).

The human metabolic pattern of cocaine, including all the main urinary and pyrolytic metabolites, is summarized in Fig. 2.1 and Table 2.1.

After ingestion, cocaine is primarily metabolized by three esterases: pseudocholinesterase-2, human carboxylesterase-1 (hCE-1), and human carboxylesterase-2 (hCE-2) (Maurer et al., 2006). Cocaine is hydrolyzed in liver mainly by hCE-1 to benzoylecgonine, the primary metabolite excreted in urine, or in blood or plasma by pseudocholinesterase-2 and hCE-2 to ecgonine methyl ester. In urine, benzoylecgonine and ecgonine methyl ester represent, respectively, 45% and 40% of the administered dose (Baselt, 2004). Moreover, cocaine is excreted as the unchanged drug in urine for 1% to 9% of the administered dose, depending on the pH of the urine, and is detectable in this form up to 24−36 h (Goldstein et al., 2009). The elimination half-life of cocaine and benzoylecgonine from plasma are 1 and 6 h, respectively (Goldstein et al., 2009); benzoylecgonine is found in urine for 1 to 2 days after an IV (intravenous) injection of a 20-mg dose, while after a higher dose, taken intranasally, is detectable up to 2−3 days (Verstraete, 2004). Norcocaine, which is formed by hepatic *N*-demethylation of cocaine, accounts to no more than 5% of the adsorbed dose (Goldstein et al., 2009) in urine and has been detected in plasma and urine in trace amounts (Baselt, 2004). The oxidative metabolism of cocaine to norcocaine consists of two different pathways. In the first one, cocaine can be directly *N*-demethylated to norcocaine by the enzyme CYP (cytocrome P450) 3A4. In the second, it is first

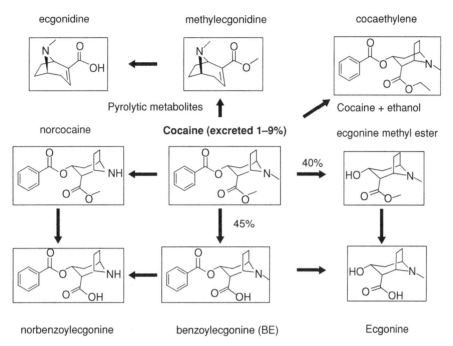

FIGURE 2.1 Metabolism of cocaine in man including all the main urinary and pyrolytic metabolites.

oxidized to *N*-oxide by a flavin-containing monooxygenase, followed by a CYP-catalyzed *N*-demethylation (Maurer et al., 2006). Norcocaine is the only metabolite that shows an activity comparable to cocaine (Hawks et al., 1974). In 1977, Williams and co-workers demonstrated that cocaine and norcocaine are equipotent inhibitors of the uptake of dopamine in the rat brain. One to three percent% of urinary metabolites are the *N*-demethylation products ecgonine, norbenzoylecgonine, and norecgonine (Goldstein et al., 2009). Fecal excretion represents a minor route of elimination of cocaine and its metabolites.

Cocaine is often used together with other substances, including alcohol. It is estimated that between 50% to 90% of cocaine users simultaneously ingest cocaine and ethanol (Goldstein et al., 2009; Farooq et al., 2009). The transesterification of these two substances catalyzed by hCE-1 produces the active metabolite cocaethylene (Maurer et al., 2006). It accounts for 0.7% of a cocaine dose in 24-hurine, with a plasma elimination half-life of about 150 min (Baselt, 2004; Goldstein et al., 2009). Cocaethylene is less active than cocaine, but has a much longer duration of action and can be neuro- and cardiotoxic (Baselt, 2004; Goldstein et al., 2009). It has a biological activity on dopaminergic neurons similar to cocaine and acts on the same receptors. Moreover, it is at least equipotent to cocaine in stimulating motor activity (Dean et al., 1992; De La Torre et al., 1995; Jatlow et al., 1996). According to Harris et al. (2003), about one-sixth of a cocaine dose is transformed into cocaethylene

TABLE 2.1 Metabolism, Pharmacokinetic, and Excretion of the Main Illicit Drugs in Humans

Drug	Principal Metabolites	Halflife (Plasma)	Percentage of a Dose excreted in Urine	Detection Time in Urine
Cocaine	• Unchanged cocaine	60–90 min[a,b]	1–9%[b]	24–36 h[b]
	• Benzoylecgonine	6 hours[a]	45%[c]	1–2 days IV, 2–3 days intranasally[a]
	• Ecgonine methyl ester		40% [c]	
	• Norcocaine		Traces[c]	
	• Ecgonine, norbenzoylecgonine, and noregonine		1–3%[b]	
	• Methyl egonidine (pyrolysis product), its metabolites ecgonidine and noregonidine	150 min[b]	0.7% 24-h Urine	
	• Cocaethylene and norcocaethylene			
Cannabinoids (THC)	• THC	18.7 h–4.1 days[h]	Traces[c]	10 h[a]
	• THC-COOH	20–50 h/ 3–13 days[a]		25 days[a]
	• 11-OH-THC		2% as a conjugate[c]	
Methamphetamine	pH 6–8:			
	• Unchanged drug	11.4 h IV and 10.7 h intranasally or smocked[i,j]	22%[d]	46–144 h[a]
	• p-Hydroxymethamphetamine		15%[d]	
	• Amphetamine		4–7%[d]	
	• p-Hydroxyamphetamine		1%[d]	
	• Phenylacetone and hydroxy products of amphetamine and p-hydroxyamphetamine (secondary products)			

Drug	Compound / Metabolite	Half-life	Excretion	Detection time
Amphetamine	pH <5:			
	• Unchanged drug		76%[e]	
	Normal condition:			
	• Unchanged drug			
	• Amphetamine			
	• Unchanged drug	7–34 h[a]	43% 24-h Urine[c]	
	• Unchanged drug	7–8 h[a]	4–7% 24-h Urine[c]	
			30-74%[d]	1–3 days[a]
MDMA	• 3-Methoxy-4-hydroxymethamphetamine		26% in 24-h Urine[c]	
	• 3,4-Di-OH-methamphetamine		65% in 3-d Urine[c]	
	• MDA		23% in 24-h Urine[c]	
MDEA	• 3-Methoxy-4-hydroxyamphetamine		20% in 24-h Urine[c]	
	• Unchanged drug		1% in 24-h Urine, 7% in 3-day Urine[c]	
	• MDA		0.9% in 24-h Urine[c]	
	• HMEA		19% in 32-h Urine[c]	
MBDB	• Unchanged drug		28% in 32-h Urine[c]	
	• BDB		32% in 32-h Urine[c]	
MDA				
Codeine	• Unchanged drug		5–17% in 24-h Urine[c]	
	• Conjugated codeine		32–46% 24-h Urine[c]	
	• Norcodeine		Traces 24-h Urine[c]	
	• Conjugated norcodeine		10–21% 24-h Urine[c]	
	• Hydrocodone		11% of Codeine concentration[g]	
	• Morphine		Traces 24-h Urine[c]	
	• Conjugated morphine		5–13% 24-h Urine[c]	

(*Continued*)

TABLE 2.1 (*Continued*)

Drug	Principal Metabolites	Halflife (Plasma)	Percentage of a Dose excreted in Urine	Detection Time in Urine
Morphine	• Unchanged drug	2–3 h[a]	10%[c]	
	• Normorphine free		1%[c]	
	• Normorphine conjugated		4%[c]	
	• M3G		75%[c]	
	• Morphine-3,6-diglucuronide, morphine-3-ethereal sulfate, M6G		Traces[c]	
	• Unchanged drug	2–7 min[a]	0.1% in 40-h Urine[c]	4.5 h (dose of 12 mg IV)[a]
	• 6-Acetylmorphine	6–25 min[a]	1.3% in 40-h Urine[c]	35.3 h (dose of 12 mg IV)[a]
Heroin	• Morphine	2–3 h[a]	4.2% + 38.3% conjugated in 40-h urine[c]	
	• M3G			
	• M6G			
	• 6-Acetylcodeine (manufacturing impurity of heroine)	237 min[l]	Trace[c]	
	• Secondary metabolites of morphine		3–5%, 5–50% in 24-h urine[c]	
	• Unchanged drug			
Methadone	• EDDP	24–36 h[k]	5%, 3–25% in 24-h urine[c]	
	• EMDP		<1% in 24-h urine[c]	
	• α-(3S6S)-Methadol and α-(3S6S)-N-desmethylmethadol			

[a]Verstraete 2004; [b]Goldstein et al., 2009; [c]Baselt 2004; [d]Schepers et al., 2003; [e]Oyler et al., 2002; [f]Smith, 2009; [g]Smith, 2009; [h]Cone and Huestis, 1993; [i]Kim et al., 2004; [j]Cruickshank and Dyer, 2009; [k]Gonzalez et al., 2002; [l]Musshoff et al., 2009.

during human metabolism; the rate of biotransformation of cocaine to cocaethylene increases proportionally to the dosage. Farrè et al. (1997) reported that the amount of ethanol involved in cocaethylene synthesis was not proportional to the level of alcohol in plasma and that the concentration of cocaethylene increased slowly. Similar to cocaine, cocaethylene is *N*-demethylated to norcocaethylene and the rate of this reaction is threefold greater than the rate of the cocaine to norcocaine reaction (Farrè et al., 1997; Harris et al., 2003).

When cocaine is smoked as crack, it is typically put on a screen in a pipe and a flame is drawn across it as the smoker inhales. The pyrolytic degradation starts at 170°C and, since cocaine has a melting point of 98°C, it travels away from the flame and cools. The vapor then condenses, forming smoke, which is an aerosol composed of cocaine base droplets and associated pyrolysis products (Wood et al., 1995). The main pyrolytic product of cocaine is methylecgonidine (anhydroecgonine methyl ester) and it is consumed together with cocaine. The amount formed during pyrolysis depends on the pyrolytic conditions and composition of cocaine (Maurer et al., 2006). Methylecgonidine has three main metabolites: ecgonidine (anhydroecgonine), its *N*-demethylation product, norecgonidine (noranhydroecgonine), and ethylecgonidine (anhydroecgonine ethyl ester). The presence of these metabolites in urine might be a potential marker indicating the smoking route of administration of cocaine. An experiment comparing various routes of administration showed that subjects receiving cocaine by smoking excrete methylecgonidine in the urine at about one-half the amount of cocaine (Peterson et al., 1995). On the contrary, those subjects who were administered cocaine by intravenous or intranasal routes had much lower amounts of methylecgonidine than cocaine in their urine (Peterson et al., 1995).

Crack smocking is generally preferred to intravenous administration, although the rapid penetration into bloodstream is similar to intravenous administration, because of the easy and convenient smoking route and the avoidance of needles and associated diseases. On the other hand, intranasal administration produces lower blood concentrations, but over a more prolonged time, because of a slower absorption rates. The bioavailability of smocked cocaine is high (approximately 70%) considering the possible loss of cocaine during smoking and its pyrolysis to methylecgonidine (Cone, 1998). Intranasal administration also results in high bioavailability (approximately 94%), but peak concentration are considerably lower than from IV or smoked administration (Cone, 1998).

Cocaine and its main metabolite benzoylecgonine were initially selected as target residues to be measured in environmental media (waste- and surface water). Later, several other metabolites were investigated, such as ecgonine methyl ester, norcocaine, norbenzoylecgonine, cocaethylene, and the pyrolytic metabolites, methylecgonidine and ecgonidine. The presence of these substances in urban wastewater can be used to measure cocaine consumption and to identify consumption routes (presence of pyrolytic metabolites when cocaine is smoked as crack) and different consumption patterns (consumption of cocaine together with alcohol, indicated by the presence of cocaethylene).

2.2.1 Cannabinoids

The United Nation Office of Drug and Crime (UNODC) estimated that between 143 and 190 million people globally used cannabis at least once in 2007 (UNODC, 2009). The most relevant species of cannabis are *Cannabis sativa* (the largest variety), *C. indica,* and *C. ruderalis.* There are three main types of cannabis products: herb (marijuana), resin (hashish), and oil (hashish oil). The plant of *Cannabis* contains more than 460 chemicals, among which more than 60 are grouped under the name of cannabinoids. Δ^9-Tetrahydrocannabinol (THC) is the primary hallucinogenic constituent of *Cannabis* mainly responsible for its psychotropic effects (Ben Amar, 2006). THC is a lipophilic compound and can accumulate in fat tissues as a function of the amount and frequency of smoking and cannabis potency (Cone and Huestis, 1993). The redistribution of THC from fat tissues to blood is the rate-limiting step in its metabolism.

THC is administered either orally or by smocking in approximate dose of 5 to 20 mg (Table 2, Baselt, 2004) or higher. Only about 22% of the THC contained in a cigarette may be recovered from the smoke; however, during the smocking process, a portion of inactive constituent, such as tetrahydrocannabinol carboxylic acid and cannabidiol, can be converted to THC (Baselt, 2004).

The human metabolism of THC, including all its main metabolites, is reported in Fig. 2.2 and Table 2.1.

After consumption, THC is rapidly metabolized by hepatic metabolism (cytocrome P450-CYP enzymes) to 11-hydroxy-THC (11-OH-THC), a pharmacologically active metabolite, the inactive metabolite 11-nor-9-carboxy-Δ^9-tetrahydrocannabinol (THC-COOH), and to various other cannabinoids (Huestis and Cone, 1998). 11-OH-THC is formed by CYP2C9 and further oxidized to the intermediate aldehyde

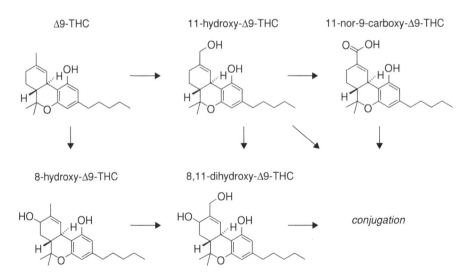

FIGURE 2.2   Metabolism of Δ^9-tetrahydrocannabinol (THC) in man.

11-oxo-THC (Fig. 2.2), followed by oxidation to THC-COOH catalyzed by a microsomal aldehyde oxygenase of the CYP2C subfamily (Maurer et al., 2006). 11-OH-THC may contribute to the psychoactive effects of THC when cannabis is administered orally, as a result of the amounts of 11-OH-THC produced by first-passage metabolism (Cone and Huestis, 1993). About 70% of a dose of THC is excreted within 72 h in feces and urine, mainly as metabolites. THC-COOH is the primary metabolite excreted in urine, 11-OH-THC (as a conjugate) accounts for only 2% of a dose, and unchanged THC is present only in traces in urine (Baselt, 2004). Most of the THC-COOH is conjugated and excreted as the water-soluble glucuronic acid (Huestis and Cone, 1998). However, after excretion, it is readily hydrolyzed back in wastewater to the free acid by the β-glucuronidases of fecal bacteria, as shown for glucuronide conjugates of estrogens in untreated wastewater (D'Ascenzo et al., 2003) and during wastewater treatment (Ternes et al., 1999). Smoking cannabis results in the rapid appearance of THC in blood. A THC concentration of 7 and 18.1 ng/mL was observed after a first inhalation of smoke from 1.75% and 3.55% THC cannabis cigarettes, respectively (Cone, 1998).

THC concentrations in blood reached a peak generally before the end of the smoking period and the bioavailability of THC for smoking cannabis was from 14% to 27%; following oral ingestion bioavailabilty was from 6% to 19% (Cone and Huestis, 1993). Estimates of the plasma half-life of THC ranged from 18.7 h to 4.1 days and the half-life of THC among occasional users was twice as long as that found among experienced cannabis users (Cone and Huestis, 1993). During the smoking period, 11-OH-THC concentration remained low relative to THC and its maximum concentration was achieved 15 min after initiation of smocking. THC-COOH concentration increased slowly and remained elevated for an extended period of time (Cone, 1998); peak THC-COOH concentrations occurred between 5.6 and 28 h after smoking (Huestis and Cone, 1998). Generally, plasma peak THC concentrations are approximately 3 times greater than peak THC-COOH concentration and 20 times greater than peak 11-OH-THC concentration (Cone, 1998).

After having smoked cannabis, THC-COOH presents a long plasma half-life, about 20 to 50 h in occasional users and 3–13 days in regular users (Verstraete, 2004). In urine, THC is detectable for 10 h [limit of detection (LOD) 10 ng/mL] and THC-COOH can be detected for 25 days (LOD 5 ng/mL) (Verstraete, 2004). Approximately 15% −30% of a THC dose has been reported to be eliminated in urine and 30%− 65% in feces as 11-OH-THC and THC-COOH (Huestis and Cone, 1998). Cannabis herb comprises the dried and crushed flower heads and surrounding leaves and often contains up to 5% THC (UNODC, 2009). Cannabis resin can contain up to 20% THC. However, the most potent form of cannabis is cannabis oil, obtained from the concentrated resin extract. It may contain more than 60% THC (UNODC, 2009). Acute cannabis effects are highly variable among individuals, also depending on their health conditions. Cannabis often induces mild euphoria, talkativeness, intensification of sensory experiences, difficulty in concentrating, altered time perception, relaxation, and drowsiness (Ben Amar, 2006; Schierenbeck et al., 2008). Cannabis also induces an increased heart rate, a lowering of blood pressure, because of vasodilatation, appetite stimulation, dry mouth, and dizziness (Baker et al., 2003).

THC is also available as a prescription drug, generally as the synthetic form, dronabinol. Some studies reported that treatment with dronabinol leads to improvement in both appetite ratings and food consumption levels (Kirkham, 2005). Consequently, it is used in stimulating appetite and preventing weight loss in cancer and AIDS patients. However, its use in these diseases is still debated, because some authors reported immunosuppressive properties of cannabinoids (Ben Amar, 2006). Nabilone, another synthetic analog of THC, has also shown an antiemetic effect. It was used for the treatment of nausea and vomiting associated with chemotherapy. Levonantradol, a synthetic cannabinoid administered intravenously, has also shown antiemetic property, but its adverse effects on central nervous system has limited its use (Hutcheon et al., 1983; Ben Amar, 2006). Cannabis use as illicit drug seems to be increasing in several countries in Latin America and Africa, whereas in the established markets of North America and western Europe the levels of use are slightly declining, particularly among young adults (UNODC, 2009). Among users, it is possible to identify: occasional users (45%), who share cannabis cigarettes four times per year on average (0.6 g/year); regular users (41%) who use more advanced inhaling techniques with higher frequency, on average, 100 times per year (15 g/ year); daily users (9%) who use one to four cannabis cigarettes per day (320 g/year); and chronic users (4%) who smoke ten cigarettes per day (1825 g/year) (UNODC, 2009).

The main urinary metabolite of THC, THC-COOH, was selected as a target compound to estimate cannabis consumption by wastewater analysis. OH-THC and THC were also used, but because of their lipophilic characteristics, they were not easily found in wastewater.

2.2.2 Amphetamine-Type Stimulants (ATS)

ATS include the amphetamine-group substances, consisting of methamphetamine and amphetamine, and the ecstasy-group substances, consisting mainly of 3,4-methylene-dioxymethamphetamine (ecstasy or MDMA), 3,4-methylene-dioxyamphetamine (MDA), 3,4-methylenedioxy-N-ethylamphetamine (MDEA), and N-methyl-1-(3,4 methylenedioxyphenyl)-2-butanamine (MBDB)(UNODC, 2009).

In 2007, UNODC estimated that between 230 and 640 tons of ATS were manufactured. The ecstasy group manufactured was between 72 and 137 tons. These substances can be produced anywhere at relative low cost and the locations of production can rapidly change. Worldwide, between 16 and 51 million people aged 15−64 years have used amphetamine-group substances at least once in 2007 (annual prevalence 0.4%−1.2%), whereas the number who used ecstasy-group drugs is between 12 and 24 million (annual prevalence 0.3%−0.5%) (UNODC, 2009). It is possible to identify different regional patterns of use for ATS: users primarily consume methamphetamine in east and southeast Asia; amphetamines are consumed in Europe (UNODC, 2009). In the United States, about one-half of stimulant users consume methamphetamine. Moreover, methamphetamine use levels are currently higher in the United States than Europe. This was also confirmed by wastewater analysis (Banta-Green et al., 2009).

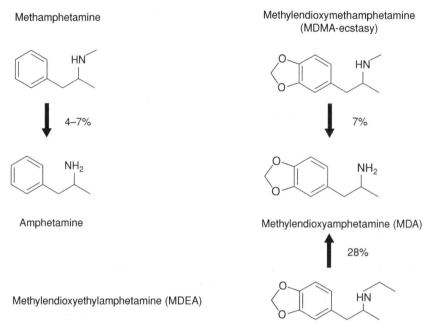

FIGURE 2.3 Metabolism of amphetamine-type stimulants (ATS) in man.

Amphetamine and related substances show symphaticomimetic and CNS stimulant activity. Amphetamines are indirect monoamine agonists and interact with the membrane transporters involved in neurotransmitter reuptake and vesicular storage systems. Therefore, they stimulate the release of norepinephrine, dopamine, and serotonin from presynaptic terminals in the CNS and at the peripheral level (De La Torre et al., 2004). Methamphetamine and the methylenedioxy derivatives (MDA, MDMA, MDEA, MBDB) can inhibit the activity of enzymes of dopamine or serotonin biosynthesis (De La Torre et al., 2004).

The main ATS substances are listed in Table 2.1 and their human metabolism is reported in Fig. 2.3.

2.2.3 Amphetamine-Group Substances

The amphetamine group is one of the most abused classes of psychotropic drugs. The usual dose of amphetamine and methamphetamine is $10-30$ mg (Table 2.2), but the doses can be also much higher, up to 2000 mg/die for amphetamine (Verstraete, 2004).

Amphetamine is a sympathomimetic phenethylamine derivative that has central stimulant activity. It is available as the *d*- or *dl*-isomeric form, having the *d*-isomeric form $3-4$ times the central activity of the *l*-form (Baselt, 2004). It has been used in the treatment of obesity, narcolepsy, and hypotension, but now it is mainly used as

TABLE 2.2   Route and Average Dose of Administrations of the Drugs

Drug	Route of Administration	Mean Dose
Cocaine	Snorted, injected, smocked	20−100 mg
Cannabinoids (THC)	Oral ingested or smocked	5−20 mg THC
		125 mg THC
Heroin	Smocked, injected, subcutaneous injection, snorted	30 mg
Methamphetamine	Smocked, snorted, oral ingested, injected	10−30 mg/die
Amphetamine	Oral ingested	10−30 mg/die
MDMA	Oral ingested	100−150 mg
MDA	Oral ingested	50−250 mg
MDEA	Oral ingested	50−250 mg

illegal substance for its stimulant activity (Baselt, 2004). Amphetamine is most often consumed orally, either as (S)-(+)-enantiomer (d-isomer) or the racemic mixture. After an oral dose, amphetamine is rapidly adsorbed, with a mean maximum plasma concentration within 4 h after ingestion (De La Torre et al., 2004). Amphetamine is oxidatively deaminated by monoamine oxidase (MAO) and this reaction is catalyzed by the cytocrome P450 (CYP) isoenzyme CYP2C (Kraemer and Maurer, 2002). Moreover, the ring hydroxylation of amphetamine and methamphetamine is mediated mainly by CYP2D6 (Kraemer and Maurer, 2002). It is normally excreted in unchanged form (30%−74% of a dose) in urine (Baselt, 2004). Davis et al. (1971) have demonstrated that the rate of decline of plasma levels of amphetamine and the rate of its excretion into urine are markedly influenced by urinary pH. This has a relevant clinical application in the treatment of overdoses, since the acidification of the urine is useful to accelerate its excretion. Amphetamine is also a metabolite of other drugs such as fenethylline, fenproporex, and methamphetamine (Baselt, 2004).

Methamphetamine, the N-methyl derivative of amphetamine, is a popular illicit drug worldwide with an annual prevalence estimated at 0.4%. Its use among adults is 14% in the Philippines, 3.2% in Australia, and 0.8% in the United States (Cruickshank and Dyer, 2009). Most synthetic procedures for illicit manufacture lead to production of relatively pure d-methamphetamine, but some methods result in d/l mixtures. The l-isomer is less potent and can be used as nasal inhaler decongestant product (Huestis and Cone, 2007). At low doses used in clinical experiments (5−30 mg), methamphetamine responses include reduced fatigue, positive mood, euphoria, accelerated heart rate and cardiovascular effects, elevated blood pressure, reduced appetite, pupil dilatation, increase of temperature, behavioral disinhibition, and short-term improvement in cognitive domains, including sustained attention and anxiety (Cruickshank and Dyer, 2009). Methamphetamine also enhances performance on some psychomotor/cognitive tasks, such as visuospatial processing, vigilance, and reaction time (Hart et al., 2008). Currently, there are only few accepted medical treatments for pharmaceutically derived d-methamphetamine, such as attention deficit disorder and short-term treatment of obesity (Huestis and Cone, 2007).

d-Methamphetamine as an illicit drug is available in different forms and can be smocked (bioavailability 67%−90% in function of smocking technique), snorted (bioavailability 79%), orally ingested (bioavailability 67%), or injected (Cruickshank and Dyer, 2009). Methamphetamine is rapidly absorbed by all routes, but the onset is faster by the IV and smocked routes. The primary site of metabolism is the liver by aromatic hydroxylation, *N*-demethylation (to form the metabolite amphetamine), and deamination (Huestis and Cone, 2007; Kim et al., 2004). Maximum methamphetamine concentrations in plasma were shown to occur at 2.7 and 2.5 h after intranasal and smocked doses. The elimination half-life was 11.4 h for an intravenous dose and 10.7 h for an intranasal and a smocked dose (Kim et al., 2004; Cruickshank and Dyer, 2009). Methamphetamine is almost entirely (70%−90%) eliminated in urine, depending on pH (Schepers et al., 2003; Cruickshank and Dyer, 2009). It is excreted mainly unchanged (43% of a dose) and partially as amphetamine (4%−7% of a dose) in the 24-h urine (Baselt, 2004). However, the percentage of excretion of methamphetamine as an unchanged drug is affected by urine pH and, when it is alkaline, only 2% of methamphetamine is excreted unchanged. When the pH is acid (pH <5) the percentage excreted as unchanged drug is up to 76% (Oyler et al., 2002). In urine at pH 6−8, 22% of a dose is eliminated as unchanged drug, 15% as *p*-hydroxymethamphetamine, 4%−7% as amphetamine, and 1% as *p*-hydroxyamphetamine (Schepers et al., 2003). Moreover, the fraction of methamphetamine excreted unchanged in urine decreases with the increased dose, which is not proportional to the percentage of drug absorbed (De La Torre et al., 2004). During phase-II biotransformation, the majority of *p*-hydroxymethamphetamine and *p*-hydroxyamphetamine is conjugated with glucuronic acid. Quantitatively minor products are phenylacetone and the hydroxylation products of amphetamine and *p*-hydroxyamphetamine (Schepers et al., 2003).

2.2.4 Ecstasy-Group Substances

This group includes the ring-substituted derivative of methamphetamine, (MDMA or ecstasy), and amphetamine, (MDA, MDEA, and MBDB). These substances are illicit recreational drugs in Europe and the United States. They are stimulatory drugs and have all been reported to produce similar central and peripheral effects in humans, but with different potency, time of onset, and duration of action (Hegadoren et al., 1999). The positive effects of MDMA are euphoria and increased energy. This drug was used as an adjunct to psychotherapy for its ability to encourage openness of emotional expression and to facilitate interpersonal communication and intimacy (Hegadoren et al., 1999). MDA induces marked sympathomimetic effects, CNS stimulatory activity, and convulsions at high doses. It was used in clinical practice for its antitussive, ataractic, and appetite suppressant properties (Hegadoren et al., 1999).

The typical doses (Table 2.2) are 100−150 mg for MDMA and 50−250 mg for MDA and MDEA (Baselt, 2004). The onset time is 30−60 min for MDA and within 30 min for MDMA and MDEA. Duration of action is longer for MDA (about 8 h) than MDMA (about 6 h) and MDEA (about 3−4 h) (Hegadoren et al., 1999). The half-life of MDMA in plasma is approximately 7 or 8 h and it is detectable in urine for more than 48 h after administration of 100 mg. For MDEA, the urinary detection time after administration of 140 mg varied between 1.4 and 2.6 days. Unchanged MBDB was excreted in urine after a single dose of 100 mg. The drug and its *N*-demethylated metabolite, BDB, were detectable in urine during the first 36 h (De La Torre et al., 2004; Verstraete, 2004).

These substances undergo predominantly two metabolic pathways: *O*-demethylenation to dihydroxy derivatives (catechol) followed by methylation and/or glucuronidation/sulfatation of one of the hydroxy groups (catalyzed by catechol-*O*-methyltransferase-COMT) and *N*-dealkylation, deamination, or oxidation of the side chain to *N*-dealkyl and deaminooxo metabolites (Maurer et al., 2000). These metabolic pathways are mainly regulated by cytocrome P450 (CYP) isoenzymes, but also by CYP-independent mechanisms (Maurer et al., 2000). Demethylenation of all these compounds is mainly catalyzed by CYP2D6/1 and CYP3A2/4 isoenzymes. *N*-Demethylation of MDMA and MBDB is catalyzed in humans and rats by CYP1A2 and only to a minor extent by the polymorphic CYP2D6/1. In addition, MBDB is *N*-demethylated by CYP3A2/4. *N*-Deethylation of MDEA is predominantly catalyzed in rats and humans by CYP3A2/4 (Maurer et al., 2000; Kraemer and Maurer, 2002). *N*-Dealkylation in humans is catalyzed mainly by CYP1A2 for MDMA and by CYP3A4 for MDEA (Maurer et al., 2000).

MDMA and MDA are both *O*-demethylenated, to 3,4- dihydroxymethamphetamine (HHMA) and 3,4-dihydroxyamphetamine (HHA), respectively (Perfetti et al., 2009). MDMA is metabolized for 80% in the liver and about 20% of the dose is excreted unaltered in urine (De La Torre et al., 2004). A single oral dose (100−125 mg) is eliminated in 24-h urine as MDMA (26%), MDA (1%), 3-methoxy-4-OH-methamphetamine (23%), 3,4-di-OH-methamphetamine (20%), and 3-methoxy-4-OH-amphetamine (0.9%). Within 3 days, urinary excretion accounted for 65% of the dose as parent drug and 7% as MDA (Baselt, 2004). MDEA, after a single oral dose, is largely eliminated in 32-h urine as parent drug (19% of the dose), MDA (28%), 4-hydroxy-3-methoxyethylamphetamine (HMEA) (32%), and trace amounts of at least eight other metabolites (Baselt, 2004). MDA is a minor metabolite of MDMA and seems to be excreted mainly unchanged (Baselt, 2004).

All the substances listed above have been measured in untreated wastewater and other environmental matrices. Amphetamine, methamphetamine, and MDMA parent compounds have been used as target residues for wastewater analysis.

2.3 OPIOIDS

The number of persons who used opiates in 2007 was between 15.2 and 21.1 million people worldwide (UNODC, 2009). More than one-half of the world's opiate users live in Asia. The highest level of use (in term of proportion of the population aged 15−64) are found along the main drug trafficking routes close to Afghanistan. Afghanistan is the world's largest producer of opium, followed by Myanmar and

Lao People's Democratic Republic. Opium poppy cultivations were also reported in Bangladesh, India, Nepal, Thailand, Vietnam, Colombia, Mexico, Peru, Guatemala, Venezuela, Baltic countries, Balkan countries, Egypt, Iraq, Lebanon, Russian Federation, Ukraine, central Asia, and Caucasus regions.

Europe has an estimated 3.4−4 million opiate users (around 0.6%−0.7% of the population aged 15−64), between 1.2−1.5 million of them live in western and central Europe and between 2.2−2.5 million in eastern and southeastern Europe. The majority of users in western Europe are estimated to be in the UK, followed by Italy, France, Germany, and Spain.

Opioids have been the mainstay of pain treatment for thousands of years. In ancient times, opium, which is derived from the juice of the opium poppy, *Papaver somniferum*, was used to fight cough and diarrhea, and to relieve pain through its euphoria-inducing properties (Nicholson, 2003). Opium contains more than 20 active alkaloid compounds, of which morphine is the most potent and was the first to be isolated in 1806 (Nicholson, 2003; Smith, 2009). *Papaver setigerum* is also able to produce morphine. Currently, opioids are used in the management of cancer and postoperative pain (Smith, 2009). Buprenorphine and methadone, are used in maintenance treatment programs for opioid dependence and opioid antagonists (such as, naloxone and naltrexone) are used to reverse the symptoms of opioid intoxication and as adjuncts in the treatment of alcohol dependence (Coller et al., 2009). Long-term use of opioids can be problematic because of the rapid development of profound tolerance to analgesic effects coupled with a slow development of tolerance to the numerous side effects that eventually limits the dose increase and analgesic efficacy (Dumas and Pollack, 2008).

These drugs exert their pharmacological effects by acting on the opioid receptor (MOP), a group of G-protein-coupled receptors. There are three types of opioid receptors, mu, delta, and kappa (Suzuki and Misawa, 1997). Opioids have quite different affinities for MOP and potency is sometimes classified as low or high (Coller et al., 2009). Side effects include respiratory depression, constipation, sedation, tolerance, nausea, vomiting, itch, dry mouth, and addiction (Thorn et al., 2009).

Opioids can be classified according to structural similarity with morphine in semisynthetic and synthetic derivatives. Structurally similar, semisynthetic morphine-like derivatives as well as structurally distinct opioids have been synthesized to search for compounds able to improve analgesic effects that minimize side effects. Semisynthetic derivatives include morphine-related agonists (hydromorphone, hydrocodone, oxycodone, oxymorphone, and codeine), and morphine-related partial agonist and antagonists (buprenorphine, naloxone, and naltrexone). Synthetic derivatives include phenylpiperidines (meperidine and loperamide), diphenylpropylamines (methadone and propoxyphene), and piperidines (fentanyl, alfentanyl, sufentanil, and remifentanil) (Dumas and Pollack, 2008).

Most opioids undergo extensive first-step metabolism in the liver before entering the systemic circulation and this step reduces the bioavailability of these drugs. All opioids are metabolized through two major enzyme systems, CYP450 during the phase-I metabolism and uridine diphosphate glucuronosyltransferases (UGT) (mainly UGT2B7) during the phase-II metabolism (Smith, 2009).

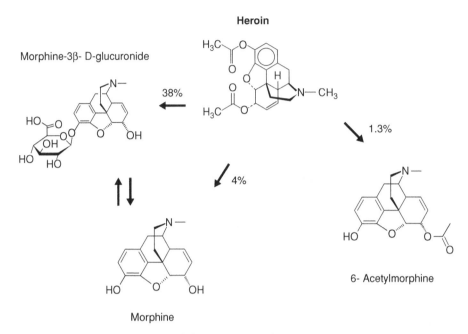

FIGURE 2.4   Metabolism of heroin in man.

2.3.1   Heroin

Heroin (3,6-diacetylmorphine or diamorphine) was originally synthesized from morphine and is regarded as one of the most dangerous drugs of abuse. The human metabolism of heroin is presented in Fig. 2.4; its main metabolites are reported in Table 2.1. It is rapidly deacetylated to 6-acetylmorphine, which is further hydrolyzed to morphine in the liver. The first reaction is catalyzed by blood esterases, while the second occurs in the liver (Baselt, 2004).

Heroin is usually administered as the hydrochloride salt by intravenous or subcutaneous injection, nasal insufflation, or inhalation (Baselt, 2004). Bioavailability of inhaled heroin, following sublimation, was estimated at 53% (Rook et al., 2006). The highest concentration of 6-acetylmorphine in plasma occurred immediately after heroin administration, but heroin and 6-acetylmorphine plasma concentrations declined rapidly and became undetectable after 10−40 min and 2−3 h, respectively. According to Rook et al. (2006) peak concentrations of heroin and 6-acetylmorphine were two- to fourfold lower after inhalation than following IV administration. After nasal insufflation, heroin was immediately absorbed through the lungs and was rapidly metabolized following absorption. Following intravenous infusion of 70 mg of heroin, approximately 45% of the dose was recovered as urinary metabolites over a 40-h period: 4.2% of morphine, 38.3% of conjugated morphine, 1.3% 6-acetylmorphine, and 0.1% unchanged (Baselt, 2004). The elimination half-life of heroin and its metabolites 6-acetylmorphine and morphine in plasma were 2−7 min,

6−25 min, and 2−3 h, respectively (Verstraete, 2004). After administration of 12 mg IV, 6-acetylmorphine was detectable in urine for 4.5 h and total morphine for 35.3 h (Verstraete, 2004).

The dose used at the beginning is approximately 10 mg, but addicted subjects can use up to 1 or 2 g, with an estimated average dose of 30 mg (Table 2.2). Heroin has very little affinity for the opiate receptors in brain and exerts its effects only after metabolism to 6-acetylmorphine and morphine catalyzed in the liver by human carboxylesterase (hCE-1 and hCE-2), in serum by pseudocholinesterase, and also nonenzymatically (Maurer et al., 2006).

6-Acetylmorphine is considered to be responsible for the immediate euphoric effect after heroin administration. The other metabolites morphine and morphine-6β-glucuronide, that remain in the human body longer, are related to feelings of well-being and calmness (Rook et al., 2006). Morphine-3β-glucuronide, the major metabolite of heroin, has no opioid activity but seems to be associated with neurotoxic effects (Rook et al., 2006). A small portion of 6-acetylmorphine is glucuronidated to 6-acetylmorphine-3-glucuronide, which is not active (Coller et al., 2009).

The use of heroin as illicit drug is normally identified by detecting its main metabolites in human body fluids. Therefore, analysis in wastewater also focused on morphine, a metabolic residue common to heroin, codeine, and morphine (Baselt, 2004), and on 6-acetylmorphine (1%−3% of a dose), a specific metabolite of heroin (Baselt, 2004). Heroin and morphine are also partially excreted as glucuronide conjugates (Baselt, 2004), which are, however, readily hydrolyzed back in wastewater to morphine by β-glucuronidases of fecal bacteria, as reported for estrogens in untreated wastewater (D'Ascenzo et al., 2003) and during wastewater treatment (Ternes et al., 1999). To calculate the amount of morphine coming from heroin consumption, we subtracted the amounts derived from the use of morphine in therapeutic treatments. Morphine is converted to metabolites up to 87% of a dose, with 75% consisting of morphine 3β-glucuronide (Baselt, 2004). Morphine is mainly metabolized by UGT2B7 to the inactive metabolite morphine-3β-glucuronide (approximately 60%), and to a lesser extent to morphine-6β-glucuronide (5%−10%) (Maurer et al., 2006). Morphine is further N-demethylated to normorphine by hepatic CYP3A4 and to a lesser extent by CYP2C8 (Maurer et al., 2006). A small proportion of morphine is metabolized to hydromorphone that can be found in the urine at low levels (less than 2.5% of a morphine dose) (Smith, 2009).

2.3.2 Codeine

Morphine is also a metabolite of codeine, a narcotic analgesic that is mostly used therapeutically as a potent antitussive (Baselt, 2004). Codeine is found naturally in the poppy plant *Papaver somniferum*, but for commercial use it is usually synthesized by 3-*O*-methylation of morphine, which is most abundant in nature (Baselt, 2004). Codeine is normally used in single doses of 15 to 60 mg given orally or by subcutaneous injection and is excreted in 24-h urine for 5%−17% as free codeine, 32%−46% conjugated codeine, 10%−21% conjugated norcodeine, and 5%−13%

conjugated morphine (Baselt, 2004). Codeine is principally metabolized in liver, where 50%−70% of the dose is converted in codeine-6-glucuronide by UGT2B7 and UGT2B4 (Maurer et al., 2006; Thorn et al., 2009). Approximately 10%−15% of codeine is N-demethylated to norcodeine by CYP3A4. The O-dealkylation of 0%−15 of codeine to morphine is catalyzed mainly by CYP2D6 (Maurer, 2006; Thorn et al., 2009). Finally, codeine is metabolized by an unknown mechanism to hydrocodone that is found in quantities up to 11% of the codeine dose (Smith, 2009).

Acetylcodeine, which is an impurity (1%−15%) of illicit heroin synthesis, was suggested as an additional biomarker of illicit heroin use (Staub et al., 2001). This is particularly useful in detecting illicit heroin use in heroine maintenance programs (Cone and Preston, 2002).

2.3.3 Methadone

Substitution treatment for opioids is largely used to reduce drug dependence. Methadone is a synthetic potent opiate receptor agonist and is the most commonly prescribed drug for this purpose in Europe in more than 90% of treatments. The substitution therapy is based on the principle of substituting a long-acting for a short-acting opiate and is actually the most effective treatment for opioid dependence (Gonzalez et al., 2002). This therapy reduces the fluctuations of opiate receptor stimulation, permits the elimination of needle use, decreases the withdrawal symptoms, and thereby decreases the probability of relapse. Methadone used clinically is a racemic mixture of (R) and (S) enantiomers of which the (R) isomer has the major activity (Cone and Preston, 2002). The affinity of (R)-methadone for opiate receptor and its analgesic activity are 10- to 50-fold higher than the (S)-methadone (Holmquist, 2009). Methadone is a synthetic, potent opiate μ-receptor agonist and is well absorbed when taken orally with 80% bounded to blood proteins. It presents a long elimination half-life of 24 to 36 h (Gonzales et al., 2002). Both enantiomers are metabolized by CYP3A4, CYP2C8, CYP2D6, CYP2C9, CYP2C19, CYP2B6, and CYP1A2 (Holmquist, 2009). Human metabolism of methadone is reported in Fig. 2.5. It is mainly excreted unchanged in urine (5%−50% of a dose) or is metabolized by

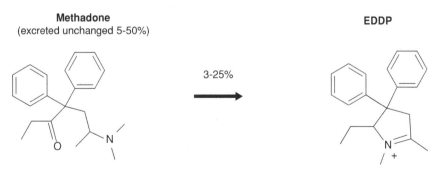

FIGURE 2.5 Metabolism of methadone in man.

N-demethylation with spontaneous cyclization to inactive metabolite 2-ethylidene-1,5-dimethyl-3,3-diphenylpyrrolidine (EDDP) (3%−25% of a dose) and to 2-ethyl-5-methyl-3,3-diphenylpyrroline (EMDP). Following a 5-mg oral dose, methadone and EDDP each accounted for 5% of the dose in the 24-h urine, with less than 1% as EMDP. In maintenance subjects, 24-h urinary methadone accounted for 5%−50% of the dose and EDDP for 3%−25%, with large individual variations because of urinary pH, volume, dose, and rate of metabolism (Baselt, 2004). In a group of 12 former heroin addicts being stabilized on daily oral methadone for 26 days, the percentage of a daily dose eliminated in the daily urine as methadone and EDDP increased from an average 19%− 42% (Baselt, 2004). Methadone may be cleared via urine and feces and the pH of urine has a large impact on its percentage of excretion. Below pH 6.0, 30% of methadone may be excreted in urine unchanged, while at higher pH, less than 5% will be eliminated in this form (Trafton and Ramani, 2009).

Methadone and EDDP were generally the molecules selected as target residues to estimate methadone consumption by wastewater analysis.

REFERENCES

Baker, D., G. Pryce, G. Giovannoni, and A. J. Thompson. 2003. The therapeutic potential of cannabis. *The Lancet* **2**:291–298.

Banta-Green, C. J., J. A. Field, A. C. Chiaia, D. L. Sudakin, L. Power, and L. de Montigny. 2009. The spatial epidemiology of cocaine, methamphetamine and 3,4-methylenedioxymethamphetamine (MDMA) use: A demonstration using a population measure of community drug load derived from municipal wastewater. *Addiction* **104**:1874–1880.

Baselt, R. C. 2004. *Disposition of Toxic Drugs and Chemicals in Man*, seventh edition. Biomedical Publications, Foster City, CA, 2004.

Ben Amar, M. 2006. Cannabinoids in medicine: a review of their therapeutic potential. *J. Ethopharmacol.* **105**:1–25.

Coller, J. K., L. L. Christrup and A. A. Somogyi. 2009. Role of active metabolites in the use of opioids. *Europ. J. Clin. Pharmacol.* **65**:121–139.

Cone, E. J. 1998. Recent discoveries in pharmacokinetics of drugs of abuse. *Toxicol. Lett.* **102−103**: 97–101.

Cone, E. J., and M. A. Huestis. 1993. Relating blood concentrations of tetrahydrocannabinol and metabolites to pharmacologic effects and time of marijuana usage. *Therap. Drug Monit.* **15**: 527–532.

Cone, E. J., and K. L. Preston. 2002. Toxicologic aspects of heroin substitution treatment. *Therap. Drug Monit.* **24**:193–198.

Cruickshank, C. C., and K. R. Dyer. 2009. A review of the clinical pharmacology of methamphetamine. *Addiction* **104**: 1085–1099.

D' Ascenzo, G., A. Di Corcia, A. Gentili, R. Mancini, R. Mastropasqua, M. Nazzari, and R. Samperi. 2003. Fate of natural estrogen conjugates in municipal sewage transport and treatment facilities. *Sci. Total Environ.* **302**: 199–209.

Davis, J. M., I. J. Kopin, L. Lemberger, and J. Axelrod. 1971. Effects of urinary pH on amphetamine metabolism. *Ann. N.Y. Acad. Sci.* **179**: 493–501.

De La Torre, R., J. Ortuño, M. L., Gonzàlez, M. Farrè, J. Camì, and J. Segua. 1995. Determination of cocaine and its metabolites in human urine by gas-chromatography/mass spectrometry after simultaneous use of cocaine and ethanol. *J Pharmaceut. Biomed. Anal.* **13**: 305–312.

De La Torre, R., M. Farrè, M. Navarro, R. Pacifici, P. Zuccaro, and S. Pichini. 2004. Clinical pharmacokinetics of amphetamine and related substances. Monitoring in conventional and non-conventional matrices. *Clin. Pharmaco.* **43**: 158–185.

Dean, R. A., E. T. Harper, N. Dumaual, D. A. Stoeckel, and W. F. Bosron. 1992. Effects of ethanol on cocaine metabolism: Formation of cocaethylene and norcocaethylene. *Toxicol. Appl. Pharmacol.* **117**:1–8.

Dumas, E. O., and G. M. Pollack. 2008. Opioid tolerance development: a pharmacokinetic/pharmacodynamic perspective. *APPS J.* **10**: 537–551.

European Monitoring Centre for Drugs and Drug Addiction (EMCDDA), 2008. Insights **9**. http://www.emcdda.europa.eu/publications/insights/wastewater. Accessed 15 July 2010.

Farooq, M. U., A. Bhatt, and M. B. Patel. 2009. Neurotoxic and cardiotoxic effects of cocaine and ethanol. *J. Med. Toxicol.* **5**: 134–138.

Farrè, M., R. De La Torre, M. L. Gonzalez, M. T. Teràn, P. N. Roset, E. Menomo, and J. Camì. 1997. Cocaine and alcohol interactions in humans: Neuroendocrine effects and cocaethylene metabolism. *J. Pharmacol. Exp. Therapeut.* **283**: 164–176.

Goldstein, R. A., C. DesLauries, and A. M. Burda. 2009. Cocaine: history, socialimplications, and toxicity-a review. *Disease Month* **55**: 6–38.

Gonzalez, G., A. Oliveto, and T. R. Kosten. 2002. Treatment of heroin (diamorphine) addiction. Current approaches and future prospects. *Drugs* **62**: 1331–1343.

Harris, D. S., E. T. Everhart, J. Mendelson, and R. T. Jones. 2003. The pharmacology of cocaethylene in humans following cocaine and ethanol administration. *Drug Alcohol Dependence* **72**: 169–182.

Hart, C. L., E. W. Gunderson, A. Perez, M, G. Kirkpatrick, A. Thurmond, S. D. Comer, and R. W. Foltin. 2008. Acute physiological and behavioural effects of intranasal methamphetamine in humans. *Neuropsycopharmacology* **33**:1847–1855.

Hawks, R. L., I. J. Kopin, R. W. Colburn, and N. B. Thoa. 1974. Norcocaine: A pharmacologically active metabolite of cocaine found in brain. *Life Sci.* **15**:2189–2195.

Hegadoren, K. M., G. B. Baker, and M. Bourin. 1999. 3,4-Methylendioxy analogues of amphetamine: Defining the risks to humans. *Neurosci. Biobehav. Rev.* **23**:539–553.

Holmquist, G. L. 2009. Opioid metabolism and effects of cytochrome P450. *Pain Med.* **10**:S20–S29.

Huestis, M. A., and E. J. Cone. 1998. Urinary excretion half-life of 11-nor-9-carboxy-delta9-tetrahydrocannabinol in humans. *Therapeut. Drug Monit.* **20**:570–576.

Huestis, M. A., and E. J. Cone. 2007. Methamphetamine disposition in oral fluid, plasma, and urine. *Ann. N.Y. Acad. Sci.* **1098**:104–121.

Hutcheon, A. W., J. B. D. Palmer, M. Soukop, D. Cunnigham, C. McArdle, J. Welsh, F. Stuart, G. Sangster, S. Kaye, D. Charlton, and H. Cash. 1983. A randomised multicentre single blind comparison of a cannabinoid antiemetic (levonantradol) with chlorpromazine in patients receiving their first cytotoxic chemioterapy. *Europ. J. Cancer Clin. Oncol.* **19**:1087–1090.

Jatlow, P., E. F. McCance, C. W. Bradberry, J. D. Elsworth, J. R. Taylor, and R. H. Roth. 1996. Alcohol plus cocaine: the whole is more than the sum of its parts. *Therapeut. Drug Monit.* **14**:460–464.

Kim, I., J. M. Oyler, E. T. Moolchan, E. J. Cone, and M. A. Huestis. 2004. Urinary pharmacokinetics of methamphetamine and its metabolite, amphetamine following controlled oral administration to humans. *Drug Monit.* **26**:664–672.

Kirkham, T. C. 2005. Endocannabinoids in the regulation of appetite and body weight. *Behav. Pharmacol.* **16**:297–313.

Kraemer, T., and H. H. Maurer. 2002 Toxicokinetics of amphetamines: metabolism and toxicokinetic. Data of designer drugs, amphetamine, methamphetamine, and their N-alkyl derivatives. *Therapeut. Drug Monit.* **24**:277–289.

Maurer, H. H., J. Bickeboeller-Friedrich, T. Kraemer, and F. T. Peters. 2000. Toxicokinetics and analytical toxicology of amphetamine-derived designer drugs (Ecstasy). *Toxicol. Lett.* **112–113**:133–142.

Maurer, H. H., C. Sauer, and D. S. Theobald. 2006. Toxicokinetics of drugs of abuse: Current knowledge of isomers involved in the human metabolism of tetrahydrocannabinol, cocaine, heroin, morphine, and codeine. *Therapeut. Drug Monit.* **28**:447–453.

Nicholson, B. 2003. Responsible prescribing of opioids for the management of chronic pain. *Drugs* **63**:17–32.

Oyler, J. M., J. E. Cone, R. E. Joseph, Jr., E. T. Moolchan, and M. A. Huestis. 2002. Duration of detectable methamphetamine and amphetamine excretion in urine after controlled oral administration of methamphetamine to humans. *Clin. Chem.* **48**:1703–1714.

Perfetti, X., B. Mathùna, N. Pizarro, E. Cuyàs, O. Khymenets, B., Almeida, M. Pellegrini, S. Pichini, S. S. Lau, T. J. Monks, M. Farrè, A. Pascual, J. Joglar, and R. De La Torre. 2009. Neurotoxic thioether adducts of 3,4-methylenedioxymethamphetamine identified in human urine after ecstasy ingestion. *Drug Metab. Disposition* **37**:1448–1455.

Peterson, K. L., K. L. Barry, and G. D. Christian. 1995. Detection of cocaine and its polar transformation products and metabolites in human urine. *Forensic Sci. Int.* **73**:183–196.

Rook, E. J., A. D. R. Huitema, W. Van der Brink, J. M. Van Ree, and J. H. Beijnen. 2006. Population pharmacokinetics of heroin and its major metabolites. *Clin. Pharmaco.* **45**:401–417.

Schepers, R. J. F., J. M. Oyler, R. E. Joseph, Jr., E. J. Cone, E. T. Moolchan, and M. A. Huestis. 2003. Methamphetamine and amphetamine pharmacokinetics in oral fluid and plasma after controlled oral methamphetamine administration to human volunteers. *Clin. Chem.* **49**:121–132.

Schierenbeck, T., D. Riemann, M. Berger, and M. Hornyak. Effect of illicit recreational drugs upon sleep: cocaine, ecstasy and marijuana. *Sleep Med. Rev.* **12**:381–389.

Smith, H. S. Opioid metabolism. *Mayo Clinic Proc.* 2009; **84**:613–624.

Suzuki, T., and M. Misawa. 1997. Opioid receptor types and dependence. *Nippon Yakurigaku Zasshi* **109**:165–74.

Staub, C., M. Marset, A. Mino, and P. Mangin. 2001. Detection of acetylcodeine in urine as an indicator of illicit heroin use: Method validation and results of a pilot study. *Clin. Chem.* **47**:301–307.

Ternes, T. A., P. Kreckel, and J. Mueller. 1999. Behaviour and occurrence of estrogens in municipal sewage treatment plants-II. Aerobic Batch experiments with active sludge. *Sci. Total Environ.* **225**:91–99.

Thorn, C. F., T. E. Klein, R. B. Altman. 2009. Codeine and morphine pathway. *Pharmacogenet. Genomics* **19**:556–558.

Trafton, J. A., and Ramani, A. 2009. Methadone: a new old drug with promises and pitfalls. *Current Pain Headache Rep.* **13**:24–30.

United Nations Office of Drug and Crime (UNODC) World Drug Report 2004. Volume 2. Statistics. http://www.unodc.org/pdf/WDR_2004/methodology.pdf

United Nation Office of Drug and Crime (UNODC) 2009. World Drug Report; 2009. http://www.unodc.org/documents/wdr/WDR_2009/WDR2009_eng_web.pdf

Verstraete, A. G. 2004. Detection times of drugs of abuse in blood, urine, and oral fluid. *Drug Monit.* **26**:200–205.

Williams, N., D. H. Clouet, A. L. Misra, and S. Mulè. 1977. Cocaine and metabolites: relationship between pharmacological activity and inhibitory action on dopamine uptake into striatal synaptosomes. *Progr. Neuro-Psycopharmacol.* **1**:265–269.

Wood, R. W., J. Shojaie, C. Ping Fang, J. F. Graefe. 1995. Methylecgonidine coats the crack particle. *Pharmacol. Biochem. Behav.* **53**:57–66.

Zuccato E., C. Chiabrando, S. Castiglioni, R. Bagnati, R. Fanelli. 2008. Estimating community drug abuse by wastewater analysis. *Environ. Health Perspect.* **116**:1027–1032.

MASS SPECTROMETRY IN ILLICIT DRUGS DETECTION AND MEASUREMENT – CURRENT AND NOVEL ENVIRONMENTAL APPLICATIONS

CHAPTER 3

ANALYTICAL METHODS FOR THE DETECTION OF ILLICIT DRUGS IN WASTEWATERS AND SURFACE WATERS

RENZO BAGNATI and ENRICO DAVOLI

3.1 INTRODUCTION

Drugs and metabolites are present in wastewater and surface waters at concentrations in the nanogram per liter range and, therefore, their determination requires the use of analytical techniques capable of high sensitivity and specificity (van Nuijs et al., 2010). Although the determination of drugs in different research fields and in forensic analyses has also been performed using gas chromatography—mass spectrometry (GC—MS) (Mari et al., 2009), the technique of choice for wastewater samples, in most of the cases reported in the literature, is high-performance liquid chromatography—mass spectrometry (HPLC—MS) and high-performance liquid chromatography—tandem mass spectrometry (HPLC—MS/MS).

The instrumental analysis is generally preceded by preconcentration and clean-up steps, which are necessary for minimizing matrix effects and chromatographic problems, which may prevent the achievement of the required selectivities and sensitivities.

As new instrumentations are available, new analytical approaches will be available. An example using new, high-resolution and high-accuracy mass spectrometry, is presented. Sewage inlets and surface waters have been analyzed by conventional HPLC—MS/MS approach and by high-resolution MS. Results indicate that, at least for benzoylecgonine and cocaine, analysis could be performed for these samples with new different strategies, each having advantages and limitations.

Illicit Drugs in the Environment: Occurrence, Analysis, and Fate Using Mass Spectrometry, Edited by Sara Castiglioni, Ettore Zuccato, and Roberto Fanelli
Copyright © 2011 John Wiley & Sons, Inc.

3.2 SAMPLE PREPARATION

Collected and stored water samples are generally filtered and then extracted using solid-phase extraction techniques (SPE).

The most used SPE sorbent materials are in the form of cartridges for off-line extraction of water samples: Oasis MCX, Oasis HLB, Strata XC, Bond-Elut Certify LRC and PLPR-s (Castiglioni et al., 2006; Hummel et al., 2006; Boleda et al., 2007; Huerta-Fontela et al., 2007; Bones et al., 2007; Gheorghe et al., 2008; van Nuijs et al., 2009; Bijlsma et al., 2009). However, there are also reports of on-line preconcentration methods (Postigo et al., 2008) or direct injection of water samples (Chiaia et al., 2008). In general, off-line extraction is considered the most versatile and reliable technique, as it allows the adjustment of the volume of extracted water (50–1000 mL) and of the elution conditions to obtain the desired sensitivities and to maximize the recoveries of different compounds. On the other side, the advantages of on-line extraction or direct-injection methods (with large volume injection) are the minimal manipulation of samples and the shorter time for the total extraction analysis procedure; however, these methods may suffer from lower sensitivities (because of the minor amount of sample that can be loaded), higher matrix effects, or worse chromatographic performances. New and more sensitive HPLC−MS instruments may help to improve the performances of on-line methods (Kasprzyk-Hordern et al., 2008).

Recovery efficiencies of several drugs and metabolites for off-line extraction methods are generally higher than 50%, except for the more polar substances (morphine, ecgonine methyl ester), for which values as low as 5% are reported.

The use of stable isotope-labeled internal standards is considered mandatory for the achievement of reliable quantitative results (Castiglioni et al., 2006). The normal analytical procedure is that of adding known amounts of internal standards at the beginning of sample extraction, as this permits all the problems deriving from variations of the recoveries and ionization matrix effects to be overcome. Currently, these standards are commercially available for most of the drugs and metabolites, which have been considered for detection in wastewater samples.

3.3 CHROMATOGRAPHY AND MASS SPECTROMETRY ANALYSIS

Quantitative determination of drugs and metabolites may be performed using a variety of chromatographic and mass spectrometric conditions.

Chromatographic separation for most of the drugs and their metabolites has been obtained with reversed-phase columns, both using high-performance liquid chromatography (HPLC) and ultra-performance liquid chromatography (UPLC) methods. Some of the more polar drug metabolites, as ecgonine methyl ester, were better resolved using direct-phase or HILIC columns (Gheorghe et al., 2008; van Nuijs et al., 2009, 2010). The use of UPLC has been reported to increase the sensitivity of the analytical methods and shorten the analysis time, however, because of the narrowness of chromatographic peaks, it requires MS instruments capable of fast scanning rates (Bijlsma et al., 2009).

The composition of the mobile phases used for separation was influenced by the MS technique used for detection. Most of the drug substances were better detected with positive ionization (except for the metabolite 11-nor-9-carboxy-Δ^9-tetrahydrocannabinol, which was also detected with negative ions) and, thus, the mobile phases were mainly constituted by mixtures of aqueous acids (formic acid or acetic acid) and organic solvents (acetonitrile or methanol).

The MS ion source which has been almost exclusively used for the ionization of drugs and metabolites is electrospray (ESI). This source gives better analytical performances, however it suffers the problem of matrix effect, a phenomenon in which the presence of coeluting substances causes suppression or enhancements of the ionization signal of analytes. This is particularly evident with more complex samples, such as wastewater influents and effluents, than with "cleaner" samples, such as river or lake water samples. The atmospheric pressure chemical ionization source (APCI) is considered to be less susceptible to matrix interferences, however, it is not suitable for all substances, in particular, the more polar ones (morphine, morphine-3β-D-glucuronide, and ecgonine methyl ester).

Most of the analytical methods so far developed make of use triple quadrupoles or quadrupole-linear ion-trap instruments for the quantitative determination of analytes (Castiglioni et al., 2006; van Nuijs et al., 2010). In these cases, the selected ion monitoring (SRM) acquisition mode provides the best sensitivity and selectivity, when at least two specific transitions are recorded. The use of isotope-stable internal standards, as already explained in the previous section, allows accurate quantitative determinations for each substance.

Thus far, the use of high-resolution instruments, like the Ion Cyclotron Resonance (ICR) or the Orbitrap analyzer for the analysis of illicit drugs in wastewaters has not been extensively reported in the literature. Hogenboom et al. (2009) used a linear ion trap–Orbitrap instrument, coupled to liquid chromatography, to analyze drugs and metabolites in waste- and surface waters. Other possible applications of the Orbitrap analyzer, with the advantages of the use of high resolution and high mass accuracy, will be discussed in the following sections.

3.4 ADVANCED HIGH-RESOLUTION APPROACHES

As mentioned before, with new instrumentation available, new analytical approaches will be available. LC−MS/MS has emerged as the technique of choice for the quantitation of small molecules in environmental sciences and in several other different fields. The striking advantage of this technique is that it brings sensitivity, selectivity, accuracy (especially with triple quadrupoles), precision, and reproducibility, all cornerstones for trace and ultratrace analysis. However, some of these factors also pose limitations for this approach. The extreme specificity, because of the multiple separations (chromatography, mass spectrometry to isolate parent ions and mass spectrometry to detect product ions), often results in chromatograms showing the single-target analytes' peaks. In this way, a postacquisition data mining in search for other contaminants reveals, at maximum, (unknown) isomers, if present. Also, as mentioned earlier, the matrix effect can be a serious problem (Mortier et al., 2002; Kjell et al., 2004; Postigo et al., 2008; Huestis and Smith, 2008; Prem et al., 2009). This phenomenon can be observed by comparing the response of the analyte in its matrix versus in solvent. The direct influence of the sample matrix on the mass spectrometer absolute response is observed at different levels in water samples like drinking water, surface water, and wastewater treatment plants (WWTP) influents and effluents. It can be caused by several factors, like the make and the model of the ion source (Bakhtiar and Majumdar, 2007), the presence of nonvolatile substances (King et al., 2000) and, mainly, the coelution of different, unexpected, substances with our target analyte. It is clear that although the LC−MS/MS is a highly specific technique, when running different environmental samples, the user has no clues, during chromatography, about sample contamination and, therefore, no ideas about possible different matrix effects in different samples. This effect is observed even at high resolution and reported both with TOF−MS (Nielena et al., 2007) and Orbitrap FT−MS (Davoli et al., 2008).

New instrumentation, like high-resolution mass spectrometers, brings higher data quality. Preliminary results on small-molecule analysis by high-resolution mass spectrometry, show that the quality of data increases with resolution and it has been proposed to add this point to the performance characteristics described in the Commission Decision 2002/657/EC, that establishes criteria and procedures for the validation of analytical methods for official laboratories of the EU.

For these reasons, we have investigated the possibilities of analyzing sewage inlets and surface water samples to measure cocaine and its main metabolite, benzoylecgonine, using new, high-resolution and high-accuracy mass spectrometers (HRMS). Samples were analyzed both by the conventional HPLC−MS/MS approach and by HPLC−HRMS, in full-scan mode, to maintain a postacquisition data mining possibility and to verify sample contamination. As an example of possible new approaches, direct infusion of the same environmental samples have been analyzed by HRMS to investigate limitations and possibilities of direct analysis of environmental drug residues.

3.4.1 High-Resolution HPLC−MS and HPLC−MS/MS

Analyses have been performed on a Thermo LTQ Orbitrap equipped with a Prosolia's Omni Spray DESI ion source, used as a nano ES source. The instrument was operated in full-scan mode (60,000 resolution) and in MS/MS mode (30,000 resolution), to confirm drug identities.

Results from HPLC−HR MS analysis show that the instrumentation is sensitive (Fig. 3.1), with a 100-fg injection limit of quantitation and a linear range over at least three orders of magnitude (Fig. 3.2). Because of these levels of sensitivity, sample preparation for waste- and surface water samples, where benzoylecgonine and cocaine residues concentration are in the parts per billion (ppb) range, can be limited to filtration and addition of internal standards, while more diluted samples can still be extracted by SPE. HPLC separation was performed on Agilent 1200 series

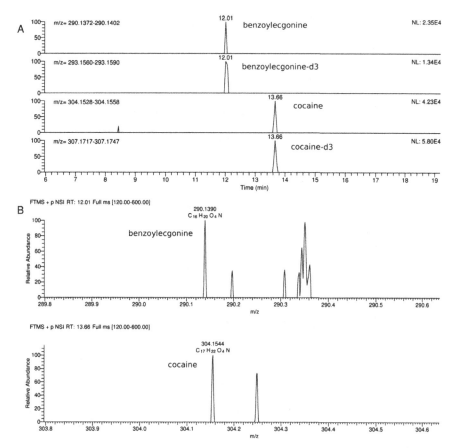

FIGURE 3.1 Benzoylecgonine and cocaine analysis. HPLC – high-resolution Orbitrap MS of benzoylecgonine, cocaine, and their deuterated analogs. A 100-fg injection of standard in full scan mode allows detection (A) and identification, based on the accurate mass assignment on the mass spectra (B).

capillary and nano pumps, using a 150×0.5 mm ID, 5 μm particles, Zorbax SB C18 column (Agilent Technologies), at a flow rate of 10 μL/min.

The wastewater sample, collected at the WWTP influent, extracted by SPE and analyzed by HPLC−MS/MS, reveals cocaine presence at 318 ppb concentration levels and it can be readily analyzed by HPLC−HR MS/MS (Fig. 3.3). Samples collected from surface water, in rivers like the Lambro River (Fig. 3.4) and in Po River (Fig. 3.5) can still be detected with this approach, even if concentration decreases down to approximately 1 ppb.

3.4.2 High-Resolution MS/MS Analysis

Because of the instrumental sensitivity and selectivity, the same wastewater sample discussed above, can be analyzed directly, with no chromatographic separation.

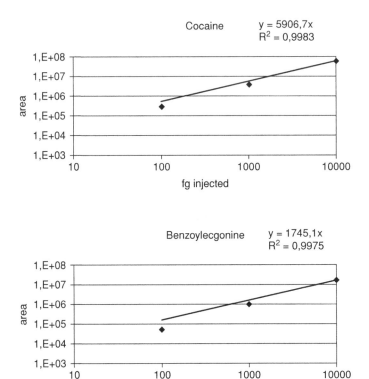

FIGURE 3.2   Linearity with HPLC–HR MS/MS approach. The approach is linear within three orders of magnitude for benzoylecgonine and cocaine. The 100-fg injection appears to be the limit of quantitation for these compounds.

Sample purification may be limited to the SPE extraction or instrumental approach only: the precursor molecular ion is selected in the ion trap (LTQ), fragmented, and the product ions are acquired in full scan at high resolution (100,000). Instrumental analysis is carried out by direct sample infusion, after dilution of the SPE extract or direct filtration of the wastewater sample and internal standard addition. In the upper part of Fig. 3.6, the HR full-scan mass spectrum of a diluted wastewater SPE extract is reported. Here, a zoom over the protonated molecular ion region of the drugs of interest, reveals several other ions at the same nominal mass, with different exact masses. Ion isolation, for the subsequent collision-induced dissociation, occurs at low resolution in the ion trap. Therefore, in the product ion scans, several peaks are present, more than those of the analytes of interest, resulting in much more contaminated MS/MS spectra. Still at this level of resolution, compounds like cocaine, benzoylecgonine, and carbamazepine can be identified by accurate mass assignment of the main product ions (Fig. 3.6). In these cases, mass assignments were accurate within 1 millimass unit (mmu) on the exact mass.

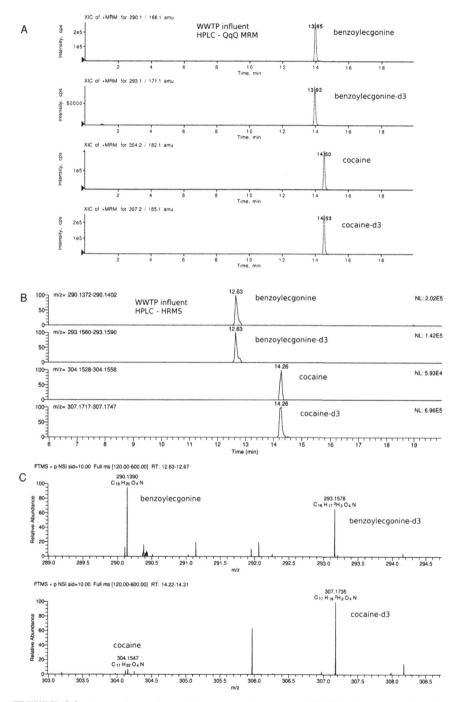

FIGURE 3.3 Benzoylecgonine and cocaine analysis of a WWTP influent. Analysis by HPLC–MS/MS, with a triple quadrupole, revealed the presence of benzoylecgonine at 2450 and cocaine at 318 ppb concentration (A). The same wastewater sample was analyzed by HPLC–HRMS (B), giving accurate mass molecular weight confirmation (C).

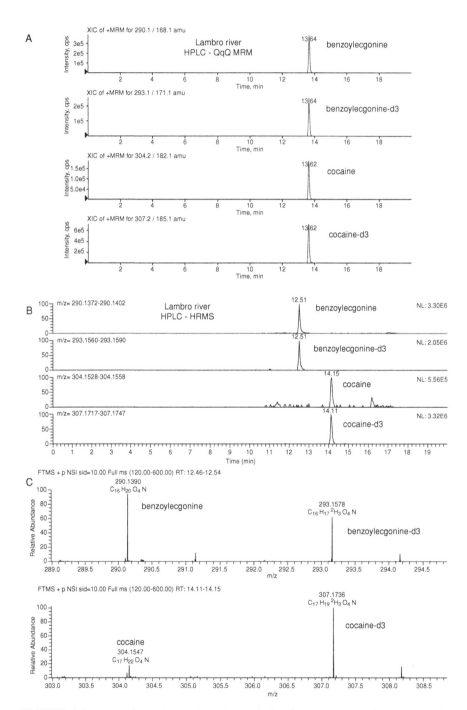

FIGURE 3.4 Benzoylecgonine and cocaine analysis of a water sample from the Lambro River (Northern Italy). Analysis by HPLC−MS/MS, with a triple quadrupole, revealed the presence of benzoylecgonine at 50 and cocaine at 15 ppb concentration (A). The same wastewater sample was analyzed by HPLC−HRMS (B), giving accurate mass molecular weight confirmation (C).

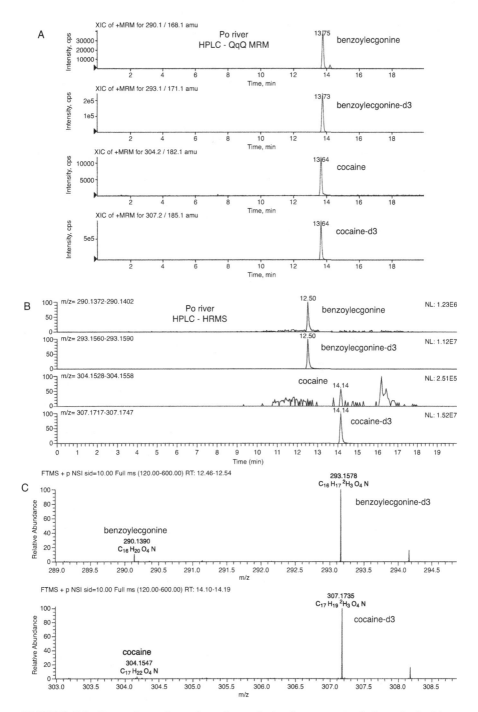

FIGURE 3.5 Benzoylecgonine and cocaine analysis of a water sample from the Po River (Northern Italy). Analysis by HPLC−MS/MS, with a triple quadrupole, revealed the presence of benzoylecgonine at 12 and cocaine at 1 ppb concentration (A). The same wastewater sample was analyzed by HPLC−HRMS (B), giving accurate mass molecular weight confirmation (C).

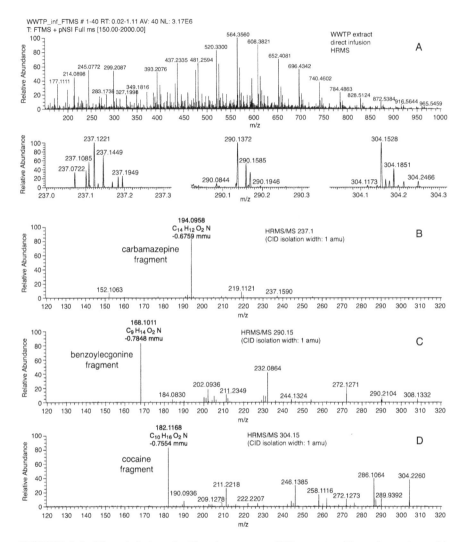

FIGURE 3.6   Direct infusion of a diluted wastewater SPE extract, without chromatographic separation. (A) Full-scan HRMS of the sample, with small boxes showing a zoom view of the ions around the pseudomolecular weights of carbamazepine ($m/z = 237.1022$), benzoylecgonine ($m/z = 290.1387$), and cocaine ($m/z = 304.1543$). (B) High-resolution MS/MS spectrum of the ion 237.1 (with CID isolation width of 1 amu), showing the identification of the main fragment of carbamazepine ($m/z = 194.0964$) with an error less than 1 mmu. (C) High-resolution MS/MS spectrum of the ion 290.15 (with CID isolation width of 1 amu), showing the identification of the main fragment of benzoylecgonine ($m/z = 168.1019$) with an error less than 1 mmu. (D) High-resolution MS/MS spectrum of the ion 304.15 (with CID isolation width of 1 amu), showing the identification of the main fragment of cocaine ($m/z = 182.1176$) with an error less than 1 mmu.

This approach can be useful at least for qualitative analyses and for the search of new contaminants in water samples.

3.5 CONCLUDING REMARKS

New mass spectrometry instruments are available and new analytical approaches are possible in this field. In this work, we have investigated new strategies deriving from the use of the Orbitrap MS, a newly developed mass spectrometer, capable of high resolution (up to 100,000 resolving power), with routine use. Results show that direct detection of illicit drugs in surface waters and in wastewater samples, is possible at low parts per billion level. Samples can be analyzed by direct infusion after internal standard addition or after SPE extraction, with no chromatographic separation. Purification is obtained by tandem mass spectrometry and high-resolution analysis. Positive samples are then validated by HPLC—MS/MS. Results show that this approach can be used for qualitative and quantitative analysis of residues of illicit drugs and their metabolites, pharmaceuticals, and pollutants in waste- and surface waters. While this method is useful for target compounds, the screening of unknowns, theoretically possible, is still limited by lack of high-resolution libraries and software limitations.

REFERENCES

Bakhtiar, R., and T. K. Majumdar. 2007. Tracking problems and possible solutions in the quantitative determination of small molecule drugs and metabolites in biological fluids using liquid chromatography–mass spectrometry. *J. Pharmacol. Toxicol. Methods* **55**:227–243.

Bijlsma, L., J. V. Sancho, E. Pitarch, M. Ibanez, and F. Hernandez, 2009. Simultaneous ultra-high-pressure liquid chromatography—tandem mass spectrometry determination of amphetamine and amphetamine-like stimulants, cocaine and its metabolites, and a cannabis metabolite in surface water and urban wastewater. *J. Chromatogr. A* **1216**:3078–3089.

Boleda, R., T. Galceran, and F. Ventura. 2007. Trace determination of cannabinoids and opiates in wastewater and surface waters by ultra-performance liquid chromatography–tandem mass spectrometry. *J. Chromatogr. A* **1175**:38–48.

Bones, J., K. V. Thomas, and B. Paull. 2007. Using environmental analytical data to estimate levels of community consumption of illicit drugs and abused pharmaceuticals. *J. Environ. Monit.* **9**:701–707.

Castiglioni, S., E. Zuccato, E. Crisci, C. Chiabrando, R. Fanelli, and R. Bagnati. 2006. Identification and measurement of illicit drugs and their metabolites in urban wastewater by liquid chromatography-tandem mass spectrometry. *Anal. Chem.* **78**:8421–8429.

Chiaia, A. C., C. Banta-Green, and J. Field. 2008. Eliminating solid phase extraction with large-volume injection LC/MS/MS: analysis of illicit and legal drugs and human urine indicators in US wastewaters. *Environ. Sci. Technol.* **42**:8841–8848.

Davoli, E., S. Castiglioni, R. Bagnati, G. Bianchi, and R. Fanelli. 2008. Direct, trace level, environmental drug residues analysis by orbitrap MS. In *Proceedings of the 56th ASMS Conference on Mass Spectrometry and Allied Topics,* Denver, CO.

Gheorghe, A., A. van Nuijs, B. Pecceu, L. Bervoets, P. G. Jorens, R. Blust, H. Neels, and A. Covaci. 2008. Analysis of cocaine and its principal metabolites in waste and surface water using solid-phase extraction and liquid chromatography-ion trap tandem mass spectrometry. *Anal. Bioanal. Chem.* **391**:1309−1319.

Gupta, P. K., M. Hubbard, B. Gurley, and H. P. Hendrickson. 2009. Validation of a liquid chromatography–tandem mass spectrometric assay for the quantitative determination of hydrastine and berberine in human serum. *J. Pharm. Biometric Anal.* **49**:1021−1026.

Hogenboom, A. C., J. A. van Leerdam, and P. de Voogt. Accurate mass screening and identification of emerging contaminants in environmental samples by liquid chromatography-hybrid linear ion trap Orbitrap mass spectrometry. *J. Chromatogr. A* **1216**:510−519.

Huerta-Fontela, M., M. T. Galceran, J. Martin-Alonso, and F. Ventura. 2007. Ultra-performance liquid chromatography-tandem mass spectrometry analysis of stimulatory drugs of abuse in wastewater and surface waters. *Anal. Chem.* **79**:3821−3829.

Huestis, M. A., and M. L. Smith. 2006. Modern analytical technologies for the detection of drug abuse and doping. *DDT* **3**:49−57.

Hummel, D., D. Loffler, G. Fink, and T. A. Ternes. 2006. Simultaneous determination of psychoactive drugs and their metabolites in aqueous matrices by liquid chromatography mass spectrometry. *Environ. Sci. Technol.* **40**:7321−7328.

Kasprzyk-Hordern, B., R. M. Dinsdale, and A. J. Guwy. 2008. Multiresidue methods for the analysis of pharmaceuticals, personal care products and illicit drugs in surface water and wastewater by solid-phase extraction and ultra performance liquid chromatography-electrospray tandem mass spectrometry. *Anal. Bioanal. Chem.* **391**:1293−1308.

King, R., R. Bonfiglio, C. F. Metzler, C. Miller-Stein, and T. Olah. 2000. Mechanistic investigation of ionization suppression in electrospray ionization. *J. Am. Soc. Mass Spectrom.* **11**:942−950.

Kjell, A. M., G. V. Alain, Z. Guo-Fang, and E. L. Willy. 2004. Enhanced method performance due to a shorter chromatographic run-time in a liquid chromatography–tandem mass spectrometry assay for paclitaxel. *J. Chromatogr. A* **1041**:235−238

Mari, F., L. Politi, A. Biggeri, G. Accetta, C. Trignano, M. Di Padua, and E. Bertol. 2009. Cocaine and heroin in waste water plants: a 1-year study in the city of Florence. *Forensic Sci. Int.* **198**:88−92.

Mortier, K. A., K. E. Maudens, W. E. Lambert, K. M. Clauwaert, J. F. Van Bocxlaer, D. L. Deforce, C. H. Van Peteghem, and A. P. De Leenheera. 2002. Simultaneous, quantitative determination of opiates, amphetamines, cocaine and benzoylecgonine in oral fluid by liquid chromatography quadrupole-time-of-flight mass spectrometry. *J. Chromatogr. B* **779**:321–330.

Nielena, M. W. F., M. C. van Engelena, R. Zuiderentb, and R. Ramakerc. 2007. Screening and confirmation criteria for hormone residue analysis using liquid chromatography accurate mass time-of-flight, Fourier transform ion cyclotron resonance and orbitrap mass spectrometry techniques. *Anal. Chim. Acta* **586**:122−129.

Postigo, C., M. J. Lopez de Alda, and D. Barceló D. 2008. Fully automated determination in the low nanogram per liter level of different classes of drugs of abuse in sewage water by on-line solid-phase extraction-liquid chromatography-electrospray-tandem mass spectrometry. *Anal. Chem.* **80**:3123−3134.

Postigo, C., M. J. Lopez de Alda, and D. Barceló. 2008. Analysis of drugs of abuse and their human metabolites in water by LC-MS2: A non-intrusive tool for drug abuse estimation at the community level. *TrAC* **27**:1053−1069.

van Nuijs, A. L. N., I. Tarcomnicu, P. G. Jorens, L. Bervoets, R. Blust, H. Neels, and A. Covaci. 2009. Analysis of drugs of abuse in wastewater by hydrophilic interaction liquid chromatography–tandem mass spectrometry. *Anal. Bioanal. Chem.* **395**:819–828.

van Nuijs, A. L. N., S. Castiglioni, I. Tarcomnicu, C. Postigo, M. L. de Alda, H. Neels, E. Zuccato, D. Barceló, and A. Covaci A. 2010. Illicit drug consumption estimations derived from wastewater analysis: a critical review. *Sci. Total Environ.* 2010; in press. doi:10.1016/j.scitotenv.2010.05.030

CHAPTER 4

WIDE-SCOPE SCREENING OF ILLICIT DRUGS IN URBAN WASTEWATER BY UHPLC–QTOF MS

FÉLIX HERNÁNDEZ, JUAN V. SANCHO, and LUBERTUS BIJLSMA

4.1 TOF AND QTOF MS ANALYSIS

The application of time-of-flight (TOF) mass spectrometry (MS) to environmental analyses has tremendously increased over the last decade. The sensitive full-spectrum data acquisition and high-mass accuracy of new generation TOF instruments together with their high-mass resolving power provide a high degree confidence for identification of compounds detected in samples. The application of orthogonal acceleration (oa-)TOF MS instruments minimizes the possibility of reporting false positives because of the accurate mass measurement of (typically) the (de)protonated molecules. This is especially relevant when facing too highly complex environmental samples like urban wastewater. Confirmatory analysis, even with greater reliability, can be performed by using a hybrid quadrupole–TOF (QTOF) MS, where an analyte m/z ion can be preselected in the first quadrupole as precursor ion. When carrying out MS/MS experiments, the fragmentation of the precursor ion results in full-product ion spectra at accurate mass giving relevant structural information and extra confidence to confirm or deny a suggested structure.

In addition to their excellent potential in the qualitative field, TOF instruments offer the possibility of a wide-scope screening once the analysis has been performed and MS data acquired, because the preselection of analytes is not compulsory and searching for contaminants is made in a subsequent step. Ideally, screening methods should be able to rapidly detect the presence of as many contaminants as possible in a sample, preferably with little sample manipulation. As expected, in environmental MS, the instrumentation available is of particular importance when selecting the screening approach. Thus, TOF instruments allow performing target screening after MS acquisition, taking advantage of their sensitivity in full acquisition mode and

Illicit Drugs in the Environment: Occurrence, Analysis, and Fate Using Mass Spectrometry, Edited by Sara Castiglioni, Ettore Zuccato, and Roberto Fanelli
Copyright © 2011 John Wiley & Sons, Inc.

mass accuracy measurements. However, when analytes have to be preselected before acquisition (as in target methods based on LC–MS/MS with triple quadrupole), there is a need for reference standards of each target compound for MS optimization and many other pollutants (nonpreselected) that might be present in samples would be ignored in the analysis.

For a wide-scope screening in environmental MS using TOF instruments, two alternatives can typically be applied (Hernández et al., 2005a; Ibáñez et al., 2008):

1. Post-target screening—after full-spectrum acquisition, selected m/z ions for the target compounds are extracted from the Total Ion Current (TIC) chromatogram, giving rise to eXtracted Ion Chromatograms (XIC) where the presence of the compound in the sample leads to the corresponding chromatographic peak. The high resolving power of TOF MS allows using narrow-window XIC (normally 10–20 mDa width), which leads to improved selectivity and sensitivity.

2. Nontarget screening—where compounds eluting from the chromatographic column may be detected and subsequently identified without any kind of previous selection. Here, the analyst is searching for actual unknown compounds, as no previous information about the compounds to be investigated is taken into account.

Identification of non-arget compounds in environmental samples is an interesting feature for analytical chemists, but it is still a challenge. Its routine application in environmental laboratories will take some time, as the possibilities of success when identifying unknown compounds at low levels in environmental complex matrices are still very scarce (Ibáñez et al., 2005, 2008). Thus, this chapter is focused on the screening of target compounds (post-target approach), taking illicit drugs of abuse in water as an illustrative example.

The main benefit of a post-target approach is the possibility to detect (almost) an unlimited number of compounds of interest, without using reference standards or performing additional analysis. Obviously, the compounds investigated should fit the requirements of sample preparation and MS analysis to be able of being detected. The accurate mass spectra can rapidly and simply be evaluated using an in-house compound database and powerful software, currently available, such as ChromaLynx XS (Waters). In subsequent steps, in order to facilitate a wider screening, earlier data can be reprocessed and reevaluated using new or extended databases.

In this chapter, the potential of QTOF MS combined with ultrahigh-pressure liquid chromatography (UHPLC) for screening and confirmation of illicit drugs in wastewater is illustrated. The work is focused on the applicability of the post-target approach, which, from the authors' point of view, is the most relevant application when using TOF in environmental MS.

4.2 ANALYTICAL STRATEGIES FOR THE DETERMINATION OF ILLICIT DRUGS IN THE ENVIRONMENT

The occurrence of emerging contaminants in the environment has been a major issue for analytical chemists during the last few years (Richardson, 2008, 2009). Emerging environmental contaminants, such as pharmaceuticals, personal care products, and veterinary drugs received much interest (Barceló and Petrovic, 2007) and studies of illicit drugs increased, since their presence were reported in surface water and urban wastewaters (Jones-Lepp et al., 2004; Zuccato et al., 2005.; Castiglioni et al., 2006). The low concentration levels of illicit drugs typically found in the aquatic environment in combination with the complexity of the matrix make their reliable determination difficult. Therefore, highly selective and sensitive analytical methods are required.

Several multiclass methods based on liquid chromatography (LC) coupled to triple quadrupole (QqQ) instruments have been developed for quantification of illicit drugs in environmental samples (Castiglioni et al., 2008). Until now, most of the methods developed have used off-line solid-phase extraction (SPE) for sample pretreatment and preconcentration. Typically, deuterated analytes are used as surrogate internal standards to compensate for possible errors resulting from matrix effects, as well as those associated with the sample preparation step. The use of QqQ analyzers in selective-reaction monitoring (SRM) mode allows the minimization, or, in some cases, the eliminatation, of interferences thereby improving selectivity and sensitivity because of the possibility of adequate precursor and product ion selection, leading to low chemical noise in the chromatograms. The excellent sensitivity together with the proved robustness of this analyzer make LC–MS/MS with triple quadrupole a powerful analytical tool for quantification purposes of illicit drugs at the low concentration levels normally present in water. In spite of the good selectivity of tandem MS, when using low-resolution instruments, such as QqQ, at least two SRM transitions should be monitored for a safe identification, but the quality of the transitions should be taken into account, avoiding nonspecific transitions (e.g., H_2O, CO_2, HCl) in order to prevent false positives or even false negatives (Hernandez et al., 2005a, 2005b; Pozo et al., 2006). Thus, when acquiring at least two transitions, together with the accomplishment of the ion ratios deviations between recorded transitions, QqQ analyzers in SRM mode are highly reliable and suitable for trace determination of drugs of abuse in water (Bijlsma et al., 2009).

As any analytical technique, LC–MS/MS with QqQ analyzer also has some limitations. The number of compounds to be included in the method is restricted, since only a limited number of transitions may be selected. Including many compounds in a multiresidue method may lead to a loss in sensitivity, as an increase in acquired transitions usually involves a decrease in either the time of acquiring the transition (dwell time) or the number of points obtained per peak. Thus, a compromise between sensitivity and peak shape has to be reached when developing multiresidue methods. However, recent developments in triple quadrupole instruments, such as dynamic or scheduled—SRMs, permits the counterbalancing of this limitation. Another issue is that reference standards of each target compound are needed for SRM optimization. Most of the illicit drugs standards are expensive and it is often time-consuming to purchase them, because of extensive administrative requirements imposed on the importation or exportation of controlled substances. However, the main limitation is that other nonselected compounds cannot be revealed, even if they are present at high levels in the sample, since the analytes are preselected before MS data acquisition. This might be considered as an important drawback, especially when wide-scope screening is the objective of analyses.

The low sensitivity in full-scan mode, the measurements in nominal mass, and need of reference standards for identification make it unadvisable to use QqQ instruments in post-target or nontarget screening of drugs of abuse in wastewater. Conversely, TOF MS instruments have shown a great potential for screening and confirmation purposes in various fields. Compounds such as pesticides, pharmaceuticals, and antibiotics have been screened in environmental samples by LC–TOF MS (Petrovic et al., 2006; Ibáñez et al., 2008, 2009). Besides, several studies have been reported on screening of compounds of toxicological interest (including illicit drugs) in forensic samples (Palander et al., 2003; Ojanperä et al., 2006; Badoud et al., 2009), and on screening of pesticides in food (Ferrer and Thurman, 2007; García-Reyes et al., 2008; Grimalt et al., 2010). Several of these methods are directed towards a relatively high number of compounds (around or even more than 100) exploiting the elevated resolution and mass-accuracy capabilities of TOF analyzers.

Because of the novelty of this approach, a wide-scope screening (300–400 analytes) has only been applied to a few studies (Grimalt et al., 2010; Mezcua et al., 2009). The ability of TOF MS to provide a notable amount of relevant chemical information in a single experiment makes this technique very attractive for a wide-scope screening of illicit drugs in water using a post-target style.

4.3 SCREENING BY UHPLC–QTOF MS

4.3.1 Methodology

As earlier stated, sensitive full-spectrum data acquisition, high-mass resolution, speed, and mass accuracy of new generation TOF MS or QTOF MS instruments allows a highly reliable confirmation and a wide-scope screening of illicit drugs in the aquatic environment.

Orthogonal acceleration TOF MS instruments permit measurements with a mass accuracy commonly better than 5 ppm and for mass resolution exceeding 10,000 full-width half maximum (FWHM). The mass of any ionizable component in different sample matrices can be measured accurately, which gives high confidence to the identification process and also implies the possibility to distinguish in between isobaric compounds. Isobars have identical nominal mass, but different elemental composition and thus different exact mass. When using QqQ instruments (which work in nominal mass), isobars can be discriminated from their retention time or fragmentation. Thus, when chromatographic separation is not fully achieved, a false positive or negative (noncompliance of ion ratio) might be reported. TOF MS allows discriminating isobaric compounds from differences in mass defect reducing the possibilities of false identification. Moreover, the elevated mass resolution of modern TOF analyzers allows obtaining narrow-window XIC. For example, a 0.02 Da mass window can be used at the exact mass of the selected compound. Decreasing the mass window leads to a reduction of interferences from isobaric compounds and an increase of the signal-to-noise ratio (Hogenboom et al., 1999; Petrovic et al., 2006; Sancho et al., 2006; Ibáñez et al., 2008) with the final result of much improved selectivity. As an example, in the analysis of a wastewater sample we observed two chromatographic peaks, corresponding to the antibiotics oxonilic acid (m/z 262.0715) and flumequine (m/z 262.0879), when a mass window of 1 Da was selected. Reducing the mass window from 1 to 0.02 Da led to only one peak per window, which corresponded to each individual antibiotic (Ibáñez et al., 2009).

An interesting feature in order to generate molecular formulas and to facilitate an extra confident identification is the use of different isotopic filters. These filters work based on the isotopic pattern deviation between the empirically measured and the theoretical spectrum. The presence of an abundant isotopic pattern in the analyte molecule helps to confirm the presence of that compound in the sample. Thus, the presence of atoms such as carbon, chlorine, bromine, or sulfur in the molecule gives a characteristic isotopic pattern that allows reducing the number of possible elemental compositions for a certain mass-accuracy window. The match between empirical and theoretical data is given by the isotope fit (i-FIT) or sigmaFIT values. These values are calculated, taking into account not only the isotopic distribution but also the accurate masses. The lower the value, the more plausible the elemental composition (Ojanperä et al., 2006; Ibáñez et al., 2008).

One of the main drawbacks of TOF or QTOF instruments compared to QqQ working in SRM mode is their lower sensitivity, since the principal difficulties for the analysis of illicit drugs in wastewater and surface water are their low concentration levels in combination with the complexity of the matrix. This disadvantage might be resolved, or at least minimized, by activating specific functions like, for example, the enhanced-duty cycle (EDC) mode in the instrument used in our work (QTOF Premier, Waters). When TOF operates in EDC mode, the ion abundance is expected to be improved, as ions are transported in packets which make it possible to synchronize the TOF pusher with each ion packet and enhance the duty cycle over a selected m/z range (Weaver et al., 2007).

On the contrary, in some occasions (e.g., during musical events) concentrations of illicit drugs in urban wastewater are higher (Bijlsma et al., 2009). This might provoke detector saturation and mass shifts that would increase mass errors, negatively affecting the identification of potential positives. Dynamic-range enhancement (DRE) is an approach that not only affects the detector saturation, but also the accuracy of mass measurements. The effectiveness of DRE has been reported in the bibliography (Weaver et al., 2007; Tiller et al., 2008; Ibáñez et al., 2009) leading to significant improvement in quantitative and qualitative analysis (i.e., to determine the accurate mass of compounds with more certainty over a large range of concentrations).

In the last few years, several methods using ultrahigh-pressure liquid chromatography (UHPLC) coupled to different MS/MS analyzers have been reported for screening of illicit drugs in water (Huerta-Fontela et al., 2007; Kasprzyk-Hordern et al., 2007; Bijlsma et al., 2009). Because of the high selectivity of tandem MS, the importance of an efficient chromatographic separation might be neglected. However, efficient chromatography is essential to avoid or minimize matrix effects, especially in trace analysis, where failure to completely separate analytes from each other and from the matrix components may result in reporting false positives or negatives (Niessen et al., 2006; Petrovic et al., 2006; Pozo et al., 2006; Bijlsma et al., 2009). UHPLC presents several advantages over conventional liquid chromatography as it generates narrow peaks (increasing peak height and improving sensitivity), facilitates resolving the analytes and matrix interferences (a crucial aspect when dealing with very complex samples), and allows multiresidue analysis to be performed with shorter chromatographic runs. Because of the better resolution and ultra-fast separation, UHPLC has become a powerful tool with significant improvement in terms of sample throughput and sensitivity. However, it requires a fast detector, adequate for narrow chromatographic peaks. Triple quadrupole instruments rely on defined dwell times to monitor analytes. This is not a limitation when a few analytes are monitored; however, it may become a potential problem for multiresidue methods where a wide-scope screening is the objective. Although modern QqQ instruments have been significantly improved, the shortening of dwell times negatively affects the sensitivity and reproducibility. TOF, being inherently faster than QqQ, might detect a huge number of analytes without compromising sensitivity and selectivity. In addition, the higher chromatographic resolution renders high-mass spectra purity. The high acquisition speed of TOF analyzers makes them fully compatible with UHPLC, which is highly beneficial for multiresidue screening methods where difficult matrices are involved (Kaufmann, 2009). As an example, UHPLC-TOF MS has been successfully applied in our group for screening of antibiotics in water (Ibáñez et al., 2009).

When screening a sample for unknown, nonselected, compounds (nontarget screening), it becomes difficult to pick out individual ions, especially when the matrix is complicated or when the concentration of the compound is low. Under these circumstances, it is necessary to use powerful software with chromatographic peak deconvolution capabilities to identify the presence of multiple components and to produce pure spectra for each individual component. As the objective of this chapter is searching for illicit drugs, a more realistic and efficient approach using TOF would be the post-target screening, acquiring the full-scan spectrum at high resolution and performing subsequent nw-XIC at selected accurate m/z. Nevertheless, this procedure is highly time-consuming, since each exact m/z needs to be typed in individually. Powerful software is currently available, which offers the possibility of applying a post-target approach using compound databases based on preselected exact masses. This permits a rapid and simple reviewing by cataloguing analytes (e.g., based on color) as a function of mass error. In addition, this software allows simultaneously visualizing the complete mass spectrum of the positive findings. In many cases, accurate mass spectra can confirm or deny the presence of a specific compound. In theory, an unlimited number of compounds can be included in a database and expensive reference standards are not compulsory, allowing the possibility of setting up a wide-scope screening (Laks et al., 2004). However, reference standards are highly valuable, as experimental data obtained from their analysis (e.g., retention time and fragmentation) provide additional confidence in the identification process.

A database does not necessarily consist of merely illicit drugs. In fact, it is desirable to add several drug-related compounds, such as metabolites and conjugates (e.g., sulfates or glucuronides), as they might produce similar or even higher ecotoxicological problems. In addition, the occurrence of these compounds in the environment is often higher than that of parent compounds. For example, higher concentrations were found for benzoylecgonine in comparison to its parent compound, cocaine (Zuccato et al., 2005.; Huerta-Fontela et al., 2007; Gheorghe et al., 2008; Bijlsma et al., 2009). In certain occasions, it can also be very interesting to include specific fragment ions in the database.

The compromised standard conditions for a wide-scope screening might lead to extensive in-source fragmentation for certain unstable compounds and only searching for (de)protonated molecules might be insufficient. Moreover, today's medicines can be tomorrow's drugs of abuse, as some prescription drugs have the potential of abuse and the inventiveness of drug addicts and dealers are high. In other words "new" drugs of abuse may be "discovered" and are of interest in a later stage. Since evaluation of TOF MS data is performed after acquisition, data can be reexamined in searching for other or new compounds. By including their theoretical mass and empirical formulas into a database, the earlier obtained data can be reprocessed and reevaluated without performing additional analysis. Without the need of reanalyzing the samples, the presence of many compounds can be investigated months or even years later, taking advantage of the abundant and useful information contained in TOF full-spectra acquisition.

As previously stated, TOF instruments normally measure the accurate mass of (de)protonated molecules, whereas QTOF MS can provide additional structural information by obtaining full spectra of products ions at accurate mass. QTOF MS is an excellent technique to perform valuable MS/MS experiments for elucidation purposes. The accurate mass together with the acquisition of the full product ion spectrum is also a powerful tool for the unequivocal confirmation of positives in target analysis. In particular, for a number of compounds (e.g., isomers), information on products ions is almost indispensable for a correct identification. Isomers share the same empirical formula and, therefore, their exact mass, and can only be discriminated either from retention times or from fragmentation pathway. For example, thanks to the use of QTOF MS/MS experiments, the hydroxyl metabolite of the insecticide buprofezin could be distinguished from the sulfoxide metabolite in a banana sample that contained the parent pesticide (Grimalt et al., 2007). Both metabolites share the same empirical formula and exact mass, but their different fragmentation when performing MS/MS experiments gave us the information required to discriminate between them and to assign the correct chemical structure to the metabolite found in the sample.

On the other hand, the potential of QTOF used in MS/MS mode for wide-scope screening purposes is limited, as this implies preselection of an analytes m/z in order to filter it in the quadrupole. However, information of both (de)protonated molecule and fragment ions from a single experiment is feasible by QTOF MS running in the so-called MS[E] acquisition mode, without the need of selecting the precursor ion or losses of isotopic information. Obviously, a QTOF instrument can be used in a TOF mode, which would be the approach recommended for wide-scope screening purposes. However, this tandem mass analyzer offers the possibility of promoting fragmentation in the collision cell, when also being used as a TOF. MS[E] experiments involve the simultaneous acquisition of exact TOF MS data at low- and high-collision energy (Castro-Perez et al., 2002; Plumb et al., 2006; Weaver et al.,

2007; Tiller et al., 2008). By applying low energy (LE) in the collision cell, compound fragmentation is minimized; consequently, information given in the accurate TOF MS spectra corresponds mainly to nonfragmented ions. At high-collision energy (HE), a more efficient fragmentation will take place, resulting in more abundant accurate mass fragments, but still maintaining, in most of cases, the (de)protonated molecule information. Although these ions are not generated in a traditional MS/MS manner, analog terms are sometimes used because of similarities in both approaches by using the same QTOF instrument. Thus, spectra at low energy are mainly dominated by "precursor" ions, while high-energy spectra are dominated by "product" ions. Although similar information might be acquired by varying the cone voltage of a single TOF MS, to produce in-source fragmentation, data obtained are of reduced signal intensity, because of the formation of adducts and the effect of neutrals in the source (Plumb et al., 2006). In addition, we also observed that mass accuracy deteriorated and less abundant fragmentation was obtained in comparison to the use of a collision cell (Díaz et al., 2010).

Hence, the MS^E approach applied in QTOF instruments to produce collision-cell fragmentation is preferable. All MS data obtained by this approach are processed after acquisition, allowing the search for many other compounds, without additional analysis, simply by reprocessing the sample data. Despite these advantages, the technique can be acknowledged as pseudo-MS/MS, the main drawback of MS^E. While the first (quadrupole) mass analyzer operates in wide-band transition mode, multiple components might simultaneously enter the mass spectrometer, making the interpretation of the high-energy mass spectra difficult. For this reason, good chromatographic separation is paramount. When working with UHPLC–QTOF MS in MS^E acquisition mode, full advantage is taken of the higher chromatographic resolution of UHPLC and both (de)protonated molecule and fragmentation data can rapidly be obtained, facilitating simultaneous screening and confirmation.

4.3.2 Application to Real Samples

To illustrate the potential of QTOF MS for confirmation of illicit drugs in the aquatic environment, we will show illustrative examples taken from effluent urban wastewater samples (EWW) of the Castellón province (Spain), which were analyzed in 2008–2009. These complex matrix samples were firstly analyzed by a target UHPLC–QqQ–MS/MS method developed for eleven drugs and metabolites (Bijlsma et al., 2009). Most of samples analyzed contained benzoylecgonine, a major cocaine metabolite. It is interesting to note that another cocaine metabolite, norcocaine, shares the same empirical formula, and thus the same exact mass ([M+H$^+$, *m/z* 290.1392). As reference standards were available to the authors and chromatographic separation was satisfactory, discrimination of both analytes could be made from their retention times (benzoylecgonine, 4.99 min and norcocaine, 5.37 min). However, reference standards were not indispensable when analysis was performed by QTOF MS, as these compounds could also be discriminated from fragmentation (Fig. 4.1). Although good chromatographic separation is important, these two

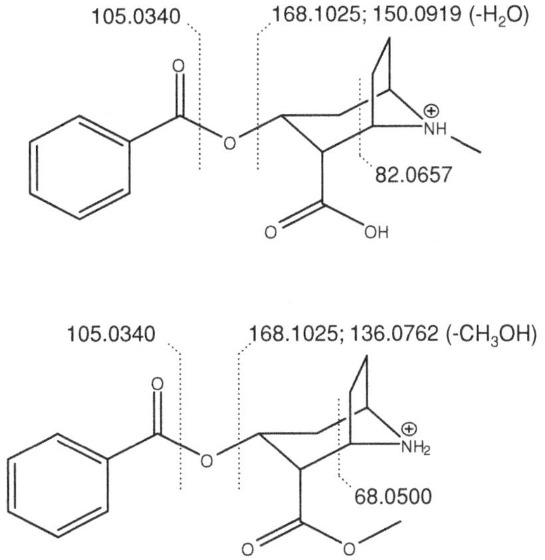

FIGURE 4.1 Top, fragmentation pattern benzoylecgonine [M+H]$^+$. Bottom, fragmentation pattern norcocaine [M+H]$^+$.

isomers could be discriminated by specific fragment ions. This is illustrated in Fig. 4.2, where full-scan spectra of an EWW sample (Fig. 4.2a) and benzoylecgonine and nor-cocaine standard (Fig. 4.2b and c, respectively) by UHPLC–QTOF MS running in both MS^E (LE and HE) and MS/MS acquisition mode are depicted. The relative intensities and accurate masses of the main ions were compared with those of a reference standard. For almost all ions, experimental accurate masses compared to theoretical exact masses (Fig. 4.1) presented errors lower than 2 mDa, giving a high degree of confidence for confirmation of benzoylecgonine. Although relative intensities of high-energy MS^E spectra versus MS/MS spectra are slightly different, because of a small change in the collision energy ramps applied to the protonated molecule, spectra were comparable, demonstrating similar capabilities for both approaches. Nevertheless, QTOF MS/MS provides better spectra with higher sensitivity, as it is more efficient because of the improved duty cycle as well as to cone voltage optimization.

Similarly, QTOF MS allowed confirming positives found by LC–MS/MS QqQ, thanks to the accurate mass measurements of the (de)protonated molecule and the most abundant fragments at LE and HE simultaneous acquisition. As an example, Table 4.1 shows the drugs of abuse confirmed by MS^E in an effluent wastewater sample.

In addition to the reliable confirmation of compounds identity, QTOF MS in MS^E acquisition mode is an excellent tool for a wide-scope screening of illicit drugs in a post-target way. Reference standards are not fully necessary and searching of

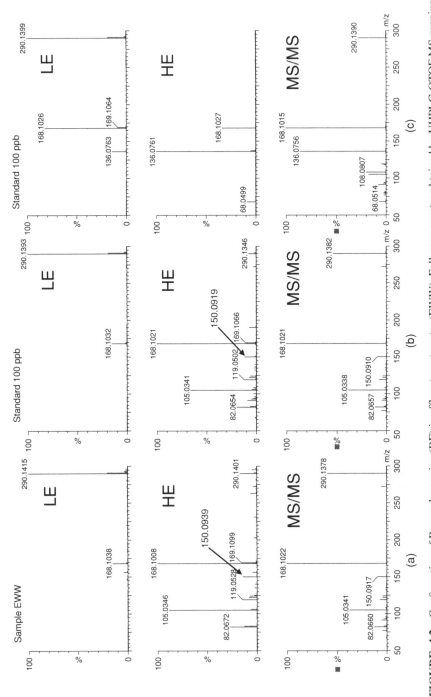

FIGURE 4.2 Confirmation of Benzoylecgonine (BE) in effluent wastewater (EWW). Full-scan spectra obtained by UHPLC-QTOF MS running in MSᴱ (low-collision energy LE, high-collision energy HE) and MS/MS acquisition mode. (a) EWW sample; (b) BE standard; (c) norcocaine standard.

TABLE 4.1   Confirmation of Different Positives of an Effluent Wastewater in MS^E Acquisition Mode at Low (LE) and High (HE) Energy

Compound	Precursor/Fragment	Exact Mass	Accurate Mass	Error) (mDa)
MDMA	$C_{11}H_{16}NO_2$ (LE)	194.1181	194.1164	−1.7
	$C_{10}H_{11}O_2$	163.0759	163.0775	1.6
	$C_8H_7O_2$	135.0446	135.0441	0.2
	C_9H_9O	133.0653	133.0661	−0.8
Cocaine	$C_{17}H_{22}NO_4$ (LE)	304.1549	304.1562	1.3
	$C_{10}H_{16}NO_2$	182.1181	182.1178	−0.3
	$C_9H_{12}NO$	150.0919	150.0923	0.4
	C_5H_8N	82.0657	82.0648	0.9
Benzoylecgonine	$C_{16}H_{20}NO_4$ (LE)	290.1392	290.1407	1.5
	$C_9H_{14}NO_2$	168.1025	168.1025	0
	$C_9H_{12}NO$	150.0919	150.0918	−0.1
	C_7H_5O	105.0340	105.0351	1.1
Norbenzoyl-ecgonine	$C_{15}H_{18}NO_4$ (LE)	276.1236	276.1260	2.4
	$C_8H_{12}NO_2$	154.0868	154.0882	1.4
	$C_8H_{10}NO$	136.0762	136.0741	−2.1
	C_7H_5O	105.0340	105.0336	−0.4

compounds can be easily extended, permitting a wider screening. In our case, a specific software (ChromaLynx XS, Waters) allowed a rapid and simple reviewing by cataloguing, based on color, the accurate mass of the analyte protonated molecule [M+H]$^+$, as a function of mass error. In this way, an effluent wastewater sample was suspected to contain ketamine (Table 4.2). This compound had not been considered in the list of target analytes included in the LC–MS/MS QqQ method applied in our previous work, but it was subsequently investigated by reexamining data obtained by LC–QTOF MS. When performing a narrow-window XIC at its exact mass, an abundant peak was observed (Fig. 4.3a). The MS^E spectra at LE and HE are shown in Fig. 4.3b. Notice from the LE spectra that a characteristic isotopic pattern, corresponding to ^{35}Cl and ^{37}Cl, can be observed. As stated before, the quadrupole operates during MS^E acquisition in wide-band transition mode in such a way that

TABLE 4.2   Ketamine and Codeine Discovered after QTOF Screening of an Effluent Wastewater by MS^E Acquisition Mode at Low (LE) and High (HE) Energy

Compound	Precursor/Fragment	Exact Mass	Accurate Mass	Error (mDa)
Ketamine	$C_{13}H_{17}ClNO$ (LE)	238.0999	238.1017	1.8
	C_7H_6Cl	125.0159	125.0166	0.7
Codeine	$C_{18}H_{22}NO_3$ (LE)	300.1600	300.1610	1.0
	$C_{13}H_9O$	181.0653	181.0636	1.7
	$C_{13}H_9$	165.0704	165.0731	2.7
	$C_{12}H_8$	152.0626	152.0641	1.5

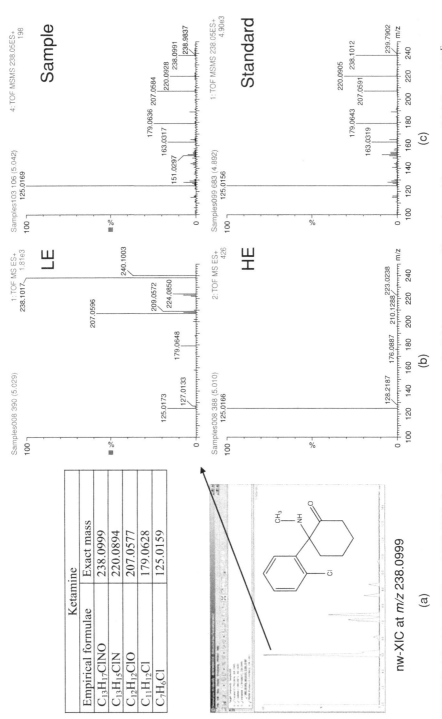

Ketamine	
Empirical formulae	Exact mass
C₁₃H₁₇ClNO	238.0999
C₁₃H₁₅ClN	220.0894
C₁₂H₁₂ClO	207.0577
C₁₁H₁₂Cl	179.0628
C₇H₆Cl	125.0159

(a)

nw-XIC at m/z 238.0999

(b)

(c)

FIGURE 4.3 (a) nw-XIC at m/z 238.0999 of an effluent wastewater sample. Table: exact mass of ketamine and fragments; (b) full-scan MS^E spectra of sample peak (Rt time: 5.0 min); (c) full-scan MS/MS spectra of the sample (top) and ketamine standard (bottom).

all ions are transmitted into the collision cell, providing useful information regarding isotopic pattern. Ketamine ([M+H]$^+$, m/z 238.0999) contains one chlorine atom, which is in agreement with the isotopic pattern mentioned above. In addition, the m/z 207.0596 fragment could be linked to m/z 238.1017 (a loss of -CH$_5$N); since there is no loss of chlorine, the isotopic peak (m/z 209.0572) corresponding to ^{37}Cl was present at around 30% of that of ^{35}Cl, as expected (Fig. 4.3b, top). Another fragment, m/z 125.0173, also shows a small peak at m/z 127.0133, however the ratio does not fit with that expected for one Cl atom. At HE, complete fragmentation toward m/z 125.0166 seems to occur, showing neither other fragment nor isotopic peak. Despite this fact, we did suspect this fragment to contain chlorine, because of the mass error of the fragment in combination with the few possibilities of ketamine to have a product ion at m/z 125 without Cl. The disappearance of the isotopic peak might be explained by the fact that the background noise was subtracted from the combined peak spectra to get cleaner spectra. To ensure the compound identity, ketamine reference standard was purchased and MS/MS experiments for both reference standard and sample were carried out (Fig. 4.3c). This allowed us to confirm the presence of ketamine in the sample, as at least five product ions were coincident with low mass errors. In addition, by using the reference standard, the identity of the analyte could also be confirmed by retention time and semiquantification could even be performed.

Another example of drugs of abuse found in the samples analyzed by MS^E approach was codeine (see Table 4.2), also discovered in an effluent wastewater sample.

The illustrative examples described here show the great potential of UHPLC–QTOF MS in MS^E acquisition mode, for (post-target) wide-scope screening and confirmation of illicit drugs in complex environmental samples.

4.4 CONCLUSIONS

Most of the existing screening methods for illicit drugs in environmental matrices are focused on (pre)target analysis, typically using LC–MS/MS with triple quadrupole analyzers in SRM mode. These methods require a reference standard of each target compound for optimization, compound identification, and quantification. This approach is highly suitable and reliable for trace determination of a limited number of compounds. However, nonpreselected compounds that might be present in samples, even at high concentrations, would not be revealed using this approach. Depending on the number of transitions monitored and on their specificity, additional confirmation is sometimes necessary in order to prevent false positives or negatives.

Other strategies would be highly useful to apply a wide-scope screening for other compounds of interest. LC–QTOF in MS^E acquisition mode is much more appropriate for this goal as most of drugs of abuse and their metabolites are LC-amenable, and QTOF offers useful relevant information for a reliable identification of compounds detected in samples. By means of specialized software and appropriate compound database, accurate mass information can be processed after MS full acquisition using mass error and retention times, when available. This approach allows a safe identification of the compound detected, because of accurate mass data and fragmentation, and is highly useful for simultaneous rapid wide-scope screening and confirmation of illicit drugs and their metabolites. Since screening and confirmation is performed after MS acquisition, data can be reprocessed and reevaluated using new or modified databases in search for other interesting compounds. In theory, a huge number of compounds could be detected and confirmed, without performing additional analysis and without the need of reference standards. Obviously, compounds subjected to investigation have to satisfy the requirements for LC–MS analysis (e.g., they have to be ionized in the commonly API sources, mostly ESI) and they have to be compatible with the sample treatment applied. Therefore, for a wide-scope screening minimum sample treatment would be desirable in order to extend the scope of the method to as many compounds as possible. New generations of QTOF instruments, with improved resolution and sensitivity, will facilitate direct injection of aqueous samples (in some cases, after centrifugation or acidification) for detection of drugs present at sub-parts per billion levels.

Although negative ionization was not taken into consideration in our work, because of the basic character of most illicit drugs, this acquisition mode would complete an even wider scope screening. In addition, unknown compounds, like some metabolites or transformation products (TPs), might also be revealed, by applying a nontarget search using the same accurate mass data and other relevant information given by QTOF MS analysis (e.g., isotopic distribution, fragment ions) together with appropriate databases.

UHPLC–QTOF MS has been successfully applied to urban wastewater in order to investigate the presence of illicit drugs. Previous positives of MDMA, cocaine, benzoylecgonine, and norbenzoylecgonine found by triple quadrupole LC–MS/MS have been confirmed and other nonpreselected compounds, like ketamine or codeine, have been detected by reexamining MS data after MS^E acquisitions.

The strategy described in this chapter for illicit drugs can also be applied to many other contaminants that could be present in the aquatic environment like antibiotics, pesticides, and transformation products, widening the searching of organic pollutants up to more than 1000 compounds. This illustrates the strong potential of UHPLC–QTOF MS for screening purposes.

REFERENCES

Badoud, F., E. Grata, L. Perrenoud, L. Avois, M. Saugy, S. Rudaz, and J.–L. Veuthey. 2009. Fast analysis of doping agents in urine by ultra-high-pressure liquid chromatography-quadrupole time-of-flight mass spectrometry I. Screening analysis. *J. Chromatogr. A* **1216**:4423–4433.

Barceló, D., and M. Petrovic. 2007. Pharmaceuticals and personal care products (PPCPs) in the environment. *Anal. Bioanal. Chem.* **387**:1141–1142.

Bijlsma, L., J. V. Sancho, E. Pitarch, M. Ibáñez, F. Hernández 2009. Simultaneous ultra-high-pressure liquid chromatography-tandem mass spectrometry determination of amphetamine and amphetamine-like stimulants, cocaine and its metabolites, and a cannabis metabolite in surface water and urban wastewater. *J. Chromatogr. A* **1216**:3078–3089.

Castiglioni, S., E. Zuccato, E. Crisci, C. Chiabrando, R. Fanelli, and R. Bagnati. 2006. Identification and measurement of illicit drugs and their metabolites in urban wastewater by liquid chromatography-tandem mass spectrometry. *Anal. Chem.* **78**:8421–8429.

Castiglioni, S., E. Zuccato, C. Chiabrando, R. Fanelli, and R. Bagnati. 2008. Mass spectrometric analysis of illicit drugs in wastewater and surface water. *Mass Spectrom. Rev.* **27**:378–394.

Castro-Perez, J., J. Hoyes, H. Major, and S. Preece. 2002. Advances in MS-based approaches for drug and metabolism studies. *Chromatographia* **55**:S59–S63.

Díaz, R., M. Ibáñez, J. V. Sancho, and F. Hernández. 2010. Building an empirical mass spectra library for screening of organic pollutants by UHPLC-QTOF MS. *Rapid Commun. Mass Spectrom.* under revision.

Ferrer, I., and E. M. Thurman. 2007. Multi-residue method for the analysis of 101 pesticides and their degradates in food and water samples by liquid chromatography/time-of-flight mass spectrometry. *J. Chromatogr. A* **1175**:24–37.

García-Reyes, J. F., B. Gilbert-López, A. Molina-Díaz, and A. R. Fernández-Alba. 2008. Determination of pesticide residues in fruit-based soft drinks. *Anal. Chem.* **80**:8966–8974.

Gheorghe, A., A. van Nuijs, B. Pecceu, L. Bervoets, P. G. Jorens, R. Blust, H. Neels, and A. Covaci. 2008. Analysis of cocaine and its principal metabolites in waste and surface water using solid-phase extraction and liquid chromatography-ion trap tandem mass spectrometry. *Anal. Bioanal. Chem.* **391**:1309–1319.

Grimalt, S., O. J. Pozo, J. V. Sancho, and F. Hernández. 2007. Use of liquid chromatography coupled to quadrupole time-of-flight mass spectrometry to investigate pesticide residues in fruits. *Anal. Chem.* **79**:2833–2843.

Grimalt, S., J. V. Sancho, O. J. Pozo, and F. Hernández. 2010. Quantification, confirmation and screening capability of UHPLC coupled to triple quadrupole and hybrid quadrupole time-of-flight mass spectrometry in pesticide residue analysis. *J. Mass Spectrom.* **45**:421–436.

Hernández, F., O. J. Pozo, J. V. Sancho, F. J. López, J. M. Marín, and M. Ibáñez. 2005a. Strategies for quantification and confirmation of multi-class polar pesticides and transformation products in water by LC-MS2 using triple quarupole and hybrid quadrupole time-of-flight analyzers. *Trends Anal. Chem.* **24**:596–612.

Hernández, F., J. V. Sancho, and O. J. Pozo. 2005b. Critical review of the application of liquid chromatography/mass spectrometry to the determination of pesticide residues in biological samples. *Anal. Bioanal. Chem.* **382**:934–946.

Hogenboom, A. C., W. M. A. Niessen, D. Little, U. A. Th. Brinkman. 1999. Accurate mass determinations for the confirmation and identification of organic microcontaminants in surface water using on-line solid-phase extraction liquid chromatography electrospray orthogonal-acceleration time-of-flight mass spectrometry. *Rapid Commun. Mass Spectrom.* **13**:125–133.

Huerta-Fontela, M., M. T. Galceran, and F. Ventura. 2007. Ultraperformance liquid chromatography-tandem mass spectrometry analysis of stimulatory drugs of abuse in wastewater and surface waters. *Anal. Chem.* **79**:3821–3829.

Ibáñez, M., J. V. Sancho, O. J. Pozo, W. Niessen, and F. Hernández. 2005. Use of quadrupole time-of-flight mass spectrometry in the elucidation of unknown compounds present in environmental water. *Rapid Commun. Mass Spectrom.* **19**:169–178.

Ibáñez, M., J. V. Sancho, D. McMillan, R. Rao, and F. Hernández. 2008. Rapid non-target screening of organic pollutants in water by ultraperformance liquid chromatography coupled to time-of-flight mass spectrometry. *Trends Anal. Chem.* **27**(5): 481–489.

Ibáñez, M., C. Guerrero, J. V. Sancho, and F. Hernández. 2009. Screening of antibiotics in surface and wastewater samples by ultra-high-pressure liquid chromatography coupled to hybrid quadrupole time-of-flight mass spectrometry. *J. Chromatogr. A* **1216**:2529–2539.

Jones-Lepp, T. L., D. A. Alvarez, J. D. Petty, and J. N. Huckins. 2004. Polar organic chemical integrative sampling and liquid chromatography-electrospray/ion-trap mass spectrometry for assessing selected prescription and illicit drugs in treated sewage effluents. *Arch. Environ. Contamination Toxicol.* **47**:427–439.

Kasprzyk-Hordern, B., R. M. Dinsdale, and A. J. Guwy. 2007. Multi-residue method for the determination of basic/neutral pharmaceuticals and illicit drugs in surface water by solid-phase extraction and ultra performance liquid chromatography-positive electrospray ionisation tandem mass spectrometry. *J. Chromatogr. A* **1161**:132–145.

Kaufmann, A. 2009. Quantitative analysis of veterinary drug residues by sub 2-μm particulate high-perfromance liquid chromatography columns and time-of-flight mass spectrometry (UPLC-TOF). In: Liquid chromatography time-of-flight mass spectrometry: principles, tools, and applications for accurate mass analysis, edited by I. Ferrer and E. M. Thurman, pp. 133–150 Wiley, Hoboken, New Jersey.

Laks, S., A. Pelander, E. Vuori, E. Ali-Tolppa, E. Sippola, and I. Ojanperä. 2004. Analysis of street drugs in seized material without primary reference standards. *Anal. Chem.* **76**:7375–7379.

Mezcua, M., O. Malato, J. F. García-Reyes, A. Molina-Díaz, and A. R. Fernández-Alba. 2009. Accurate-mass databases for comprehensive screening of pesticide residues in food by fast liquid chromatography time-of-flight mass spectrometry. *Anal. Chem.* **81**:913–929.

Niessen, W. M. A., P. Manini, and R. Anderoli. 2006. Matrix effects in quantitative pesticide analysis using liquid chromatography-mass spectrometry. *Mass Spectrom. Rev.* **25**:881–899.

Ojanperä, S., A. Pelander, M. Pelzing, I. Krebs, E. Vuori, and I. Ojanperä. 2006. Isotopic pattern and accurate mass determination in urine drug screening by liquid chromatography/time-of-flight mass spectrometry. *Rapid Commun. Mass Spectrom.* **20**:1161–1167.

Pelander, A., I. Ojanperä, S. Laks, I. Rasanen, and E. Vuori. 2003. Toxicological screening with formula-based metabolite identification by liquid chromatography/time-of-flight mass spectrometry. *Anal. Chem.* **75**:5710–5718.

Petrovic, M., M. Gros, and D. Barcelo. 2006. Multi-residue analysis of pharmaceuticals in wastewater by ultra-performance liquid chromatography-quadrupole-time-of-flight mass spectrometry. *J. Chromatogr. A* **1124**:68–81.

Plumb, R. S., K. A. Johnson, P. Rainville, B. W. Smith, I. D. Wilson, J. Castro-Perez, and J. K. Nicholson. 2006. UPLC/MSE; a new approach for generating molecular fragment information for biomarker structure elucidation. *Rapid Commun. Mass Spectrom.* **20**:1989–1994.

Pozo, O. J., J. V. Sancho, M. Ibañez, F. Hernandez, and W. M. A. Niessen. 2006. Confirmation of organic micropollutants detected in environmental samples by liquid chromatography tandem mass spectrometry: achievements and pitfalls. *Trends Anal. Chem.* **25**(10):1030–1042.

Richardson, S. D. 2008. Environmental mass spectrometry: emerging contaminants and current issues. *Anal. Chem.* **80**:4373–4402.

Richardson, S. D. 2009. Water analysis: emerging contaminants and current issues. *Anal. Chem.* **81**:4645–4677.

Sancho, J. V., O. J. Pozo, M. Ibañez, and F. Hernández. 2006. Potential of liquid chromatography/time-of-flight mass spectrometry for the determination of pesticides and transformation products in water. *Anal. Bioanal. Chem.* **386**:987–997.

Tiller, P. R., S. Yu, J. Castro-Perez, K. L. Fillgrove, and T. A. Baillie. 2008. High-throughput, accurate mass liquid chromatography/tandem mass spectrometry on a quadrupole time-of-flight system as a 'first-line' approach for metabolite identification studies. *Rapid Commun. Mass Spectrom.* **22**:1053–1061.

Weaver, P. J., A. Laures, and J.–C. Wolff. 2007. Investigation of the advanced functionalities of a hybrid quadrupole orthogonal acceleration time-of-flight mass spectrometer. *Rapid Commun. Mass Spectrom.* **21**:2415–2421.

Zuccato, E., C. Chiabrando, S. Castiglioni, D. Calamari, R. Bagnati, S. Schiarea, R. Fanelli. 2005. Cocaine in surface waters: A new evidence-based tool to monitor community drug abuse. *Environ. Health Global Access Sci. Source* **4**:14. Available at http://www.ehjournal.net/content/4/1/14.

CHAPTER 5

DETERMINATION OF ILLICIT DRUGS IN THE WATER CYCLE BY LC−ORBITRAP MS

PIM DE VOOGT, ERIK EMKE, RICK HELMUS, PAVLOS PANTELIADIS, and JAN A. VAN LEERDAM

5.1 INTRODUCTION

In the past 60 years, the social impact of drug abuse and illegal trade of drugs has become significant. As a result, a strong demand has risen for the monitoring of the amounts and trends of consumption of illicit drugs and for the tracing of new ones. Thus far, this has been done with the use of traditional surveys and inventories of data collected from specific information sources, like the police, hospitals, and drug addiction treatment centers. Only limited data are available on the consumption of illicit drugs, because their use is neither registered nor controlled. Although the Dutch Foundation for Pharmaceutical Statistics (SFK, 2010) publishes data on annual prescriptions of some pharmaceuticals/drugs that are controlled, these data do not reflect the consumption volumes. Moreover, the data do not include the uncontrolled illegal selling of drugs of abuse. Although considerable progress has been made with the collection of those data, there are still a number of limitations that seem hard to overcome. More specifically, drug use is highly stigmatized in today's society, so that users tend to hide their consumption habits. Furthermore, when the proportion surveyed is statistically extrapolated to the total of the population, uncertainty is unavoidably introduced into the reported results. Finally, these methods are time-consuming, resulting in slow response reports, usually with an annual periodicity.

Each year the Trimbos Institute (Netherlands Institute of Mental Health and Addiction) in the Netherlands, publishes the National Drug Monitor (2006). In this report, the illicit drug consumption is estimated based on criteria such as (il) legal import volumes, anonymous surveys, consumption during so-called "dance parties" and so on. However, it remains very difficult to estimate the consumption of illicit drugs.

Illicit Drugs in the Environment: Occurrence, Analysis, and Fate Using Mass Spectrometry, Edited by Sara Castiglioni, Ettore Zuccato, and Roberto Fanelli
Copyright © 2011 John Wiley & Sons, Inc.

Daughton and Jones-Lepp (2001) were the first to present a new approach to the solution of this problem. They suggested that monitoring sewage treatment plants (STPs) for illicit drugs can be used for back-calculations in order to estimate collective consumption. Since then, illicit drugs have been receiving increased interest from environmental scientists. The same group of investigators was the first to report the occurrence of illicit drugs in treated sewage effluents in 2004 (Jones-Lepp et al., 2004). In 2006, Castiglioni et al. also reported the occurrence of benzoylecgonine in the Italian Po River. Since then illicit drugs have been reported in wastewaters of several countries, including the United States, Germany, Italy, Spain, United Kingdom, Switzerland, Belgium, The Netherlands, and Poland (Vanderford and Snyder, 2006; Hummel et al., 2006; Boleda et al., 2007; Huerta-Fontela et al., 2007; Kasprzyk-Hordern et al., 2007; Postigo et al., 2008; Bijlsma et al., 2009; Van Nuijs et al., 2009; Hogenboom et al., 2009).

5.1.1 Environmental Occurrence of Illicit Drugs

Since the early work at the beginning of the 21st century, several studies have been published on the occurrence of illicit drugs in environmental waters. Illicit drugs were detected in all studies investigated, either in surface waters or wastewater. The lowest concentrations, as expected, were detected in surface waters, while the highest ones occurred in wastewater influents. Analysis of results from surface water and wastewater influent published in the literature is presented in Tables 5.1–5.3.

Concentrations in surface waters of cocaine and its metabolites, which are the compounds studied most frequently, range from <1 up to 115 ng/L of cocaine, with values usually between the limit of detection (LOD) of each method and 35 ng/L. The higher value was reported for the river Zenne, 115 ng/L, downstream of Brussels, which was explained by the untreated, industrial and domestic wastewater discharged into the river (Van Nuijs et al., 2009). High values (44 ng/L) were also observed in the river Olona (Zuccato et al., 2008) a small heavily polluted river close to Milan and in the river Liffey (33 ng/L) close to Dublin (Bones et al., 2007).

Similar results were reported for benzoylecgonine (BE), cocaine's basic metabolite, but with somewhat higher levels, since most of the cocaine used is metabolized in the human body, with up to 55% being metabolized into BE (Cook et al., 1985). The levels of BE ranged from the LOD up to 520 ng/L, with the highest value reported again from the river Zenne (Van Nuijs et al., 2009). For the river Olona, a concentration of 180 ng/L was reported (Zuccato et al., 2008), while a value below the LOD was reported for the river Liffey (Bones et al., 2007).

Lower concentrations were reported for the other drugs investigated in surface waters, including the presence of 11-nor-9-carboxy-Δ^9-tetrahydrocannabinol (11-COOH-THC), 21 ng/L, in the river Llobregat, close to Barcelona (Boleda et al., 2007). Morphine (38 ng/L) was found in the river Olona (Zuccato et al., 2008) and in the river Rhine (78 ng/L), close to Koblenz (Hummel et al., 2006).

More data are available for wastewater influents and effluents. Cocaine in effluents ranged up to 560 ng/L, a value reported for the STP of Castellón in eastern Spain (Bijlsma et al., 2009), while most of the values reported are between 30 and 140 ng/L.

TABLE 5.1 Concentrations of Selected Illicit Drugs in Surface Waters Reported in Recent Studies[a]

Location	Cocaine	BE	THC	THC-COOH	Heroin	Methadone	Morphine	References[b]
Olona, I	44	183		<0.48		8.6	38	a
Lambro, I	15	50		3.7		3.4	3.5	a
Po, I	0.5	3.7		0.3 [0.2]		0.5 [0.2]	<0.55	a
Arno, I	1.7	22		0.5 [0.4]		4.8 [4]	3 [1.4]	a
Thames, UK	4	13		<0.48				a
Llobregat, ES	nd	nd	<LOQ-13.6	21	nd	6.4	5.4	b
Broadmeadow, UK	25	<LOD					nd	c
Liffey, UK	33	<LOD					nd	c
Rhine, DE	nd	3					10	d
Zenne, B	115	520					(<LOQ-78)	e

[a]All concentrations in ng/L. Values in square brackets refer to standard deviations; nd, not determined.
[b]References: a Zuccato et al., 2008; b Boleda et al., 2007; c Bones et al., 2007; d Hummel et al., 2006; e Van Nuijs et al., 2009.

TABLE 5.2 Concentrations of Illicit Drugs in Sewage Treatment Plants Influents as Reported in Recent Studies[a]

STP Influent	THC-COOH	THC-OH	THC	Heroin	Methadone	Morphine	LSD	LSD-OH	nor-LSD	References[b]
Castellón ES	<400									a
Barcelona, ES	4.3 [7.8]	8.4 [2.1]	nd	nd		162.9 [20]	2.8 [1.2]	5.6 [12.1]	4.3 [1.8]	b
Valencia, ES	13.8	37.2	22.2	2.3		75.1	4.7	2.6	22.1	b
Benicassim, ES	32.5	77.6	39.4	2.4		66.7	3	4.2	13	b
Gandia, ES	16.8	24.2	13.8	nd		62.6	1.1	nd	5.3	b
Catalonia, ES	<12.5–96.2		<8.3–31.5	<20	4-23.9	<7.1–96.7				c
Ringsend, UK						nd				d
Nosedo, I	62.7 [5]				11.6 [1.7]	83.3 [11.8]				e
Lugano, S	91.2 [24.7]				49.7 [9.6]	204.4 [49.9]				e
12 STPs, DE						310				f

TABLE 5.2 (*Continued*)

STP Influent	Cocaine	BE	CE	NBE	Norcocaine	Amphetamine	MDA	MDEA	MDMA	MA	References[b]
Castellón ES	370–1240	1480–10500	<150–190	150–430	nd	610–1400	500–1690	<500	500–27,500	<500	a
37 STPs, BE	10–753 (146)	33–2258 (497)									g
Antwerpen-Noord, BE	149	515									g
Antwerpen-Zuid, BE	584	1858									g
Brussel-Noord, BE	348	1291									g
Deurne, BE	753	2258									g
Leuven, BE	141	716									g
Barcelona, ES	860.9 [213.6]	42250.7 [1142.8]	77.5 [33.2]			41.1 [9.1]			133.6 [29.8]	18.2 [5.8]	b
Valencia, ES	651	1900	89.2			20.4			113	7.8	b
Benicassim, ES	540	1450	97.2			35.5			245	3.7	b
Gandia, ES	316	1020	49.2			6.5			47.4	3	b
Lier, BE	167	612.5									h
Hoboken, BE	399	1403									h
Tessenderlo, BE	22	82									h
Mol, BE	48	230									h
Lommel, BE	112	553									h
Clifynydd, UK	521	992				4310					i
Coslech, UK	207	1082				1196					i
Ringsend, UK	489 [117]	290 [11]									d
16 STP, ES	79	810				15		28	49		j
Nosedo, I	421.4 [83.3]	11320.1 [197.2]	11.5 [5.1]	36.6 [7.8]	13.7 [5.3]	14.7 [10.6]	4.6 [7.3]	1.5 [3.8]	14.2 [14.5]	16.2 [7.1]	e
Lugano, S	218.4 [58.4]	574.4 [169.4]	5.9 [2.6]	18.8 [5.6]	4.3 [0.9]	<5.4	<8.7	<4.19	13.6 [12.6]	<3.7	e
12 STPs, DE		78									f
Leuven, BE	109.2	658.8									k

[a] All concentrations in ng/L. Values in square brackets refer to standard deviations; values in parenthesis refer to mean values. nd, not determined. Abbreviations: CE, cocaethylene; NBE, nor-BE; MDA, 3,4-methylenedioxyamphetamine; MDEA, 3,4-methylenedioxy-N-ethylamphetamine; MA, methylamphetamine; others explained in text.

[b] References: a Bijlsma et al., 2009; b Postigo et al., 2008; c Boleda et al., 2007; d Bones et al., 2007; e Castiglioni et al., 2006; f Hummel et al., 2006; g Van Nuijs et al., 2009; h Gheorghe et al., 2008; i Kasprzyk-Hordern et al., 2007; j Huerta-Fontela et al., 2007; k Waumans et al., 2006.

TABLE 5.3 Concentrations of Illicit Drugs in Wastewater Treatment Plants Effluents[a]

STPs Effluent	Cocaine	BE	CE	NBE	Norcocaine	Amphetamine	MDA	MDEA	MDMA	MA	References[b]
2 STP, NL	nd	10							150		a
Castellón ES	30–560	30–6790	<30–80	<30–170	<30	110–210	410–680	<100	200–21,200	<100	b
Barcelona, ES	6.2 [3.7]	30.3 [17.6]	1.7 [1.2]			0.5 [0.1]			82.1 [22.2]	6.3 [0.6]	c
Valencia, ES	33.2	220				2.2			38.2	2.7	c
Benicassim, ES	83.6	49.9	2.1			1			376	2	c
Gandia, ES	105	318	6.8			3.3			30.3	1.5	c
Clifynydd, UK	128	1091				201					d
Coslech, UK	<1	13				2					d
Ringsed, UK	138 [20]	22 [4]									e
Swords, UK	nd	nd									e
Shanaganagh, UK	77 [25]	31 [18]									e
Leixlip, UK	47 [10]	nd									e
Navan, UK	111 [15]	67 [10]									e
16 STPs	17	216				<0.8		<0.3	41		f
Nosedo, I	<0.99	<0.92	<0.66	<0.56	<0.67	<0.67	1.1 [1.5]	<1.64	4.4 [3.7]	<1.11	g
Lugano, S	10.7 [3.2]	100.3 [28.6]	0.2 [0.5]	7.5 [2.9]	0.7 [0.5]	<0.67	0.9 [1.9]	<1.64	5.1 [3]	<1.11	g
12 STPs, DE		49									h
3 STPs, US									0.5	1.3	i

92

TABLE 5.3 (Continued)

STPs Effluent	THC-COOH	THC-OH	THC	Heroin	Methadone	Morphine	LSD	LSD-OH	nor-LSD	References[b]
2 STP, NL			30		20	40				a
Castellón ES	<80									b
Barcelona, ES	8.4 [3.8]	4.8 [1.9]	nd	nd		21.8 [3]	0.3 [0.2]	0.7 [0.3]	0.6 [0.5]	c
Valencia, ES	3.9	23	20.5	1.2		18.8	1.6	0.8	4	c
Benicassim, ES	19	14.1	13	nd		11.8	0.6	nd	1.5	c
Gandia, ES	10.8	8	nd	nd		29.7	0.2	nd	1.3	c
Catalonia, ES	14.8–71.7		<8.3	<20	4–24.7	<7.1–81.1				j
Ringsed, UK						<856				e
Swords, UK						874 [86]				e
Shanaganagh, UK						nd				e
Leixlip, UK						<856				e
Navan, UK						<856				e
Nosedo, I	<0.94				9.1 [0.5]	<3.22				g
Lugano, S	7.2 [3.7]				36.2 [2.8]	55.4 [11.1]				g
12 STPs, DE						110				h

[a] All concentrations in ng/L. Values in square brackets refer to standard deviations; values in parenthesis refer to mean values. nd, not determined.
[b] References: a Emke et al., 2009; b Bijlsma et al., 2009; c Postigo et al., 2008; d Kasprzyk-Hordern et al., 2007; e Bones et al., 2007; f Huerta-Fontela et al., 2007; g Castiglioni et al., 2006; h Hummel et al., 2006; i Jones-Lepp et al., 2004; j Boleda et al., 2007.

Relatively high values were reported for the STPs in the UK (Kaszprzyk-Hordern et al., 2007; Bones et al., 2007), with Ringsed and Clifynydd being two examples. For BE, the highest values were again reported for Castellón and Clifynydd, 6790 and 1091 ng/L, respectively. Two aggregated results should be pointed out, namely, one for 12 STPs in Germany (Hummel et al., 2006) and one for 16 STPs in Catalonia (Huerta-Fontela et al., 2007), giving means of 49 and 216 ng BE/L, respectively. 3,4-Methylenendioxy-methamphetamine (MDMA) was detected in high values in two STPs in the Netherlands (Emke et al., 2009), Castellón (Bijlsma et al., 2009), Barcelona, and Benicassim (Postigo et al., 2008). High amphetamine levels were reported for Clifynydd's effluents (Kasprzyk-Hordern et al., 2007). Morphine, THC, and its metabolites were reported in higher values from surface waters and, in some locations, heroin, LSD (lysergic acid diethylamide), and its metabolites were detected (see Table 5.1).

The highest concentrations of drugs are observed in STP influents. More specifically, for cocaine, levels of BE up to 10,500 ng/L were detected, the highest value again being reported for Castellón (Bijlsma et al., 2009). High values were also reported for Barcelona, 4200 ng/L, and in Valencia, 1900 ng/L (Postigo et al., 2008). An aggregated study of 37 STPs in Belgium (Van Nuijs et al., 2009) gave a mean of 500 ng/L while a similar one for 16 STPs in Catalonia gave 810 ng/L (Boleda et al., 2007). Similar trends, but at lower concentration levels, were reported for cocaine, except for the study by Bones et al. (2007) that reports higher cocaine values than BE ones. Amphetamine was also detected at several locations, mostly in Spain, and remarkably high average values were reported by Kasprzyk-Hordern et al. (2007) for Clifynydd and Coslech in the UK amounting to 4300 and 1200 ng/L, respectively.

Morphine was detected in 2 STPs in the Netherlands (40 ng/L), in Catalonia (up to 81 ng/L), in Lugano (55 ng/L), and in 12 STPs in Germany giving a mean of 110 ng/L (Table 5.3). Bones et al. (2007) reported a noticeably high value, 870 ng/L, which shows an 86% standard deviation. Moreover, these authors detected morphine in effluent while not in influent water, in samples coming from the same STP. An explanation given is the cleavage of glucuronide metabolites during wastewater treatment. However, even if that is the case, no similar results, thus far, were observed by any other researcher. THC and its metabolites were also detected in low concentrations: 14 and 34 ng/L for Δ^9-tetrahydrocannabinol (Δ^9-THC) and its human metabolite THC-COOH, respectively, in STPs located in Spain (Huerta-Fontela et al., 2007).

Thus far, only two studies have reported on the measurement of these compounds in drinking water: meprobamate was detected at a level of 6 ng/L level (Vanderford and Snyder, 2006) and BE at a mean level of 45 ng/L and a maximum of 130 ng/L (Huerta-Fontela et al., 2008).

5.1.2 Objectives

As a consequence of the findings of illicit drugs in the water cycle, the possible hazards for the environment and human health because of exposure to these drugs are increasingly being investigated. Acute health hazards because of illicit drugs and metabolites in drinking water are unlikely. However, long-term (chronic) effects and those because of mixtures of substances with similar mechanisms of action have not, thus far, been studied. A well-founded scientific risk assessment is, therefore, not yet possible. In order to understand the levels that have been reported and that may occur in surface waters, it is important to know whether or not the illicit drugs are difficult to remove in (drinking) water treatment processes.

The monitoring of illicit drugs in the aquatic environment began in The Netherlands in 2006. A large number of the illicit drugs, viz. ~130 out of a total of ~160 (Dutch Opium Act List I + II; List I refers to illicit drugs, whereas List II refers to legal, but addictive, drugs) have medium to high polarities, as is reflected by a log K_{ow} value of less than 3. The metabolites are generally more polar than the parent compound. As a result of high polarities and corresponding high aqueous solubilities, illicit drugs and their metabolites present in influents may pass easily through water treatment processes. On the other hand, degradation of a compound during wastewater treatment or in the environment can significantly decrease the environmental or health relevance of a compound. Whether or not such degradation occurs is dependent on the physicochemical properties of the illicit drug. In order to evaluate the hazard in the water cycle, the efficiency of removal of the drugs in treatment processes must not only include the parent compound, but also the transformation product(s).

A further question relates to the perception of the issue by consumers of drinking water. In general, consumers do not accept pharmaceuticals in their drinking water even if it is demonstrated not to be associated with a health hazard. Drinking water companies strive to ensure that concentration levels of contaminants in drinking water are as low as possible (Van der Hoek et al., 2008). Thus far, there are neither guidelines for the concentrations of illicit drugs in surface waters or drinking water (e.g., in the framework of Drinking Water Directive 98/83/EC), nor have tolerable daily intakes been issued for these substances (Schriks et al., 2010). Illicit drugs have not been regulated in the Water Framework Directive of the EU.

The work on determination of illicit drugs in surface waters in The Netherlands began with an inventory conducted in 2006–2007 by the KWR Institute. At that time, the standards were limited and KWR's application for a license to keep standards was pending. An analytical method, based on full-scan accurate mass screening employing the LTQ Orbitrap–mass spectrometer was selected and optimized. In 2008, the formal license for standards was obtained from the Dutch authorities and the existing accurate mass-based method was extended to cover a total of 30 different illicit drugs and some of their transformation products. This chapter presents details on the development and optimization of a single analytical method for a wide group of illicit drugs in water based on LC coupled to the LTQ Orbitrap–full-scan mass spectrometry. It discusses the typical characteristics of the methodology and presents some of the findings in samples collected in The Netherlands in the period 2006–2009.

5.2 ORBITRAP MASS SPECTROMETRY

An Orbitrap is a type of mass spectrometer that consists of an outer barrel-like electrode and a coaxial inner spindlelike electrode that form an electrostatic field.

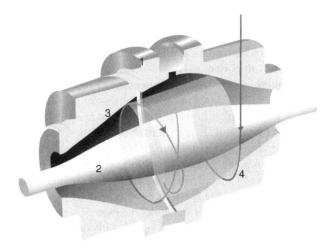

FIGURE 5.1 Ion trajectories in an Orbitrap mass spectrometer. (1) Orbitrap; (2) inner electrode; (3) outer electrodes (taken from Wikipedia).

(Makarov, 2000; Hu et al., 2005). In the Thermo Electron configuration, the Orbitrap is combined with a linear ion trap (LTQ).

5.2.1 Principle of Operation

In an Orbitrap, ions are tangentially injected into the electric field between the electrodes and trapped because their electrostatic attraction to the inner electrode is balanced by centrifugal forces. Thus, ions orbit the central electrode in rings (Fig. 5.1). In addition, the ions also move back and forth along the axis of the central electrode. Therefore, ions of a specific mass-to-charge ratio move in rings oscillating along the central spindle. The frequency, ω, of these harmonic oscillations is independent of the ion velocity and is inversely proportional to the square root of the mass-to-charge ratio.

By registering the ion oscillation, the trap can be used as a mass analyzer. Orbitraps have a high mass accuracy (typically 2–5 ppm), a high resolving power (up to 100,000), and a high dynamic range (around 5000) (Makarov et al., 2006a, b).

The resolution provided by the Orbitrap is proportional to the number of harmonic oscillations of the ions. As a result, the resolving power is inversely proportional to the square root of m/z and proportional to acquisition time. Scigelova and Makarov (2006) showed that the resolving power is 7500 per 0.1 s transient at m/z 400, a transient being the duration that the time domain signal is acquired. The resolving power decreases further as the m/z value increases so that at an m/z value of 1600 the resolving power has halved to 3750 per 0.1 s. Approximately 0.1 s per transient is required for data processing; thus, a 0.1-s transient has a cycle time of 0.2 s. These effects in combination result in a resolving power of 7500 at m/z 400 and 3750 at m/z 1600 when the Orbitrap is operating at five acquisitions/s.

The LTQ can be used as a front end for the Orbitrap. The linear form of the trap can be used either as a selective mass filter, or as an actual trap. In the latter mode, the ions entering the storage quadrupole lose their energy in collisions with gas molecules present, thereby becoming unable to escape and accumulate. Bunched ions are next extracted pulsewise by temporary supplying a negative voltage to the lens between the LTQ and the Orbitrap. The full process is described in detail by Hu et al. (2005).

The use of the LTQ FT Orbitrap−MS/MS combines the tandem mass spectrometry capability of the linear ion trap with the high-resolution and mass-accuracy capability of the FT Orbitrap. This combination allows high-quality accurate mass MS^n spectra to be acquired. Fourier transformation of the acquired transient allows wide mass range detection with high resolving power, mass accuracy, and dynamic range. High-accuracy mass measurements are within 2 parts per million (ppm) using internal standards and within 5 ppm with external calibration.

5.3   METHODS AND MATERIALS

5.3.1   Analytes

5.3.1.1   Compound Selection and Characterization   Relevant compounds/ representatives were selected from characteristic groups of List I + II Drugs of the Dutch Opium Act, including their metabolites. The selection was based on criteria that included known use, reported occurrence, and availability as standards (including deuterated standards), as well as toxicity and behavior in treatment processes. Table 5.4 presents the final selection, which consists of 32 illicit drugs that were divided into six different classes.

5.3.1.2   Preparation of Standards   Once the license was obtained by the KWR laboratory, all calibrated reference standards were based on the free drug substances, which were acquired from Lipomed AG (Arlesheim, Switzerland), except 2-ethyl-1,5-dimethyl-3,3-diphenylpyrrolidine (EDDP) and 2-ethyl-5-methyl-3,3-diphenylpyrrolidine (EDMP), which were obtained from Cerillant (Round Rock, TX, USA), and Meprobamate obtained from Sigma (St. Louis, MO, USA).

The standards from Lipomed were available in ready-to-use calibrated reference ampuls of 1 mL in either methanol, ethanol, or acetonitrile, at a concentration of 1 g/L . From the ampul, 900 μL was diluted to a concentration of 36 mg/L in methanol according to the groups mentioned in the table above. The final mixture was made up by diluting aliquots from every group to a concentration of 3.6 mg/L. Working solutions for the calibration curves were made in methanol concentrations ranging from: 7 pg/μL to 3 ng/μL. Before each analytical run, the standards were diluted 10 times with ultrapure water (Millipore, MA, USA) resulting in a mix of 90% water and 10% methanol. For the deuterated standards, the same procedure was followed, except that they were added to all the calibration standards at a concentration level of 72 pg/μL. The remaining solution of the ampul was used to made up individual standards at around 10 mg/L in methanol for checking the purity of the standards.

TABLE 5.4 Selection and Grouping of Illicit Drugs Monitored in This Study

Cannabinoids	Opiates	Barbiturates	Amphetamines	Benzodiazepins	Others
11-Nor-9-Carboxy-THC	Heroin	Barbital	Amphetamine	Oxazepam	Cocaine
11-OH-Δ^9-THC	6-Monoacetyl morphine (6-MAM)	Phenobarbital	Metamphetamine	Nordazepam	Benzoylecgonine
9-COOH-Δ^9-THC	Morphine	Pentobarbital	MDA	Diazepam	Methacathinone
Δ^9-THC	Codeine		N-Ethyl-MDA (MDEA)	Temazepam	Meta-chlorophenyl-piperazine (Meta-CPP)
	Methadone		MDMA	Desalkyl- flurazepam	Ketamine
	Fentanyl				Meprobamate
					Ritalin
					EDDP, EDMP

Stock and working solutions were stored at a temperature of −20°C. The ultrapure water was obtained by purifying demineralized water in a Milli-Q-system. All organic solvents were HPLC-grade. A 30% hydrochloric acid solution was used and the quality of the formic acid was 98%. Sea sand was obtained from Mallinckrodt Baker, Deventer, The Netherlands and had been glowed and ignited by the manufacturer.

5.3.2 Sample Treatment

5.3.2.1 Solid-Phase Extraction
Samples collected in the period 2006−2007 were preconcentrated off-line employing Oasis HLB adsorbent (5-mL glass cartridges, 200 mg, Waters, Milford, MA, USA) cartridges. A 6-mL polypropylene cartridge, filled with 4 g of sea sand, was placed above the cartridge with adsorbent, in order to prevent the SPE material from clogging. The SPE cartridges were successively conditioned with 10 mL of acetonitrile, 10 mL of methanol, and 10 mL of ultrapure water (pH 2, except for the neutral pH experiments). The sea sand cartridges were conditioned in the same way, but the methanol was omitted, because of the absence of active sites. In order to examine the influence of the pH on the recoveries, spiked samples of drinking water and surface water were concentrated at pH 7 and 2.

The internal standard compounds fenuron and chloroxuron were spiked to 1 L of the water sample (200 mL in case of STP in/effluent) at a concentration level of 1 μg/L. The samples were acidified to pH 2 with 30% hydrochloric acid (except for the experiments at neutral pH) and the water samples were percolated through the cartridges at a flow of 10 mL/min. After passing the water, the sea sand cartridges were removed and the SPE cartridges were washed twice with 5 mL of acidified water (pH 2). The cartridges were dried under vacuum for 90 min until dryness and eluted three times with 2.5 mL of acetonitrile. The eluates were concentrated to a final volume of 500 μL under a gentle flow of nitrogen. Last, the extracts were diluted with 500 μL of an internal standard (atrazine-D5 and bentazon-D6) solution in ultrapure water at a concentration level of 2 mg/L. The final concentration of the deuterated internal standards in the sample extract was 1 mg/L.

Samples collected in the period 2008−2009 were extracted with the help of an automated large-volume SPE (Gilson, Australia) system using Oasis-HLB SPE stationary phase (150 mg, 60 μm). 900 mL (in the case of tapwater, surface water), 200 mL (effluent) or 100 mL (influent) of a sample was adjusted to pH 7 and 16 deuterated analogs were added. The analytes were eluted from the SPE column with 8 mL of methanol and concentrated.

5.3.2.2 Chromatographic Separation
The liquid chromatograph was equipped with a Surveyor autosampler model Plus and a Surveyor quaternary gradient HPLC-pump (Thermo Fisher Scientific, Breda, The Netherlands). The separation of the compounds was performed either on a 3-μm Omnisphere C18 (Varian-Chrompack, Middelburg, The Netherlands) column with dimensions: 100 ×2.0 mm i.d. (2006−2007 series) or a 150 × 2.1 mm i.d. Xbridge C18-column (Bester,

Amsterdam, The Netherlands) with 3.5-μm particles (2008–2009 series). The pre-column used in both series was a 4.0 × 2.0 mm i.d. Phenomenex Security Guard column (Bester). The columns were conditioned in a column thermostat that was maintained at a temperature of 21°C. The extract (20 μL) was injected into the LC system and the analytes were separated using a linear gradient elution.

For the 2006–2007 samples, an acetonitrile/water gradient was used at a flow rate of 0.3 mL/min, starting at 5% acetonitrile/95% water/0.05% formic acid, increasing to 40% acetonitrile/60% water/0.05% formic acid in 15 min, and then, finally, to 100% acetonitrile in 10 min, and subsequently held constant for 10 min. Between the runs, the analytical column was reequilibrated for 15 min. For the 2008–2009 samples, a methanol/water gradient was used at 0.3 mL/min: (A) ultrapure water, 0.05% formic acid; (B) methanol 0.05% formic acid; a linear gradient of 95% A to 0% in 20 min was employed. The mass spectrometer was operated in positive ionization mode using an electrospray interface.

5.3.3 Mass Spectrometry

5.3.3.1 Mass Spectrometric Conditions In the present work, a linear ion trap–FT Orbitrap mass spectrometer (Thermo Electron, Bremen, Germany) was used. The ion trap part of this system was equipped with an ion max electrospray ionization (ESI) probe that can be applied in the positive- and negative-ion mode for the compounds investigated. The present chapter only presents results of the positive-ion mode.

From the LC column effluent, 125 μL/min (2006–2007 series) or 300 μL/min (2008–2009 series) was introduced into the mass spectrometer. The full-scan accurate mass spectra from 100 to 600 Da, which were obtained at a resolution of 30,000 FWHM, were processed using the Xcalibur version 2.0 software. The total cycle time depends upon the resolution; at a resolution of 30,000 FWHM the total cycle time is about 0.55 s. The following conditions were employed in the ESI positive-ion mode: source voltage 3.6 kV, heated capillary temperature 300°C, capillary voltage 30 V, and tube lens 70 V. In the LTQ component of the instrument, the FT part was set to 26°C and helium was used as damping gas. All measurements were done using the automatic gain control (AGC) of the LTQ to adjust the number of ions entering the trap. Product ions were generated in the LTQ trap at a normalized collision energy setting of 35%, using an isolation width of 2 Da.

5.3.3.2 Mass Accuracy Before measuring the samples, accurate mass calibration was performed using flow injection of a 1,3,6-polytyrosine solution (*m/z* 181, 507, and 997) in methanol/water (50/50; v/v) with 0.1% formic acid at a flow rate of 10 μL/min. No recalibration was performed during the experiments. While for identification of unknowns the mass accuracy is of greatest importance (Hogenboom et al., 2009), this is not a strict prerequisite for target analysis, such as in this study, as the mass error could be accounted for from the injection of calibration solutions. Still, the mass accuracy should be stable over a whole analytical sequence to ensure accurate peak detection and integration. To study the mass accuracy, the drift in mass error of seven deuterated compounds covering the mass range of target compounds was evaluated over a time period of $\sim$60 h. The accurate mass was derived from the apex of the chromatographic peak. The concentration of the compounds in the test solution was 72 μg/L.

5.3.3.3 Identification and Confirmation
Identification of the compounds was performed using the accurate mass of the protonated molecule within a mass window of 5 ppm together with one product ion (nominal mass). The retention times of the compounds were compared to those of the compounds in the calibration standard solution of the final analysis.

For confirmation of target compounds, LC relative retention time criteria (retention time window <2.5%) and mass spectrometric identification criteria need to be fulfilled. The latter are based on the concept of identification points (EC, 2002). For accurate mass screening using TOF or Orbitrap MS instruments, no criteria are described. Recently, some propositions for these types of instruments were made by Nielen et al. (2007). For high-resolution screening (resolution ≥20,000 and a mass accuracy ≤5 mDa), these authors proposed two identification points. Each product ion (low-mass- resolution MS) also contributes two points. Thus, acquiring a high-resolution precursor ion in combination with at least one product ion and the LC relative retention time meets the minimum requirement of four identification points.

5.3.3.4 Matrix Effects
The matrix effects in the mass spectrometer (that is, signal suppression or enhancement) were assessed using a method described by Matuszewski et al. (2003). To that end, the peak area of the analyte in the sample extract spiked after extraction (B) was compared with the peak area of the corresponding analyte in the standard solution (A). The matrix effects (ME) were calculated as follows:

$$ME(\%) = B/A \times 100$$

Consequently, a value higher than 100% means that signal enhancement occurs in the mass spectrometer because of the coextracted and coeluted matrix. A value lower than 100% implies that signal suppression occurs in the mass spectrometer. This matrix effect was measured twofold and calculated for ultrapure water, drinking water, and surface water at a concentration level of 0.2 mg/L for each compound. In order to investigate the effect of the matrix on the response of the mass spectrometer, the signal suppression was calculated for 15 deuterated standards in sample extracts of surface water and wastewater compared to a standard in drinking water. The results were corrected for the blank values.

5.3.3.5 Quality Assurance
Recovery experiments were conducted by spiking drinking water and river water samples, respectively, with known amounts of the analytes at neutral pH and pH 2. Quantification of the compounds was obtained by using a nine-point external calibration curve ranging between $\sim$0.001 and 0.3 mg/L,

corresponding to ~1 - 300 μg/L water. The solutions for the calibration curve were prepared in acetonitrile/water or methanol/water (50/50; v/v). Because the noise level in the Orbitrap–MS, especially at m/z values >200 is virtually absent, a standard signal-to-noise approach is not feasible at these m/z values. LOD were determined based on the lowest concentration of a deuterated standard in pure water producing an appreciable signal. The LOD in other matrices was subsequently derived by correcting for the matrix suppression observed in the Orbitrap–MS.

For confirmation purposes, the samples collected in the period 2006–2007 were also analyzed by an independent laboratory that had a license for standards of illicit drugs at their disposal.

5.4 RESULTS

5.4.1 Method Characteristics

5.4.1.1 Chromatographic Separation For the present study, 29 illicit drugs were selected. For 15 of these compounds, the corresponding deuterated standards were obtained. Their CAS numbers, bruto formulas, and n-octanol/water partition coefficients (K_{ow}) are tabulated in Table 5.5. Group 3 substances (barbiturates) were not analyzed in the 2006–2007 monitoring, and, although included in the 2009 monitoring studies, could not be evaluated because of optimization problems that have not yet been solved.

5.4.1.2 Identification and Confirmation In Table 5.6, the accurate mass of the precursor ions are tabulated together with the two most abundant (nominal mass) product ions of the analytes generated in the LTQ part of the instrument by a normalized collision energy of 35%. One product ion was used to check the identity of the compounds.

5.4.1.3 Mass Accuracy The accurate masses of seven deuterated standards during a period of 60 h immediately after mass calibration of the Orbitrap instrument are presented in Table 5.7. In general, the mass error ranged from −5.7 to +2.66 ppm for the seven deuterated compounds evaluated (Table 5.7). The average mass accuracy of the compound with the lowest molecular mass ($m/z = 147$) was −3.73 ppm, that is, the actual mass observed was on average 3.73 ppm less than the theoretical mass. The mass accuracy of the compound with the highest molecular mass (m/z 334) was +0.26 ppm. Typically, the mass error (drift) invariably was negative and did not exceed a rate of −0.05 ppm/h over the course of a sequence of 70 to 80 injections, corresponding to a total analysis time of ~60 h. The drift appeared to be independent of the m/z ratio (cf. slopes given in Table 5.7). No effect of the signal intensity on mass accuracy was noticed, indicating the large dynamic range of the Orbitrap–MS with respect to mass accuracy (Van Leerdam et al., 2009). Similar results were observed by Krauss and Hollender (2008), who used the Orbitrap for the trace level determination of low molecular weight nitrosamines.

TABLE 5.5 Names, CAS numbers, Bruto Formulas, and Calculated log K_{ow} Values of Illicit Drugs Reported in This Study

Group	Name	CAS[a]	Bruto Formula	log K_{ow}[b]
	Δ^9-THC	1972−08-3	$C_{21}H_{30}O_2$	7.68
1	Δ^9-THC-D3	81586-39-2	$C_{21}H_{27}D_3O_2$	
1	11-Nor-9-Carboxy-THC	56354-06-4	$C_{21}H_{28}O_4$	6.21
1	11-Nor-9-Carboxy-THC-D3	136844-96-7	$C_{21}H_{25}D_3O_4$	
1	11-OH-Δ^9-THC	2/36557-05-8	$C_{21}H_{30}O_3$	6.58
1	11-OH-Δ^9-THC-D3	n.a	$C_{21}H_{27}D_3O_3$	
1	9-COOH-Δ^9-THC	23978-85-0	$C_{22}H_{30}O_4$	8.41
2	Codeine	76-57-3	$C_{18}H_{21}NO_3$	1.20
2	Heroin	561-27-3	$C_{21}H_{23}NO_5$	1.52
2	6-Monoacetylmorphine	2784-73-8	$C_{19}H_{21}NO_4$	1.32
2	6-Monoacetylmorphine-D3	136765-25-8	$C_{19}H_{18}D_3NO_4$	
2	Morphine	57-27-2	$C_{17}H_{19}NO_3$	0.43
2	Morphine-D3	67293-88-3	$C_{17}H_{16}D_3NO_3$	
2	Methadone	76-99-3	$C_{21}H_{27}NO$	4.20
2	EDDP	66729-78-0	$C_{20}H_{23}N$	5.51
2	EMDP	n.a.	$C_{19}H_{21}N$	5.14
4	Amphetamine	300-62-9	$C_9H_{13}N$	1.81
4	Amphetamine-D11	73758-24-4	$C_9H_2D_{11}N$	
4	Methamphetamine	51-57-0	$C_{10}H_{15}N$	1.94
4	Methamphetamine-D5	60124-88-1	$C_{10}H_{10}D_5N$	
4	MDA	4764-17-4	$C_{10}H_{13}NO_2$	1.67
4	MDA-D2	n.a.	$C_{10}H_{11}D_2NO_2$	
4	N-Ethyl-MDA, MDEA	82801-81-8	$C_{12}H_{17}NO_2$	2.34
4	MDEA-D5	160227-43-0	$C_{12}H_{12}D_5NO_2$	
4	MDMA	42542-10-9	$C_{11}H_{15}NO_2$	1.81
4	MDMA-D5	136765-43-0	$C_{11}H_{10}D_5NO_2$	
5	Diazepam	439-14-5	$C_{16}H_{13}ClN_2O$	2.90
5	Diazepam-D5	65854-76-4	$C_{16}H_8D_5ClN_2O$	
5	Nordazepam	1088-11-5	$C_{15}H_{11}ClN_2O$	3.15
5	Nordazepam-D5	65891-80-7	$C_{15}H_6D_5ClN_2O$	
5	Oxazepam	604-75-1	$C_{15}H_{11}ClN_2O_2$	2.31
5	Oxazepam-D5	65854-78-6	$C_{15}H_6D_5ClN_2O_2$	
5	Temazepam	846-50-4	$C_{16}H_{13}ClN_2O_2$	2.15
5	Desalk-flurazepam	2886-65-9	$C_{15}H_{10}ClFN2O$	3.02
6	Cocaine	50-36-2	$C_{17}H_{21}NO_4$	3.08
6	Cocaine-D3	65266-73-1	$C_{17}H_{18}D_3NO_4$	
6	Benzoylecgonine	519-09-5	$C_{16}H_{19}NO_4$	2.72
6	Benzoylecgonine-D3	115732-68-8	$C_{16}H_{16}D_3NO_4$	
6	Methcathinone	5650-44-2	$C_{10}H_{13}NO$	1.40
6	Meta-CCP	6640-24-0	$C_{10}H_{13}ClN_2$	2.07
6	Ketamine	1867-66-9	$C_{13}H_{16}ClNO$	2.28
6	Fentanyl	437-38-7	$C_{22}H_{28}N2O$	3.89
6	Meprobamate	57-53-4	$C_9H_{18}N2O_4$	0.70
6	Ritalin	113-45-1	$C_{14}H_{19}NO_2$	2.55

[a]n.a., CAS number not available.
[b]Partition coefficient n-octanol/water, calculated by ClogP.

TABLE 5.6 Accurate Masses of Precursor Ion, Nominal Masses and Relative Abundance of Product Ions of Nondeuterated Illicit Drugs by LTQ Orbitrap MS[a]

Component	Precursor m/z [M+H][+]	Ion 1 m/z	Ion 2 m/z	Abundance (%)	RSD %[a] (n = 16-24)
Morphine	286.14334	201.1	229.1	51.9	8.9
Methcathinone	164.10699	146.1	133.1	1.7	20.9
Codeine	300.15942	215.2	243.1	47.7	6.1
Amphetamine	136.11208	119.1	91.1	0.5	14.6
6-Monoacetylmorphine	328.15433	211.2	268.2	73.7	6.9
Metamphetamine	150.12773	119.0	91.1	9.0	5.2
MDA	180.10191	163.2	–[b]		
MDMA	194.11755	163.1	58.0	1.0	21.3
MDEA	208.13321	163.1	72.0	2.7	7.3
Ketamine	238.09932	220.1	207.1	23.9	5.3
Benzoylecgonine	290.13868	168.2	272.2	4.8	12.9
Heroin	370.16490	328.2	268.2	99.1	2.1
Cocaine	304.15433	182.1	150.2	2.6	14.6
Nordazepam(desmethyldiazepam)	271.06327	243.1	208.1	37.7	9.2
Ritalin	234.14886	84.0	174.2	0.3	60.7
metaCPP	197.08400	154.0	119.1	6.9	18.7
Fentanyl	337.22744	188.2	216.3	5.6	9.4
Meprobamate	219.13393	158.1	–		
Methadone	310.21654	265.1	247.2	0.1	54.3
Oxazepam	287.05818	269.1	241.1	3.9	8.6
desalk-flurazepam	289.05385	261.1	140.0	44.5	22.2
Temazepam	301.07383	283.0	255.2	9.2	5.0
Diazepam	285.07892	257.1	222.2	30.4	8.7
EDDP	278.19033	249.1	234.1	13.0	12.5
EDMP	264.17468	235.1	–		
11-OH-Δ^9-THC	331.22677	313.3	–		
11-nor-9-Carboxy-THC	345.20604	327.2	299.3	6.1	8.2
Δ^9-THC	315.23186	259.2	193.2	76.7	8.7
9-COOH-Δ^9-THC	359.22169	-	-		

[a]RSD, relative standard deviation in abundance of second product ion; –, no stable product ion observed.

On the basis of these results, peak integration for quantitative analysis was done from ion chromatograms extracted for each ion at a range of ±5 ppm around the theoretical m/z value. If a better mass accuracy at the low m/z range would be needed, a lower calibration mass would be required than that of m/z 181 of polytyrosine in the routine calibration solution.

5.4.1.4 Repeatability The repeatability of the method was evaluated using a mixture solution of 14 standards in ultrapure water (milli-Q water)/acetonitrile (50/50) at a concentration of 50 μg/L and injecting each into the MS seven times. The

TABLE 5.7   Drift of the Accurate Mass of Seven Deuterated Standards Recorded during a 60-h Period

Substance	m/z	Range Mass Deviation (ppm) (n = 70−80)	Slope (ppm/h)
Amphetamine-D11	147.18112	−1.97 to −5.71	−0.05
MDA-D2	182.11446	0.00 to −3.79	−0.04
MDEA-D5	213.16459	+0.47 to −3.33	−0.04
Desmethyldiazepam-D5	276.09465	+2.50 to −2.03	−0.05
Diazepam-D5	290.11030	+2.27 to −1.31	−0.05
Cocaine-D3	307.17316	+2.18 to −1.79	−0.04
11-Hydroxy-Δ^9-THC-D3	334.24560	+2.66 to −1.62	−0.05

concentration was then calculated against a nine-point calibration curve covering a concentration range of two orders of magnitude. The coefficients of variation thus obtained were satisfactory and ranged from 3% (heroin) to 13% (Δ^9-THC).

5.4.1.5   *Recovery* Recovery experiments were conducted by spiking drinking water and river water samples, respectively, with known amounts of the analytes at two pH values. The results are shown in Table 5.8. Acceptable recoveries were observed for the diazepines, XTC, codeine, heroin, cocaine, BE, and methadone. In

TABLE 5.8   Recoveries of Selected Illicit Drugs from Spiked Water Samples[a]

Substance	Spiked dw at pH 2.3 (n = 3)	Spiked dw at pH 7.0 (n = 2)	Spiked rw at pH 2.3 (n = 3)	Spiked rw at pH 7.0 (n = 2)
Diazepam	65.7	103.5	48.0	73
Nordazepam	70.0	48.0	68.0	66.5
Oxazepam	80.0	0.0	83.0	14
Δ^9-THC	4.0	3.0	19.5	6.5
Amphetamine	6.0	3.0	9.3	4.5
Methamphetamine	22.3	0.0	32.0	0
XTC	65.7	0.0	75.0	0
Codeine	53.7	61.0	57.0	56.5
Heroin[b] (isomer 1)	9.0	9.0	42.5	9
Heroin[b] (isomer 2)	53.7	68.0	64.7	64
Morphine	0.0	0.0	1.7	0.5
6-MAM	20.5	5.0	27.5	21
Cocaine	77.7	83.5	66.3	74
Benzoylecgonine	91.7	0.0	88.0	0
Methadone	88.3	73.5	80.7	67.5
Average	47.2	30.5	50.9	30.5

[a]dw, drinking water; rw, river water.
[b]Standard supplied reveals two peaks in chromatogram.

general, low recoveries were observed for Δ^9-THC, metamphetamine, amphetamine, 6-MAM, and morphine. Lowering the pH significantly improved the recoveries of nordazepam, oxazepam, the amphetamines, and BE.

Low recoveries of morphine and 6-MAM can be explained by their high polarities (see log K_{ow} values in Table 5.5; both compounds elute very early in the chromatograms); therefore, both are likely to get poorly trapped at the SPE conditions used. Low recoveries of amphetamines may also be because of thermal instabilities of some of the compounds (Delta Lab, personal communication). The low recovery of Δ^9-THC is because of its relatively high hydrophobicity (Table 5.5) and its corresponding tendency to stick to glassware and tube walls.

5.4.1.6 Matrix Effects

The suppression or enhancement of drug response in the Orbitrap–MS because of matrix effects were evaluated by comparing responses of deuterated standards added to raw drinking water, wastewater, and surface water samples to the response of the same standards dissolved in tap water. Tap water is used as a reference matrix in drinking water laboratories. As Table 5.9 shows, matrix suppression increases in the order raw drinking water $\approx$ surface water $<$ effluent $<$ influent. Matrix suppression is strongest for amphetamine, metamphetamine, MDA, MDMA, MDEA, Δ^9-THC and nor-9-COOH-THC with signal reductions of more than 90% for each of these compounds. For several substances, signal enhancement can be observed relative to the tap water matrix.

5.4.1.7 Limits of Detection (LODs)

The LODs are substance and matrix dependent as a result of matrix suppression or enhancement, as can be seen in

TABLE 5.9 Suppression or Enhancement (%) of the Orbitrap-MS Response of Deuterated Standards of Illicit Drugs in Different Matrices Relative to Tap Water

Deuterated Standard	Drinking Water (raw)	Surface Water	STP Effluent	STP Influent
Morphine-d3	77.1	59.5	30.5	19.3
Amphetamine-D11	167.3	171.1	147.9	7.4
6-Monoacetylmorphine-D3	135.4	137.5	117.9	62.1
Metamphetamine-D5	161.3	156.1	119.0	8.5
MDA-D2	114.5	106.7	86.0	2.7
MDMA-D5	95.5	81.5	53.7	3.3
MDEA-D5	97.1	84.7	50.8	1.2
Benzoylecgonine-D3	77.2	65.7	43.5	15.7
Cocaine-D3	114.8	105.4	75.2	38.0
Oxazepam-D5	101.8	102.3	75.8	28.2
Desmethyldiazepam-D5	96.9	96.2	75.8	35.2
Diazepam-D5	91.6	90.4	76.2	41.7
11-OH-Δ^9-THC-D3	167.9	202.2	141.7	14.6
11-Nor-9-COOH-THC-D3	103.5	108.0	31.4	5.8
Δ^9-THC-D3	194.2	242.2	147.4	0.0

TABLE 5.10   Limits of Detection for Determination of Selected Illicit Drugs in Various Matrices by Orbitrap-MS (in ng/L)

Substance	Drinking Water	Surface Water	Effluent	Influent
Amphetamine	2	2	2	42
Metamphetamine	1	1	1	19
MDA	2	2	2	63
MDMA	2	2	3	48
MDEA	1	1	2	63
Nordazepam	1	1	2	4
Oxazepam	1	1	1	2
Diazepam	1	1	1	2
Morphine	1	1	3	5
11-OH-Δ^9-THC	22	10	13	131
11-Nor-9-carboxy-THC	10	10	28	152
Benzoylecgonine	0.1−1	0.1−1	2	5
Cocaine	1	1	2	3

Table 5.10. Matrix effects are, of course, influenced by the sample intake, which varied depending on the type of matrix. The data in Table 5.10 show that the LODs range from 1 to 22 ng/L for drinking water and from <1 to 10 ng/L in surface waters. In wastewaters, LODs can be much higher because of stronger matrix suppression of the signal. In effluents, LODs varied from 1 to 28 ng/L and, in influents, from 2 to 152 ng/L.

5.4.1.8 Quantitation and Linearity   In order to evaluate the linear range, the regression coefficient (r^2) of the concentration-response curve was calculated for the analytes in ultrapure water and in extracts of ultrapure, drinking water, and surface water, spiked before extraction. In ultrapure water, the nine-point calibration curves for all 32 standards (Table 5.4) had regression coefficients (r^2) better than 0.99. This means that the optimized LC−MS method can be used for the quantitative analyses of the analytes in the range 1−1500 ng/L in drinking water, surface waters, and wastewater.

5.4.2 Occurrence of Illicit Drugs in Dutch Environmental Waters

5.4.2.1 Period 2006−2007   In 2006 and 2007 samples of surface water were collected at five locations in the river Meuse catchment. In addition, tap water was collected from three locations, and a single STP effluent was also collected. As discussed in the materials and methods section, in the period 2006−2007 our laboratory did not possess a license to have in stock the standard compounds. Nevertheless, by employing the Orbitrap−high-resolution MS several of the analytes could be unambiguously identified and a (semi-)quantitative analysis could be performed. The results are shown in Table 5.11. As can be seen from Table 5.11, the STP effluent

TABLE 5.11 Concentrations (ng/L) of Illicit Drugs in Dutch Surface Waters, in Drinking and Bottled Water in the Period 2006–2007[a]

Compound	dw 1	dw 2	Bottled Water	River Meuse A[b] 2 May	River Meuse A[b] 2 May	River Meuse A 22 May	River Meuse B	River Meuse C	River Meuse D	Estuary Rivers Rhine/Meuse	STP Effluent	dw 3	River Meuse D (2006)	River Meuse E (2006)
Codeine	–	–	–	–	–	–	–	–	<10	–	90	–	<10	–
Heroin, isomer 1	–	–	–	–	–	–	–	–	–	–	10	–	–	–
Heroin, isomer 2	–	–	–	–	–	–	–	–	–	–	20	–	–	–
Morphine	–	<10	–	–	–	–	–	–	–	–	30	na	na	na
6-MAM	–	–	–	–	–	–	–	–	–	–	240	na	na	na
Cocaine	–	–	–	–	–	–	–	–	–	–	–	–	–	<10
Benzoylecgonine	–	–	–	10	10	10	10	10	10	–	–	–	20	30
Methadone	–	–	–	<10	<10	<10	<10	–	<10	<10	10	–	–	<10
Amphetamine	–	–	–	–	–	–	–	–	–	–	–	–	–	–
Methamphetamine	–	–	–	–	–	–	–	–	–	–	–	–	–	–
MDMA (XTC)	–	–	–	<10	–	–	10	<10	–	–	110	–	–	–
Diazepam	–	–	–	–	–	<10	<10	–	–	–	<10	–	–	–
Nordazepam	–	–	–	–	10	–	<10	<10	–	–	10	–	–	–
Oxazepam	–	–	–	10	10	10	<10	<10	10	<10	800	–	20	10
Desalkylflurazepam[c]	–	–	–	+	+	+	+	+	+	+	+	na	na	na
Δ9-THC	–	–	–	10	20	30	–	–	–	10	<10	–	–	–
11-hydroxy-Δ9-THC[c]	–	–	–	–	–	–	–	–	–	–	–	na	na	na
9-COOH-Δ9-THC[c]	–	–	–	+	+	+	+	+	–	+	–	na	na	na

[a]Only data above LOD reported; dw, drinking water; <10, below limit of quantification; + present, not quantified; – below LOD; na, not analyzed.

[b]Replicates.

[c]Qualitative analysis only.

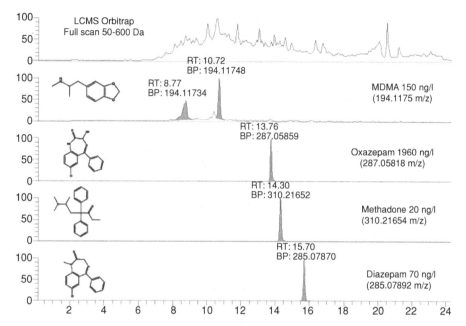

FIGURE 5.2 LC–MS (+ mode) extracted-ion chromatograms of illicit drugs in a Dutch STP effluent sample collected in 2007.

contained the largest number of identified illicit drugs and levels ranged from $<$ LOQ to up to 800 ng/L for oxazepam. Benzoylecgonine, Δ^9-THC, and some of the diazepines were observed at levels of several tens of nanograms per liter at several locations along the river Meuse. Some transformation products of Δ^9-THC and the diazepines could also be detected qualitatively. Neither in tap water nor in bottled water were any of the analytes present at levels above the LOQ.

Figure 5.2 shows LC–MS (positive-ion mode) extracted ion chromatograms for the identification of some illicit drugs and metabolites in the STP effluent sample. The presence of codeine, MDMA, nordazepam, and oxazepam in this STP effluent water was confirmed by an independent laboratory (with a license to store and analyze drug reference standards).

5.4.2.2 Period 2008 In 2008 samples were again collected, this time from the two major rivers in the Netherlands: Meuse and Rhine. Samples were collected on four different dates at two locations from the river Meuse and on six different days (in consecutive months) from the river Rhine. Full quantitative analysis could be performed in the 2008 campaign since the KWR laboratory had been accredited and a license obtained by that time. Figure 5.3 presents the main results from the 2008 campaign. As can be seen, the major components were invariably two of the diazepines (temazepam and oxazepam) (concentrations ranged from 0.5 to 10 and 3 to 35 ng/L, respectively) and benzoylecgonine (0.1–9 ng/L). In general, in so far as

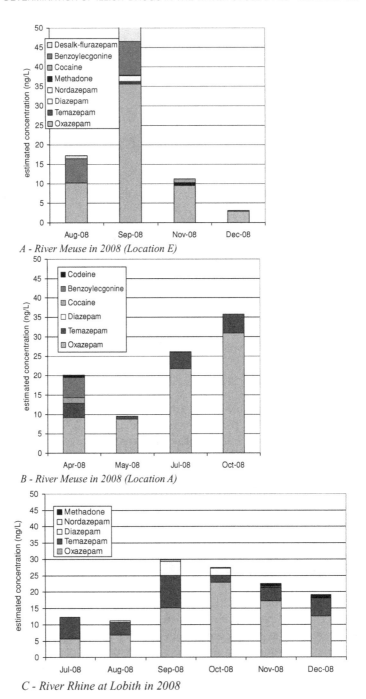

A - River Meuse in 2008 (Location E)

B - River Meuse in 2008 (Location A)

C - River Rhine at Lobith in 2008

FIGURE 5.3 Occurrence of illicit drugs in rivers Meuse (upper and middle panel: two locations, same as 2006, cf. Table 5.8) and Rhine (bottom panel) sampled during a one-half year period in 2008.

seasonal changes could be interpreted (no samples were collected in January–March), the concentrations appear to be highest in the samples collected in the postsummer period (September–October).

5.4.2.3 Screening for Illicit Drugs with a Validated Method in the Year 2009

In 2009, KWR Watercycle Research Institute in collaboration with RIVM conducted a monitoring campaign in The Netherlands. A total of 64 samples of surface waters and wastewaters were collected and analyzed. A selection of sewage treatment plants (STPs) were monitored, from small rural site STPs to relatively large urban STPs. From eight sewage treatment plants both influent and effluents were analyzed by LC–LTQ FT Orbitrap–MS, according to the conditions described in the methods section and in more detail by Hogenboom et al. (2009). In order to compare results, several of the samples were sent to two collaborating laboratories.

In Dutch surface waters collected in 2009, 10 out of 30 illicit drugs were detected. The quantitative results of this campaign were not available at the time of writing in so far as surface waters and drinking water was concerned. Table 5.12 presents the major illicit drugs observed in the influents and effluents of eight Dutch STPs. The results shown in Table 5.12 demonstrate that illicit drugs are present in domestic wastewater influents and effluents at levels similar to those reported elsewhere in Europe. Diazepines, codeine, morphine (which are all partly available on medical prescriptions and partly used illegally), MDMA, and benzoylecgonine are the most abundant illicit drugs observed, with concentrations amounting to several micrograms per liter in STP effluents. The significance of the present findings for the drinking water cycle in The Netherlands, where almost 40% of the total amount of drinking water consumed is produced from surface waters, needs further investigation.

TABLE 5.12   Presence of Major Illicit Drugs in Influents and Effluents from Eight Dutch STP Collected During the 2009 Sampling Campaign (concentrations in ng/L)

Compound	Influents	Effluents
Morphine	400–1600	4–75
Codeine	45–370	2–230
Amphetamine	+[a]	+
Metamphetamine	+	+
MDMA	22–760	<LOQ – 143
Benzoylecgonine	600–3600	7–98
Cocaine	+	+
Ritalin	–[a]	+
Nordazepam	<LOD –27	9–34
Oxazepam	800–2200	800–1900
Temazepam	100–330	200–650
Desalk-fluazepam	40–100	40–180
Diazepam	0.2–56	<LOQ – 56

[a]+, present, not quantified; – below LOD; na, not analyzed.

5.5 CONCLUSIONS

In Dutch surface waters sampled in the period 2006–2008, nine out of thirty compounds tested were actually observed in concentration levels ranging between 10 and 100 ng/L. These included four benzodiazepines + one metabolite (maximum concentration 40 ng/L), cocaine + benzoylecgonine (maximum 30 ng/L), and codeine + methadone (<10 ng/L). In effluents from STPs, levels were higher, typically up to one order of magnitude. In Dutch drinking water samples collected in the period 2006–2008, no drugs were found to be present. Apparently, the treatment steps used in producing drinking water efficiently remove most of the illicit drugs present in surface water used as source water.

The broad screening method is suitable for the analysis of illicit drugs in water at low concentration levels. In Dutch surface waters collected in 2009, 10 out of 30 illicit drugs were detected. Influents and effluents of STPs again contained much higher levels, and the same compounds that were observed in the 2006–2008 campaigns appeared to be the most prominent in the 2009 campaign: diazepines, codeine, methadone, BE, and MDMA. It can be concluded that illicit drugs are present in the aquatic environment and constitute emerging contaminants, the importance of which for the Dutch water cycle needs further study.

High-resolution mass spectrometry coupled to LC offers a powerful combination for screening purposes and identifying new illicit drug and their metabolites in the environment. The application of accurate mass screening described in this chapter demonstrates that current analytical instrumentation is well equipped to meet the challenges posed by newly emerging polar chemicals.

5.6 ACKNOWLEDGMENT

Parts of the work presented in this chapter were supported financially by the Joint Research Programme of the Dutch Water utilities (BTO). We thank Pascal Kooij (KWR) for his valuable assistance during the analytical optimizations.

REFERENCES

Bijlsma, L., J. V. Sancho, E. Pitarch, M. Ibáñez, and F. Hernández. 2009. Simultaneous ultra-high-pressure liquid chromatography–tandem ms determination of amphetamine and amphetamine-like stimulants, cocaine and its metabolites, and a cannabis metabolite in surface water and urban wastewater. *J. Chromatogr. A* **1216**:3078–3089.

Boleda, R., M. T. Galceran, and F. Ventura. 2007. Trace determination of cannabinoids and opiates in wastewater and surface waters by ultra-performance liquid chromatography-tandem mass spectrometry. *J Chromatogr A* **1175**:38–48.

Bones, J., K. V. Thomas, and B. Paull 2007. Using environmental analytical data to estimate community consumprion of illicit drugs and abused pharmaceuticals. *J. Environ. Monit.* **9**:701–707.

Castiglioni, S., E. Zuccato, E. Crisci, C. Chiabrando, R. Fanelly, and R. Bagnati. 2006. Identification and measurement of illicit drugs and their metabolites in urban wastewater by liquid chromatography-tandem ms. *Anal. Chem.* **78**:8421−8429.

Cook, C. E., A. R. Jeffcoat, and M. Perez-Reyes. 1985. Pharmakokinetic studies of cocaine and phencyclidine in man. In *Pharmacokinetics and Pharmacodynamics of Psychoactive Drugs*, edited by G. Barnett, and Chiang CN, pp. 48−74. Foster City, CA: Biomedical Publ.

Daughton, C. G., and T. Jones-Lepp Eds. 2001. Pharmaceuticals and Personal Care Products in the Environment: Scientific and Regulatory Issues, pp. 348−363. Washington, DC: American Chemical Society.

EC. 2002. Commission Decision 2002/657/EC of 12 August 2002 implementing Council Directive 96/23/EC concerning performance of analytical methods and the interpretation of results. *Office J. Eur. Commun.* **2002**;L221/8.

Emke, E., T. van Leerdam, and P. de Voogt 2009. The analysis of illicit drugs in water by HPLC-LTQ-Orbitrap MS. Nieuwegein, The Netherlands: KWR Watercycle Research Institute.

Gheorghe, A., A. L. N. van Nuijs, B. Pecceu, L. Bervoets, P. G. Jorens, R. Blust, H. Neels, and A. Covaci. 2008. Analysis of cocaine and its principal metabolites in waste and surface water using solid phase extraction and liquid chromatography—ion trap tandem mass spectrometry. *Anal. Bioanal. Chem.* **391**:1175−1181.

Hogenboom, A. C., J. A. van Leerdam, and P. de Voogt. 2009. Accurate mass screening and identification of emerging contaminants in environmental samples by liquid chromatography-LTQ FT Orbitrap ms. *J. Chromatogr. A* **1216**:510−519.

Hu, Q., R. J. Noll, H. Li, A. Makarov, M. Hardman, and R. G. Cooks. 2005. The Orbitrap: a new mass spectrometer. *J. Mass Spectrom.* **40**:430–443.

Huerta-Fontela, M., M. T. Galceran, and F. Ventura. 2007. Ultraperformance liquid chromatography-tandem mass spectrometry analysis of stimulatory drugs of abuse in wastewater and surface waters. *Anal. Chem.* **79**:3821−3829.

Huerta-Fontela, M., M. T. Galceran, and F. Ventura. 2008. Stimulatory drugs of abuse in surface waters and their removal in a conventional drinking water treatment plant. *Environ. Sci. Technol.* **42**:6809−6816.

Hummel, D., D. Löffler, G. Fink, and T. A. Ternes. 2006. Simultaneous determination of psychoactive drugs and their metabolites in aqueous matrices by liquid chromatography mass spectrometry. *Environ. Sci. Technol.* **40**:7321−7328.

Jones-Lepp, T. L., D. A. Alvarez, J. D. Petty, and J. N. Huckins. 2004. Polar organic chemical integrative sampling (POCIS) and LC–ES/ITMS for assessing selected prescription and illicit drugs in treated sewage effluents. *Arch. Environ. Contam. Toxicol.* **47**:427–439.

Kasprzyk-Hordern, B., R. M. Dinsdale, and A. J. Guwy. 2007. Multi-residue method for the determination of basic/neutral pharmaceuticals and illicit drugs in surface water by solid-phase extraction and ultra performance liquid chromatography-positive electrospray ionization tandem mass spectrometry. *J. Chromatogr. A* **1161**:132−145.

Krauss, M., and Hollender J. 2008. Analysis of nitrosamines in wastewater: exploring the trace level quantification capabilities of a hybrid linear ion trap/Orbitrap mass spectrometer. *Anal. Chem.* **80**:834−842.

Makarov, A. 2000. Electrostatic axially harmonic orbital trapping: A high-performance technique of mass analysis. *Anal.Chem.* **72**:1156–1162.

Makarov, A, E. Denisov, O. Lange, and S. Horning. 2006a. Dynamic range of mass accuracy in LTQ Orbitrap hybrid mass spectrometer. *J. Am. Soc. Mass Spectrom.* **17**:977–982.

Makarov, A., E. Denisov, A. Kholomeev, W. Balschun, O. Lange, K. Strupat, and S. Horning 2006b. Performance evaluation of a hybrid linear ion trap/orbitrap mass spectrometer. *Anal. Chem.* **78**:2113–2120.

Matuszewski, B. K., M. L. Constanzer, and C. M. Chavez-Eng. 2003. Strategies for the assessment of matrix effect in quantitative bioanalytical methods based on HPLC-MS/MS. *Anal. Chem.* **75**:3019–3030.

National Drug Monitor 2006. Jaarbericht, Trimbos Institute (Netherlands Institute of Mental Health and Addiction), Utrecht

Nielen, M. W. F., M. C. van Engelen, R. Zuiderent, and R. Ramaker. 2007. Screening and confirmation criteria for hormone residue analysis using liquid chromatography accurate mass time-of-flight, Fourier transform ion cyclotron resonance and orbitrap mass spectrometry techniques. *Anal. Chim. Acta* **586**:122–129.

Postigo, C., M. J. López De Alda, and D. Barceló. 2008. Fully automated determination in the low nanogram per liter level of different classes of drugs of abuse in sewage water by on-line solid-phase extraction-liquid chromatography-electrospray tandem mass spectrometry. *Anal. Chem.* **80**:3123–3134.

Schriks, M., M. C. Heringa, M. M. E. van der Kooi,, P. de Voogt, and A. P. van Wezel. 2010. Toxicological relevance of emerging contaminants for drinking water quality. *Water Res.* **44**:461–476.

Scigelova, M., and A. Makarov. 2006. Orbitrap mass analyzer—overview and applications in proteomics. *Proteomics* **6**(Suppl 2): 16–21.

SFK 2010. Stichting Farmaceutische Kengetallen http://www.sfk.nl/publicaties/data_en_ feiten.html.

Vanderford, B. J., and S. A. Snyder. 2006. Analysis of pharmaceuticals in water by isotope dilution liquid chromatography/tandem mass spectrometry. *Environ. Sci. Technol.* **40**:7312–7320.

Van der Hoek, J. P., P. Stoks, M. Mons, and D. van der Kooij. 2008. Visie en streefwaarden voor milieuvreemde stoffen in drinkwater. H_2O **4**:33–35 (in Dutch).

Van Leerdam, J. A., A. C. Hogenboom, M. M. E. van der Kooi, and P. de Voogt. 2009. Determination of 1H-benzotriazoles and benzothiazoles in water by solid-phase extraction and liquid chromatography LTQ FT Orbitrap mass spectrometry. *Int. J. Mass Spectrom.* **282**:99–107.

Van Nuijs, A. L. N., B. Pecceu, L. Theunis, N. Dubois, C. Charlier, P. G. Jorens, L. Bervoets, R. Blust, H. Neels, and A. Covaci. 2009. Cocaine and metabolites in waste and surface water across Belgium. *Environ. Pollut.* **157**:123–129.

Waumans, D., S. Pauwels, N. Bruneel, and J. Tytgat. 2006. Quantification of cocaine and benzoylecgonine in wastewater by gas chromatography-mass spectrometry: mapping cocaine in the Belgian university city of Leuven. The International Association of Forensic Toxicologists, In *Proceedings 44th Meeting Ljubljana*

Zuccato, E., S, Castiglioni, R. Bagnati, C. Chiabrando, P. Grassi, and R, Fanelli. 2008. Illicit drugs, a novel group of environmental contaminants. *Water Res.* **42**:961–968.

SECTION IVA

MASS SPECTROMETRIC ANALYSIS OF ILLICIT DRUGS IN THE ENVIRONMENT

Occurrence and Fate in Wastewater and Surface Water

CHAPTER 6

OCCURRENCE OF ILLICIT DRUGS IN WASTEWATER IN SPAIN

CRISTINA POSTIGO, MIREN LÓPEZ DE ALDA, and DAMIA BARCELÒ

6.1 INTRODUCTION

Between 172 and 250 million people consume illicit drugs annually worldwide. Most of them are occasional consumers and only about 10%–15% are heavy drug users (United Nations Office on Drugs and Crime, 2009). These consumption figures are sustained by high volumes of production, e.g., more than 800 tons of cocaine per year, up to 774 tons of heroin and morphine per year, and between 13,300 and 66,100 tons of cannabis herb per year (United Nations Office on Drugs and Crime, 2009).

As a consequence, illicit drugs and their metabolites have become pseudopersistent in the environment and, at the moment, they constitute a recognized group of emerging contaminants of growing concern (Richardson, 2008). Discharged treated wastewaters constitute their main source in the environment as direct disposal is less likely. Like prescribed pharmaceuticals, illicit drugs and metabolites that end up in wastewaters after human consumption and excretion may not be completely removed during wastewater treatment. Thus, certain amounts of these compounds may be released into the aquatic environment.

The presence of illicit drugs in surface waters may have toxicological implications on biota that are still unknown. Data on environmental levels of these substances are required to assess their ecotoxicological risk. Moreover, levels of drug residues determined in aqueous environmental samples have been also used to estimate illicit drug use at the community level (Zuccato et al., 2005; Huerta-Fontela et al., 2008a; Zuccato et al., 2008; Boleda et al., 2009; Kasprzyk-Hordern et al., 2009b; Van Nuijs et al., 2009a; Postigo et al., 2010). Thus far, drug consumption estimations have been made from influent drug residues levels, because back-calculations in this matrix are straightforward and less subjected to bias than those performed in effluent or surface

Illicit Drugs in the Environment: Occurrence, Analysis, and Fate Using Mass Spectrometry, Edited by Sara Castiglioni, Ettore Zuccato, and Roberto Fanelli
Copyright © 2011 John Wiley & Sons, Inc.

waters (Postigo et al., 2008a; Zuccato et al., 2008) where the drugs residues may undergo degradation and dispersion processes.

Several peer-reviewed publications have reported the occurrence of various illicit drugs and metabolites in waste and surface waters in the United States (Jones-Lepp et al., 2004; Chiaia et al.; 2008; Bartelt-Hunt et al., 2009; Loganathan et al., 2009) and in European countries, such as Italy (Zuccato et al., 2005; Castiglioni et al., 2006; Zuccato et al., 2008; Mari et al., 2009), Switzerland (Zuccato et al., 2008), the United Kingdom (Kasprzyk-Hordern et al., 2008, 2009c; Zuccato et al., 2008), Belgium (Van Nuijs et al., 2009b, c), Ireland (Bones et al., 2007), Germany (Hummel et al., 2006), and Spain (Boleda et al., 2007; Huerta-Fontela et al., 2007, 2008a, 2008b; Postigo et al., 2008b, 2010; Boleda et al., 2009).

According to data estimated by means of official methods, e.g., statistics and medical and crime indicators, the most abused drug in Europe is cannabis followed by cocaine (European Monitoring Centre for Drugs and Drug Addiction, 2009). Despite the fact that Spain follows this pattern, that is, higher use of cannabis than cocaine, this country currently presents the highest prevalence on cocaine use among the adult population, aged 15–64, in Europe (European Monitoring Centre for Drugs and Drug Addiction, 2009). In order to assess the environmental impact of drug consumption, various analytical methodologies have been developed in Spain to determine the presence of illicit drugs and metabolites residues belonging to several chemical classes, namely, opioids, cannabinoids, cocaine-like, amphetamine-like, and lysergic compounds in waters (Boleda et al., 2007; Huerta-Fontela et al., 2007; Postigo et al., 2008b; Bijlsma et al., 2009; González-Mariño et al., 2009) and in airborne particles (Postigo et al., 2009).

The occurrence of this type of emerging contaminants in Spain has been monitored mostly in waste- and surface waters from northeastern (Boleda et al., 2007, 2009; Huerta-Fontela et al., 2007, 2008a, 2008b; Postigo et al., 2008b, 2010) and eastern regions of Spain (Postigo et al., 2008b; Bijlsma et al., 2009). The presence of these substances has been more thoroughly studied in wastewaters than in surface waters. Nevertheless, both matrices have been investigated in the Ebro River basin (Postigo et al., 2010) and in the Llobregat River basin (Huerta-Fontela et al., 2008a, 2008b; Boleda et al., 2009), both located in northeast Spain. The presence of amphetamine-like compounds has also been studied in wastewaters from the northwestern region of Spain (González-Mariño et al., 2009).

The following sections of the present manuscript will review the analytical methodologies developed in Spain to detect illicit drugs and metabolites in wastewater treatment plant (WWTP) influent and effluent water samples, the removal efficiency of the investigated WWTPs, and the impact that the discharged treated waters have on the levels of these compounds in surface waters in Spain.

6.2 WASTEWATER ANALYSIS

The analytical methodologies developed to investigate the presence of illicit drugs and metabolites in environmental waters in Spain are summarized in Table 6.1.

TABLE 6.1 Analytical Methodologies Used to Measure Illicit Drugs and Their Metabolites in Wastewaters in Spain[a]

Drug Classes	Sample Pretreatment	Type (Volume)	Cartridge	Type	Chromatographic Column	Mobile Phase	Analyzer (Interface)	Acquisition Mode	Rec. (%)	LOD (ng/L)	Reference
		Sample Preconcentration			Liquid Chromatography		Mass Spectrometry		Influent WW		
Cocaine and its metabolites, amphetamine-like and lysergic compounds	Filtration (1.6 µm) Surrogate std (12.5 ng/L)	SPE (100 mL)	Oasis HLB (200 mg)	RP-UPLC	Acquity BEH C18 (100×2.1mm, 1.7 µm)	A: MeOH B: H$_2$O: 5mM CH$_3$COONH$_4$: 0.1% FAc	QqQ (ESI +)	2 SRM	70–101	0.2–2.1	Huerta-Fontela et al., 2007
Opioids, cannabinoids	Filtration (1.6 µm) Surrogate std (50 ng/L)	SPE (200 mL)	Oasis HLB (200 mg)	RP-UPLC	Acquity BEH C18 (100×2.1mm, 1.7 µm)	A: MeOH B: H$_2$O: 50mM NH$_4$HCO$_2$ (pH 3.8 with FAc)	QqQ (ESI +)	2 SRM	42–96	0.3–25	Boleda et al., 2007
Cocaine and its metabolites, amphetaminelike and lysergic compounds, opioids, cannabinoids	Filtration (0.45 µm) Surrogate std (20 ng/L)	On-line SPE (5 mL +5 mL)	Oasis HLB (cannabinoids) PLRPs (all rest)	RP-HPLC	Purospher Star RP18 (125×2mm, 5 µm)	A: AcN; B: H$_2$O	QqLIT (ESI +/−)	2 SRM	5–59	0.07–1.94	Postigo et al., 2008b
Cocaine and its metabolites, amphetamine-like compounds, cannabinoids	Centrifugation Surrogate std (100 ng/L) FAc (pH 2)	SPE (50 mL)	Oasis MCX, (60 mg)	RP-UPLC	Acquity BEH C18 (50×2.1mm, 1.7 µm)	A: AcN: 0.1% FAc B: H$_2$O: 30 mM NH$_4$HCO$_2$ (pH = 3.5)	QqQ (ESI +)	3 SRM	50–116	1–2500	Bijlsma et al., 2009
Amphetamine-like compounds	Filtration (0.45 µm) Surrogate std (100 ng/L)	SPE (50 mL) MIPs (50 mL)	Oasis HLB (60 mg) Oasis MCX (60 mg) Supel-MIP amphetamine (25 mg)	RP-HPLC	Halo C18 (100×2.1mm, 2.7 µm)	A: MeOH: 5mM CH$_3$COONH$_4$ B: H$_2$O: 5mM CH$_3$COONH$_4$	QqQ (ESI +)	2 SRM	112–451 101–127 92–111	4.8–7.2 1.5–5.2 0.5–2.7	González-Mariño et al., 2009

[a]FAc: formic acid, SPE: solid-phase extraction. MIPs: molecular imprinted polymers, RP-HPLC: reverse-phase high-performance liquid chromatography, RP-UPLC: reverse-phase ultraperformance liquid chromatography, AcN: acetonitrile, MeOH: methanol, ESI: electrospray, QqQ: triple quadrupole, QqLIT: quadrupole-linear ion trap, SRM: selective-reaction monitoring mode, Rec: Analyte recovery, LOD: limit of detection.

They are all based on detection by liquid chromatography coupled to tandem mass spectrometry (LC–MS/MS) (Boleda et al., 2007; Huerta-Fontela et al., 2007; Postigo et al., 2008b; Bijlsma et al., 2009; González-Mariño et al., 2009).

Water samples were usually vacuum filtered through glass microfiber (1.6 μm) (Boleda et al., 2007; Huerta-Fontela et al., 2007), nylon membrane (0.45 μm) (Postigo et al., 2008b) or nitrocellulose (0.45 μm) (González-Mariño et al., 2009) filters prior to extraction. Sample centrifugation (5 min at 4500 rpm) was also applied to separate suspended solids from the aqueous phase (Bijlsma et al., 2009). In all the cases, filtered or centrifuged samples were spiked with a known concentration of the isotopically labeled compounds used as surrogate standards. Acidification of the sample was also performed when mixed-mode cation-exchange sorbents were used for sample extraction (Bijlsma et al., 2009; González-Mariño et al., 2009).

Samples were always preconcentrated prior to analysis in order to increase method sensitivity and to reduce matrix interferences. Most of the reported methodologies used solid-phase extraction (SPE) to isolate the target analytes from the matrix. Polymeric SPE sorbents, such as Oasis HLB and PLRPs (Boleda et al., 2007; Huerta-Fontela et al., 2007; Postigo et al., 2008b; González-Mariño et al., 2009), as well as mixed-mode cation-exchange sorbents, such as Oasis MCX (Bijlsma et al., 2009; González-Mariño et al., 2009), were used for this purpose. Molecular imprinted polymers (MIPs) were also tested for the extraction and concentration of amphetamine-like drugs by González-Mariño et al. (2009), showing better analytical performance than some SPE sorbents, such as Oasis HLB and Oasis MCX.

On-line coupling of the SPE procedure to the LC–MS/MS analysis has also allowed the development of fully automated methodologies (Postigo et al., 2008b). Automated approaches are less labor and time-consuming than off-line SPE procedures and offer a series of other advantages, such as lower sample volume requirements and minimal sample handling, which improve reproducibility, accuracy, and sensitivity of the method. Conversely, the main disadvantages are the absence of extracts for further analyses and potentially higher matrix effects because of less flexibility in the selection of washing and eluting solvents.

Overall, method recoveries in complex aqueous matrices, e.g., wastewaters, were lower than those achieved in cleaner matrices. Several additional matrix components are concentrated during SPE and can compete with the target analytes in the ionization process, thus interfering with their MS signal. Analytical methods developed to determine illicit drugs and metabolites in wastewaters in Spain used isotopically labeled compounds as surrogate standards to correct for the presence of matrix effects, which vary from one sample to another and affect negatively analytes recoveries.

The chromatographic separation, carried out always using reverse-phase columns, was achieved indistinctly by means of high-performance liquid chromatography (HPLC) (Postigo et al., 2008b; González-Mariño et al., 2009) and ultraperformance liquid chromatography (UPLC) (Boleda et al., 2007; Huerta-Fontela et al., 2007; Bijlsma et al., 2009) (see Table 6.1). The latter provides a more efficient separation (higher peak capacity) in shorter time of analysis and, hence, higher throughput. As shown in Table 6.1, elution of the analytes from the chromatographic column was achieved with organic gradients using acetonitrile (Huerta-Fontela et al., 2007;

Postigo et al., 2008b) or methanol (Boleda et al., 2007; Bijlsma et al., 2009; González-Mariño et al., 2009) as organic solvents. Aqueous phases were often acidified with formic acid to improve analyte ionization and peak shape in the positive-ionization (PI) mode (Boleda et al., 2007; Huerta-Fontela et al., 2007; Bijlsma et al., 2009). Eluted analytes were ionized in all the cases by means of electrospray (ESI). Most illicit drugs and metabolites are better ionized in the PI mode. However, cannabinoids show a better response in the negative-ionization (NI) mode (Postigo et al., 2008b). It is also important to note that none of the reported methodologies discerned between enantiomeric forms of amphetamine-like drugs.

Mass determination was accomplished by means of triple quadrupole (QqQ) (Boleda et al., 2007; Huerta-Fontela et al., 2007; Bijlsma et al., 2009; González-Mariño et al., 2009) and quadrupole-linear ion trap (QqLIT) analyzers (Postigo et al., 2008b) operating in the selected-reaction monitoring mode (SRM). In all the studies, at least two SRM were monitored for each target analyte (Table 6.1) in order to fulfill the European Union requirements for identification and confirmation of banned substances (Commission Decision 2002/657/EC of 12 August 2002).

6.3 OCCURRENCE OF ILLICIT DRUGS AND METABOLITES IN INFLUENT WASTEWATER

Illicit drugs and metabolites levels determined in WWTP influents in Spain are summarized in Table 6.2 and Fig. 6.1. In agreement with other peer-reviewed multiresidue monitoring studies reporting the presence of these substances in various European countries (Zuccato et al., 2005; Castiglioni et al., 2006; Hummel et al., 2006; Bones et al., 2007; Gheorghe et al., 2008; Kasprzyk-Hordern et al., 2008, 2009c; Van Nuijs et al., 2009b, 2009c) and in the United States (Chiaia et al., 2008), the most ubiquitous and abundant compounds in influent wastewaters in Spain were cocaine (CO) and its major metabolite benzoylecgonine (BE). These compounds were usually present at the high nanogram per liter level or even at the microgram per liter level. Maximum BE concentrations (10.5 μg/L) were reported by Bijlsma et al. (2009) in a WWTP located in the Castellón province (eastern Spain). This level is the highest registered so far in Europe for BE; however, it was strongly influenced by the occurrence of an important music event in the area served by the WWTP during the sampling campaign. In other investigated areas of Spain, maximum BE levels attained up to 7.5 μg/L (northeast Spain) (Huerta-Fontela et al., 2008a) and 3.8 μg/L (Ebro River basin area) (Postigo et al., 2010).

The study carried out by Huerta-Fontela et al. (2008a) in northeastern Spain, in which 42 WWTPs were monitored, reported very variable CO concentrations in influent waters, ranging from 4 ng/L to 4.7 μg/L (which is the highest CO level registered in Europe). CO concentrations in the seven WWTPs monitored along the Ebro River basin were less variable (195–384 ng/L). Other metabolites of cocaine, such as cocaethylene (CE), norcocaine (nor-CO), and norbenzoylecgonine (nor-BE), were also monitored in wastewaters in Spain (Postigo et al., 2008b, 2010; Bijlsma et al., 2009). CE, which is a transesterification product of CO formed when this

TABLE 6.2 Concentration Ranges of Illicit Drugs and Metabolites in Wastewater in the Monitoring Studies Carried Out in Spain

Drug	Influent WW (ng/L)	Effluent WW (ng/L)	Reference[a,b,c,d,e,f,g]
Cocaine	370–1240	<30–560	Bijlsma et al., 2009[a]
	4–4700	1–100	Huerta-Fontela et al., 2008a[b]
	268–305	20–28	
	517–1120	3–13	Postigo et al., 2008b[c]
	651/540/316	33/84/105	
	195–961	2–31	Postigo et al., 2010[d]
	79/225	17/47	Huerta-Fontela et al., 2007[e]
Norcocaine	nd	<30–30	Bijlsma et al., 2009[a]
Benzoylecgonine	1480–10500	<30–6790	Bijlsma et al., 2009[a]
	9–7500	1–1500	Huerta-Fontela et al., 2008a[b]
	2500–3900	50–96	
	2550–5980	9–61	Postigo et al., 2008b[c]
	1900/1450/1020	220/50/318	
	545–3790	4–510	Postigo et al., 2010[d]
	810/2307	216/928	Huerta-Fontela et al., 2007[e]
Norbenzoylecgonine	<150–430	<30–170	Bijlsma et al., 2009[a]
Cocaethylene	<150–190	<30–80	Bijlsma et al., 2009[a]
	44–125	0.9–4	Postigo et al., 2008b[c]
	89/97/49	4/2/7	
	6–50	0.2–3	Postigo et al., 2010[d]
Ephedrine	340–725	103–144	Postigo et al., 2008b[c]
	394/444/360	266/138/163	
	203–660	3–276	Postigo et al., 2010[d]
Amphetamine	<500–1400	110–210	Bijlsma et al., 2009[a]
	3–688	4–210	Huerta-Fontela et al., 2008a[b]
	24–101	<1	
	26–53	0.4	Postigo et al., 2008b[c]
	20/36/7	2/1/3	
	3–664	0.9–58	Postigo et al., 2010[d]
	15/15	<1	Huerta-Fontela et al., 2007[e]
	<2.7–9.1	nd	González-Mariño et al., 2009[f]
Methamphetamine	<500	<100	Bijlsma et al., 2009[a]
	3–277	3–90	Huerta-Fontela et al., 2008a[b]
	<0.4–12	<0.4–1	
	10–28	5–7	Postigo et al., 2008b[c]
	8/4/3	3/2/2	
	0.8–8	0.5–8	Postigo et al., 2010[d]
	nd	nd	Huerta-Fontela et al., 2007[e]
	nd	nd	González-Mariño et al., 2009[f]
MDA or	<500–1690	410–680	Bijlsma et al., 2009[a]
3,4-methylenedioxy-	3–266	1–200	Huerta-Fontela et al., 2008a[b]
amphetamine	<0.4–5	<0.4	
	nd	nd	Huerta-Fontela et al., 2007[e]
	13–19	6–20	González-Mariño et al., 2009[f]

TABLE 6.2 (*Continued*)

Drug	Influent WW (ng/L)	Effluent WW (ng/L)	Reference[a,b,c,d,e,f,g]
MDMA or	<500–27500	<100–21200	Bijlsma et al., 2009[a]
3,4-methylenedioxy-	2–598	2–267	Huerta-Fontela et al., 2008a[b]
metamphetamine	26–75	13–40	
	99–174	56–116	Postigo et al., 2008b[c]
	113/245/47	38/376/30	
	4–180	3–120	Postigo et al., 2010[d]
	49/91	41/67	Huerta-Fontela et al., 2007[e]
	4–11	4–9	González-Mariño et al., 2009[f]
MDEA or	<500	<100	Bijlsma et al., 2009[a]
Methylenedioxy-	6–114	12	Huerta-Fontela et al., 2008a[b]
ethylamphetamine	28/28	<1.5	(Huerta-Fontela et al., 2007[e]
	nd	nd	González-Mariño et al., 2009[f]
LSD or lysergic acid	2–4	0.3–0.8	Postigo et al., 2008b[c]
diethylamide	5/3/1	2/0.6/0.2	
	nd	nd	Postigo et al., 2010[d]
	nd	nd	Huerta-Fontela et al., 2007[e]
Nor-LSD	2–8	0.1–2	Postigo et al., 2008b[c]
	22/13/5	4/2/1	
	4	0.4	Postigo et al., 2010[d]
O-H-LSD or	4–9	0.3–1	Postigo et al., 2008b[c]
2-oxo-3-hydroxy-LSD	3/4/nd	0.8/nd/nd	
	nd	nd	Postigo et al., 2010[d]
Morphine	146–196	17–25	Postigo et al., 2008b[c]
	75/67/63	19/12/30	
	54–166	5–81	Postigo et al., 2010[d]
	26–278	12–81	Boleda et al., 2009[g]
	356	–	
	<7–97	<7–81	Boleda et al., 2007[h]
Normorphine	13–30	31–107	Boleda et al., 2009[g]
	28	–	
	<25	<25–31	Boleda et al., 2007[h]
Codeine	6–120	3–397	Boleda et al., 2009[g]
	314	–	
	18–120	3–397	Boleda et al., 2007[h]
Norcodeine	5–46	5–40	Boleda et al., 2009[g]
	8	–	
	<5–7	<5–23	Boleda et al., 2007[h]
Methadone	3–1531	3–732	Boleda et al., 2009[g]
	55	–	
	4–24	4–25	Boleda et al., 2007[h]

(*Continued*)

TABLE 6.2 *(Continued)*

Drug	Influent WW (ng/L)	Effluent WW (ng/L)	Reference[a,b,c,d,e,f,g]
EDDP or 2-ethylidene-	3–1029	3–1150	Boleda et al., 2009[g]
1,5-dimethyl-3,3-	132	–	
diphenylpyrrolidine	5–41	5–57	Boleda et al., 2007[h]
Heroin	nd-2	nd-1	Postigo et al., 2008b[c]
	nd	nd	Postigo et al., 2010[d]
	nd	nd	Boleda et al., 2009[g]
	nd	–	
	<20	<20	Boleda et al., 2007[h]
6-Acetylmorphine	10–19	3–5	Postigo et al., 2008b[c]
	11/9/6	3/2/3	
	1–4	nd	Postigo et al., 2010[d]
	nd	nd	Boleda et al., 2009[g]
	nd	–	
	<3	<3	Boleda et al., 2007[h]
THC or Δ⁹-tetrahydro-	nd	nd	Postigo et al., 2008b[c]
cannabinol	22/39/14	21/13/nd	
	48.4	nd	Postigo et al., 2010[d]
	11–127 7	21 -	Boleda et al., 2009[g]
	<8–32	<8	Boleda et al., 2007[h]
THC-COOH or	nd	5–15	Postigo et al., 2008b[c]
11-nor-9-carboxy-	14/33/17	4/19/11	
THC	11–22	5–73	Postigo et al., 2010[d]
	24–402	15–72	Boleda et al., 2009[g]
	152	–	
	<13–96	<13–15	Boleda et al., 2007[h]
OH-THC or	7–12	2–8	Postigo et al., 2008b[c]
11-hydroxy-THC	37/78/24	23/14/8	
	2–91	0.4	Postigo et al., 2010[d]

[a]One WWTP in Castellón (eastern Spain) ($n = 14$). Sampling period: 1 week in June 2008 and 1 week in July 2008.

[b]Top: 42 WWTPs in northeastern Spain ($n = 42$). Sampling period: April–January 2007.
Bottom: one WWTP in northeastern Spain ($n = 7$). Samples collected during one week.

[c] Top: one WWTP in Barcelona ($n = 7$). Sampling period: 1 week in July 2007.
Bottom: one WWTP Valencia ($n = 1$)/one WWTP Benicassim ($n = 1$)/one WWTP Gandía ($n = 1$). Sampling: 26th July 2007. Concentrations found in each plant.

[d]Seven WWTPs in the Ebro River basin ($n = 14$). Sampling period: October 2007 and July 2008.

[e]16 WWTPs in Catalonia ($n = 16$). Sampling period: April–September 2006. Average concentration/maximum concentration.

[f]Four WWTPs in NW Spain ($n = 3$ for influent waters and $n = 4$ for effluent waters). Sampling period: June 2009.

[g]Top:15 WWTPs in Catalonia ($n = 15$). Sampling period: April–May 2007.
Bottom: 1 WWTP in Barcelona ($n = 7$). Samples collected during one week. Average concentration.

[h]Five WWTP in Catalonia ($n = 5$). Sampling period: March–May 2007. nd: not detected.

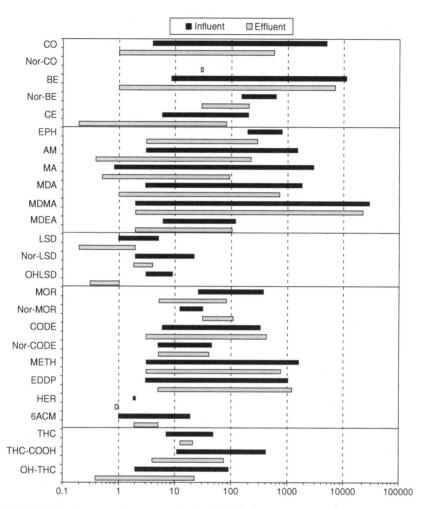

FIGURE 6.1 Concentration ranges (ng/L) of illicit drugs and metabolites in influent and effluent wastewaters in Spain. In the case of HER, MDEA, nor-BE, and nor-CO, which were only occasionally detected, concentrations range from the lowest limit of quantification of the method to the only positive value reported.

substance is consumed together with ethanol, was present at levels below 190 ng/L in most of the water samples investigated (Postigo et al., 2008b, 2010; Bijlsma et al., 2009). Nor-BE levels increased up to 430 ng/L in the waters collected at the inlet of the Castellón WWTP. Nor-CO was not detected in any of the investigated samples (Bijlsma et al., 2009).

Ephedrine (EPH), which is used mainly for therapeutic purposes, was the amphetamine-like compound detected most frequently at the inlet of the Spanish WWTPs investigated, being measured at levels between 220 and 725 ng/L (Postigo et al., 2008b, 2010). 3,4-Methylenedioxymethamphetamine (MDMA or ecstasy) was also positively identified in most of the investigated samples. The highest MDMA level (27.5 µg/L) was reported by Bijlsma et al. (2009) in the WWTP located in Castellón (eastern Spain). This is the highest MDMA value reported in wastewaters in all the peer-reviewed literature. However, as it was above that discussed for BE, the sample with such a high value was collected during an international music event that took place in the area served by the investigated WWTP. The levels of MDMA were between 2 and 598 ng/L (Huerta-Fontela et al., 2008a) in Catalonia, and between 4 and 180 ng/L in the Ebro River basin area (Postigo et al., 2010). The concentrations of MDMA in northwest Spain were somewhat lower (<11 ng/L) than in northeast Spain (González-Mariño et al., 2009). Other related drugs, such as 3,4-methylenedioxyamphetamine (MDA) and 3,4-methylenedioxyethamphetamine (MDEA), were also determined in Spanish WWTP influents, but their frequency of detection and their concentrations were lower than those of MDMA. Huerta-Fontela et al. (2008a) reported MDA and MDEA levels in Catalonia between 3 and 266 and 6 and 114 ng/L, respectively. Up to 1.7 µg/L of MDA were measured in the WWTP sampled in Castellón (Bijlsma et al., 2009) at the time of the previously mentioned music event. Conversely, MDEA was below the limit of quantification of the method (<500 ng/L) in the same samples (Bijlsma et al., 2009). In influent wastewaters from northwest Spain, MDA levels were below 20 ng/L, whereas MDEA was not detected (González-Mariño et al., 2009). Note that MDA in wastewater may have sources other than MDA consumption, as it is also a metabolite of MDEA and MDMA (Postigo et al., 2008a).

Amphetamine (AM) levels were usually in the nanogram per liter range, except in the case of the samples collected in Castellón at the time of the music event previously mentioned, where this compound reached 1.4 µg/L. Up to 688 ng/L of AM were measured in WWTPs sampled in Catalonia and a similar value, 664 ng/L, was reported by Postigo et al. (2010) as the maximum AM concentration measured in the Ebro River basin area. AM levels at the inlet of the WWTPs located in northwest Spain did not exceed 9 ng/L (González-Mariño et al., 2009). Overall, methamphetamine (MA) levels determined in influent wastewaters were below those reported for AM and also the frequency of detection of MA was lower than that registered for AM. MA reached concentrations up to 277 ng/L in Catalonia (Huerta-Fontela et al., 2008a) and below 10 ng/L in the Ebro River basin area (Postigo et al., 2010). In northwest Spain, MA was not detected in any of the investigated samples (González-Mariño et al., 2009).

The detection of lysergic compounds (LSD and its metabolites nor-LSD and 2-oxo-3-hydroxy LSD [O-H-LSD]) was rare. A few nanograms per liter were reported by Postigo et al. (2008b) in WWTPs located in Barcelona and in different cities of the Autonomous Community of Valencia (eastern Spain).

The opioids that were most frequently detected at the entrance of the investigated WWTPs in Spain were morphine (MOR) and codeine (CODE); however, the maximum levels of these compounds were one order of magnitude lower than those reported for CO and BE. The highest MOR and CODE levels reported in Spain were 356 and 314 ng/L, respectively. These values correspond to the average level detected in raw water samples collected daily during a 1-week sampling (spring 2007) in a large WWTP serving the city of Barcelona (Boleda et al., 2009). About one-half of the aforementioned MOR concentration (162 ng/L) was determined, on average, by Postigo et al. (2008b) in the same WWTP in a sampling carried out in summer 2007. MOR levels ranged from 26 to 278 ng/L in wastewater samples from Catalonia (Boleda et al., 2009) and between 54 and 166 ng/L in those from the Ebro River basin area (Postigo et al., 2010). On the other hand, CODE levels in Catalonia ranged from 6 to 120 ng/L (Boleda et al., 2009). The MOR metabolite normorphine (nor-MOR) was detected less frequently than MOR, at levels up to 30 ng/L in Catalonian WWTPs (Boleda et al., 2009). The CODE metabolite norcodeine (nor-CODE) was detected more frequently than nor-MOR in Catalonia and its levels were below 50 ng/L (Boleda et al., 2009).

Heroin (HER) and its exclusive metabolite 6-acetylmorphine (6ACM) were rarely detected in wastewaters in Spain; they were mostly measured at low levels (few nanograms per liter) (Postigo et al., 2008b, 2010). This may result from the low heroin usage in the areas served by the investigated WWTPs or, more likely, from the low rates at which these compounds are excreted. In contrast, the synthetic opioid agonist methadone (METH), used to treat heroin addiction and its main metabolite 2-ethylidine-1,5-dimethyl-3,3-diphenylpyrrolidine perchlorate (EDDP) were very ubiquitous and abundant in influent wastewaters. METH and EDDP were determined at maximum levels of 1.5 and 1.0 μg/L, respectively, in a WWTP of Catalonia (Boleda et al., 2009). Thus far, these concentrations are the highest reported worldwide.

Despite cannabinoids are hydrophobic (log $K_{ow} > 5$) (PHYSPROPdatabase 2010) they were occasionally detected in influent wastewaters. Concentrations of the main psychoactive cannabinoid, Δ^9-tetrahydrocannabinol (THC), usually accounted for few nanograms per liter. Higher levels, up to 127 ng/L, were measured only in Catalonia. The THC metabolites 11-nor-9-carboxy-THC (THC-COOH) and 11-hydroxy-THC (OH-THC), which are slightly more hydrophilic than THC, were found more frequently than the parent compound. Their concentrations ranged from few nanograms per liter to 402 ng/L for THC-COOH (Postigo et al., 2008b, 2010; Bijlsma et al., 2009, Boleda et al., 2009) and from 7 to 91 ng/L for OH-THC (Postigo et al., 2008b, 2010).

6.4   OCCURRENCE OF ILLICIT DRUGS AND METABOLITES IN EFFLUENT WASTEWATER

Levels of illicit drugs and metabolites measured in WWTP effluents in Spain are summarized in Table 6.2 and Fig. 6.1. As it was observed in the WWTPs influents, CO and its major metabolite BE were the most abundant compounds. The highest concentration of CO in treated wastewaters (560 ng/L) was reported by Bijlsma et al. (2009) in the Castellón WWTP (eastern Spain). BE was usually found at comparatively higher concentrations than CO, reaching the microgram per liter range in effluents of the Castellón WWTP (6.8 μg/L) (Bijlsma et al., 2009) and of one WWTP sampled northeast of Spain (1.5 μg/L) (Huerta-Fontela et al., 2008a). Levels of CO and BE measured in the outlet of the WWTP of Castellón were the highest ever reported in the peer-reviewed bibliography. CE was less frequently detected than CO and BE and it was present usually in the low nanogram per liter range. The maximum concentration of CE measured in effluent wastewaters (80 ng/L) is almost one-half of that found in influent wastewaters (Bijlsma et al., 2009). Levels of nor-BE in effluent wastewaters reached 170 ng/L. Nor-CO, that was absent in influent wastewaters, was detected in effluent wastewaters at levels between the limits of detection of the method and 30 ng/L, as reported by Bijlsma et al. (2009).

EPH and MDMA were the amphetamine-like compounds most frequently identified in Spanish WWTPs effluents. The maximum concentration of EPH, 276 ng/L, was registered in a sample collected in the Ebro River basin area whereas the maximum concentration of MDMA (21.2 μg/L), which was determined in the Castellón WWTP during the already mentioned music event, was two orders of magnitude higher (Bijlsma et al., 2009). In the Castellón WWTP effluent, MDA and MDEA were also measured at concentrations higher than those found in effluents of the other investigated WWTPs. These compounds were present only in samples collected at the time of the music event, at levels between 410 and 680 ng/L for MDA, and below 100 ng/L for MDEA (Bijlsma et al., 2009). MDMA, MDA, and MDEA concentrations in other areas did not exceed 376 (Postigo et al., 2008b), 200, and 12 ng/L (Huerta-Fontela et al., 2008a), respectively.

AM was always less frequently detected and at lower concentration in effluent than in influent wastewaters. The highest AM concentration (210 ng/L) was observed in treated waters from Castellón (Bijlsma et al., 2009) and from northeast of Spain (Huerta-Fontela et al., 2008a). MA showed a higher frequency of detection in effluent wastewaters compared to influent wastewaters in the monitoring studies carried out in the WWTP of Castellón and in the Ebro River basin (Bijlsma et al., 2009; Postigo et al., 2010). MA levels in WWTP effluents were always below 100 ng/L and most often they did not exceed 10 ng/L.

LSD and its metabolites were rarely identified in the WWTP effluents and their levels were usually in the picogram per liter range (Postigo et al., 2008b, 2010).

Concerning opioids, METH and its metabolite EDDP, followed by CODE, were the most ubiquitous compounds in effluent wastewaters. These compounds were measured only in Catalonia at concentrations ranging from few nanograms per liter to 732 ng/L for METH, 1.2 μg/L for EDDP, and 397 ng/L for CODE (Boleda et al., 2009). MOR levels in the Ebro River basin area (Postigo et al., 2010) and in Catalonia (Boleda et al., 2007, 2009; Postigo et al., 2008b) were below 81 ng/L. The MOR metabolite, nor-MOR, was at lower levels in influents than in effluents where it reached 107 ng/L (Boleda et al., 2009). HER and its metabolite 6ACM were usually below the limit of detection of the method in the investigated effluent wastewaters. They were measured only occasionally in the low nanogram per liter range in WWTPs in Barcelona and in the Autonomous Community of Valencia (Postigo et al., 2008b).

The most frequently detected and abundant cannabinoid in effluent wastewaters was the main metabolite of THC, THC-COOH. Its levels (up to 73 ng/L) were above those of the other investigated metabolite of THC, OH-THC (up to 23 ng/L) (Boleda et al., 2009; Postigo et al., 2010). THC was identified less frequently in effluents than in influents and at maximum levels of 21 ng/L.

6.5   REMOVAL OF ILLICIT DRUGS AND METABOLITES IN WASTEWATER TREATMENT PLANTS

Levels of illicit drugs and metabolites measured in treated wastewaters were one order of magnitude lower than those measured in raw wastewaters, indicating their partial removal during wastewater treatment processes. However, some of the investigated substances were found at higher levels in WWTP effluents than in influents. Figure 6.2 shows removal rates calculated for each compound in the WWTPs investigated in Spain. These rates, however, have to be considered with caution, since not all the studies reviewed have analyzed composite samples and the WWTP hydraulic residence time was not always taken into account when collecting samples in the inlet and outlet.

Removal of illicit drugs residues during wastewater treatment has been mostly studied in WWTPs equipped with conventional activated sludge (CAS) processes as secondary treatment (Huerta-Fontela et al., 2008a; Postigo et al., 2008b, 2010; Bijlsma et al., 2009; Boleda et al., 2009). In only one of the investigated plants, the secondary treatment consisted of biological filtration (Postigo et al., 2010). Comparison of the removal rates obtained using different wastewater treatment processes is, however, difficult because the analyses were done with different methodologies, samples were not collected and pretreated equally, and the WWTPs operational conditions were not systematically documented.

According to the reviewed data, cocaine and its metabolites were the class of illicit drugs best eliminated with CAS treatments (above 90% of removal, in most cases, for CO, BE, and CE). However, a drastic decrease in their removal efficiency was observed by Bijlsma et al. (2009) when the wastewaters entering the plant contained high levels of these compounds.

Amphetaminelike compounds were eliminated differently in the investigated WWTPs. Some of them, such as MA, MDMA, and MDA occasionally presented higher levels in samples from the WWTPs outlet than from the inlet (Huerta-Fontela et al., 2008a; Postigo et al., 2008b, 2010; Bijlsma et al., 2009; González-Mariño et al., 2009). The increase of MDA concentration in WWTPs effluents has been attributed to N-demethylation of MDMA (Huerta-Fontela et al., 2008a). On the other hand, MDMA and MA might result from desorption processes during wastewater treatment, but the processes involved in the formation of these compounds have not yet been elucidated. Among all the amphetamine-like compounds investigated, AM showed the highest removal rates (above 90% in most of the cases), followed by EPH and MA (Fig. 6.2).

LSD and its metabolites, when present, were, in general, efficiently removed in the investigated WWTPs. Only nor-LSD was apparently not removed in one of the plants (Postigo et al., 2010).

Opioids showed different behavior during wastewater treatment. As observed in similar studies performed in other European countries (Castiglioni et al., 2006; Bones et al., 2007) METH and its metabolite EDDP were poorly or not removed in the Spanish WWTPs investigated (Boleda et al., 2007, 2009). Maximum removal rates observed for METH and EDDP reached 53% and 15%, respectively (Fig. 6.2).

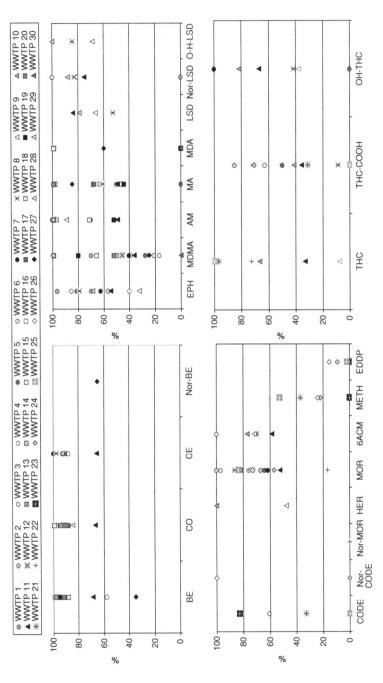

FIGURE 6.2 Removal of illicit drugs and metabolites in WWTPs in Spain. WWTP 1 –WWTP 7: Ebro River basin (average of $n = 2$ different samples), WWTP4 with biological filters as secondary treatment (Postigo et al., 2010); WWTP 8: WWTP in Barcelona (average of a week sampling, $n = 7$), WWTP 9: WWTP in Valencia ($n = 1$), WWTP10: WWTP in Benicassim ($n = 1$), WWTP 11: WWTP in Gandía ($n = 1$) (Postigo et al., 2008b); WWTP12: WWTP in NE Spain (average of a week sampling, $n = 7$) (Huerta-Fontela et al., 2008a), WWTP13 –WWTP 20: WWTPs in northeastern Spain (average of $n = 4$) (Huerta-Fontela et al., 2008a), WWTP 21: Average of WWTPs sampled in Catalonia (Boleda et al., 2009), WWTP 22 –WWTP 26: WWTPs in Catalonia ($n = 1$) (Boleda et al., 2007), WWTP 27: WWTP in Castellón (Bijlsma et al., 2009), WWTP 28 –WWTP 30: NW Spain (González-Mariño et al., 2009).

In general, MOR was efficiently removed in the investigated WWTPs with removal rates above 60% in most of the cases (Boleda et al., 2007, 2009; Postigo et al., 2008b, 2010). Only one WWTP located in Catalonia showed poor removal percentages for this compound (17%) (Boleda et al., 2007). Low MOR removals may be attributed to the cleavage of conjugated MOR during treatment process (Castiglioni et al., 2006), since it is mainly excreted in urine as glucuronide metabolites, and/or to inadequate operational conditions (e.g., low hydraulic retention time, low solids retention time). Nor-MOR levels also increased during wastewater treatment, which might be because of N-demethylation of MOR (Boleda et al., 2009). HER and its metabolite 6ACM were efficiently removed when present in wastewater (Postigo et al., 2008b, 2010).

CODE and its metabolite nor-CODE were fairly persistent (Boleda et al., 2007, 2009), since higher concentrations of these compounds in effluent compared to influent wastewaters were frequently measured. Like MOR, the lack of or poor removal of CODE may result from cleavage of the glucuronide forms of CODE during wastewater treatment. On the other hand, N-demethylation of CODE has been proposed as a possible reaction contributing to the formation of nor-CODE from CODE (Boleda et al., 2009).

Removal of cannabinoids was observed to vary widely among the investigated WWTPs. THC and its metabolite OH-THC were removed comparatively better than THC-COOH, which is often present at higher concentrations in effluent than in influent wastewaters. In this case, the cleavage of conjugated forms is also proposed as a possible cause, but it has not yet been investigated (Boleda et al., 2007, 2009; Postigo et al., 2008b, 2010).

6.6 EFFECT OF DISCHARGED TREATED WASTEWATER ON SURFACE WATER QUALITY

Incomplete removal of illicit drugs and metabolites during wastewater treatment is directly related to their presence in surface waters, as it has been shown in several works carried out in Europe (Kasprzyk-Hordern et al., 2009a) and North America (Bartelt-Hunt et al., 2009). Treated waters are usually diluted several times when discharged into the receiving natural waters; hence, concentrations of these compounds in surface waters are generally lower than in effluent wastewaters. Although not yet investigated in Spain, levels of illicit drugs and metabolites were reported elsewhere to significantly decrease with increasing distance from the discharge point (Bones et al., 2007).

The dilution of contaminants is directly related to flow and physical characteristics of the receiving surface waters. In general, rivers are turbulent enough to allow rapid mixing and dilution of WWTPs effluents. Rivers in their lower course present usually higher dilution factors than upstream. However, levels of illicit drugs and metabolites in Spanish rivers were observed to either keep fairly constant along the course of the river (Postigo et al., 2010) or increase from upstream to downstream (Huerta-Fontela et al., 2008b, Boleda et al., 2009) because of the increasing urban pressure toward the mouth.

In the Ebro River basin, calculated dilution factors, that is, the ratios between WWTP effluent and river flows, varied between 3 and 1785. The lowest values were observed in a tributary, the Arga River, and the largest ones close to the mouth of the Ebro River. Variability of dilution factors was also observed between different seasons, probably because of the continental climate conditions (sparse and irregular precipitation and high evaporation rates) that affect this water basin (Postigo et al., 2010).

Another watershed in Spain where the occurrence of illicit drugs and metabolites in surface waters was investigated is the Llobregat River basin. This water catchment is subjected to Mediterranean climate conditions, characterized by scarce precipitation concentrated in just a few days. As a consequence, the Llobregat River flow rate varies widely, which is a common feature to all Mediterranean rivers. In addition, the Mediterranean river basins are usually heavily urbanized, industrialized, and/or cultivated and, thus, surface waters are most often more contaminated than in other larger European basins. Moreover, during drought periods, effluent waters may represent almost 100% of the total flow of the river and, for this reason, very high levels of illicit drugs and metabolites have been occasionally reported, for instance, for CO, BE, CODE, and MDMA, in the Llobregat River basin (Huerta-Fontela et al., 2008b; Boleda et al., 2009).

In agreement with the discussion above, the highest total levels of illicit drugs and metabolites in Spain have been reported in small creeks, such as the Ebro tributary Huerva River (578 ng/L) (Postigo et al., 2010) and the Llobregat tributary Rubí River (330 ng/L) (Boleda et al., 2009).

Despite the fact that CO and BE were usually efficiently removed (>80%) in WWTPs, they were the two most abundant and ubiquitous substances found in surface waters in Spain. CO and BE levels in the Ebro River basin reached 59 and 346 ng/L, respectively (Postigo et al., 2010). In the Llobregat River, the maximum seasonal average concentrations of CO and BE were 60 and 150 ng/L, respectively, as reported by Huerta-Fontela et al. (2008b).

EPH was the most ubiquitous and abundant amphetamine-like compound found in the Spanish rivers. EPH levels measured in the Ebro River basin were up to 145 ng/L (Postigo et al., 2010). MDMA was the second most ubiquitous amphetamine-like compound found in surface waters. MDMA levels in Spanish rivers were below 12 ng/L in the Ebro River basin (Postigo et al., 2010), with a maximum concentration of 20 ng/L in the Llobregat River (Huerta-Fontela et al., 2008b). The other investigated amphetamine-like compounds AM, MA, and MDA were only occasionally detected in surface waters in Spain in the low nanogram per liter range.

Detection of LSD and metabolites in river waters was rare, with levels in the picogram per liter range (Postigo et al., 2010).

Among opioids, CODE, METH, and its metabolite EDDP were the most ubiquitous compounds; however, concentrations measured in river waters were usually substantially lower than those of CO and BE. In the Llobregat River, concentrations of these compounds were increasing downstream, reaching values of 40 ng/L for CODE, 4 ng/L for METH, and 17 ng/L for EDDP (Boleda et al., 2009). MOR was found at low concentrations in some surface water samples collected from the Ebro

River basin (<11 ng/L) (Postigo et al., 2010) and in Catalonia (below 31 ng/L in Rubí Creek and below 7 ng/L in the Llobregat River) (Boleda et al., 2009). HER was never detected in surface waters in Spain and its metabolite 6ACM was rarely detected.

THC-COOH was often detected in the Llobregat River, at concentrations up to 79.5 ng/L, but less frequently than opioids, CO and BE. THC was detected only in one sample collected from the Llobregat River, at a concentration of 14 ng/L (Boleda et al., 2007).

6.7 CONCLUSIONS

In recent years, several publications have reported the occurrence of illicit drugs and metabolites in waste and surface waters in Spain. The studies performed so far have shown, in general, incomplete removal of these compounds in the WWTPs investigated, and their consequent presence in the receiving river waters, although at lower concentrations than in wastewater. CO and BE seem to be the most efficiently removed compounds, and cannabinoids and amphetamine-like compounds the least. Nevertheless, removal rates varied widely between compounds and plants and the interpretation of the results must be handled with caution, since sample collection in most of the cases did not take into account the WWTP hydraulic residence time and water samples occasionally were not composite.

Overall, the most ubiquitous and abundant compounds in both treated wastewaters and surface waters in Spain were CO and its metabolite BE, the opioid CODE, METH, its metabolite EDDP, the THC metabolite THC-COOH, and the amphetamine-like compounds EPH and MDMA. Geographically, comparatively lower concentrations of these substances, and presumably lower consumption of illicit drugs, were reported in northwest than in northeast-east of Spain. The highest levels were found in samples from the Castellón WWTP (eastern Spain), but this occurred on the occasion of a music event and, therefore, are probably not representative.

The fate and transport of illicit drugs and their metabolites in the environment requires further investigation, not only in Spain but worldwide. For instance, the presence of these compounds in solid matrices, e.g., sediments and sludge, has not yet been investigated, in spite of the fact that some compounds, such as cannabinoids, are likely to partition to solid matrices. Obviously, this also requires the development of methodologies suitable for the analysis of these complex matrices. Ecotoxicological data, not yet available, are also necessary to assess the risk that the presence of these substances may pose to ecosystems. Further studies are also needed in the field of waste and drinking water treatment and to assess the stability of these substances under different environmental conditions.

The main drawback of the analytical methodologies available is that monitoring of this type of contaminant is limited to selected compounds. Methodologies based on target analysis are insufficient to identify the relevance of new illicit drugs, such as the synthetic cannabinoids that constitute the commercially available products known as "spice," or the substances with stimulant and psychotropic activity derived from cathinone and piperazine. Moreover, further photo- and biodegradation products that may be of interest to refine the environmental risk assessment of these substances cannot be monitored by means of target analysis.

ACKNOWLEDGMENTS

This work has been supported by the EU Project MODELKEY [GOCE 511237] and by the Spanish Ministry of Science and Innovation (projects CGL2007-64551/HID and Consolider-Ingenio 2010 CSD2009-00065). It reflects only the authors' views. The EU is not liable for any use that may be made of the information contained therein. Cristina Postigo acknowledges the European Social Fund and AGAUR (Generalitat de Catalunya, Spain) for their economical support through the FI predoctoral grant.

REFERENCES

Bartelt-Hunt, S. L., D. D. Snow, T. Damon, J. Shockley, and K. Hoagland. 2009. The occurrence of illicit and therapeutic pharmaceuticals in wastewater effluent and surface waters in Nebraska. *Environ. Pollut.* **157**:786–791.

Bijlsma, L., J. V. Sancho, E. Pitarch, M. Ibáñez, and F. 2009. Hernández F. Simultaneous ultra-high-pressure liquid chromatography–tandem mass spectrometry determination of amphetamine and amphetamine-like stimulants, cocaine and its metabolites, and a cannabis metabolite in surface water and urban wastewater. *J. Chromatogr. A* **1216**:3078–3089.

Boleda, M. R., M. T. Galcerán, and F. Ventura. 2007. Trace determination of cannabinoids and opiates in wastewater and surface waters by ultra-performance liquid chromatography-tandem mass spectrometry. *J. Chromatogr. A* **1175**:38–48.

Boleda, M. R., M. T. Galcerán, and F. Ventura F. 2009. Monitoring of opiates, cannabinoids and their metabolites in wastewater, surface water and finished water in Catalonia, Spain. *Water Res.* **43**:1126–1136.

Bones, J., K. V. Thomas, and B. Paull. 2007. Using environmental analytical data to estimate levels of community consumption of illicit drugs and abused pharmaceuticals. *J. Environ. Monit.* **9**:701–707.

Castiglioni, S., E. Zuccato, E. Crisci, C. Chiabrando, R. Fanelly, and R. Bagnati. 2006. Identification and measurement of illicit drugs and their metabolites in urban wastewater by liquid chromatography-tandem mass spectrometry. *Anal. Chem.* **78**:8421–8429.

Chiaia, A. C., C. Banta-Green, and J. Field. 2008. Eliminating solid phase extraction with large-volume injection LC/MS/MS: Analysis of illicit and legal drugs and human urine indicators in US wastewaters. *Environ. Sci. Technol.* **42**:8841–8848.

Commission Decision 2002/657/EC of 12 August 2002 implementing Council Directive 96/23/EC concerning the performance of analytical methods and the interpretation of results.

European Monitoring Centre for Drugs and Drug Addiction (EMCDDA) 2009. Annual report on the state of the drugs problem in Europe.

Gheorghe, A., A. Van Nuijs, B. Pecceu, L. Bervoets, P. G. Jorens, R. Blust, H. Neels, and A. Covaci. 2008. Analysis of cocaine and its principal metabolites in waste and surface water using solid-phase extraction and liquid chromatography-ion trap tandem mass spectrometry. *Anal. Bioanal. Chem.* **391**:1309–1319.

González-Mariño, I., J. B. Quintana, I. Rodríguez, R. Rodil, J. González-Peñas, and R. Cela. 2009. Comparison of molecularly imprinted, mixed-mode and hydrophilic balance sorbents performance in the solid-phase extraction of amphetamine drugs from wastewater samples for liquid chromatography-tandem mass spectrometry determination. *J. Chromatogr. A* **216**:8435–8441.

Huerta-Fontela, M., M. T. Galcerán, and F. Ventura. 2007. Ultraperformance Liquid Chromatography-tandem mass spectrometry analysis of stimulatory drugs of abuse in wastewater and surface waters. *Anal. Chem.* **79**:3821–3829.

Huerta-Fontela, M., M. T. Galcerán, J. Martin-Alonso, and F. Ventura. 2008a. Occurrence of psychoactive stimulatory drugs in wastewaters in north-eastern Spain. *Sci. Total Environ.* **397**:31–40.

Huerta-Fontela, M., M. T. Galcerán, and F. Ventura. 2008b. Stimulatory drugs of abuse in surface waters and their removal in a conventional drinking water treatment plant. *Environ. Sci. Technol.* **42**:6809–6816.

Hummel, D., D. Löffler, G. Fink, and T. A. Ternes. 2006. Simultaneous determination of psychoactive drugs and their metabolites in aqueous matrices by liquid chromatography mass spectrometry. *Environ. Sci. Technol.* **40**:7321–7328.

Jones-Lepp, T. L., D. A. Alvarez, J. D. Petty, and J. N. Huckins. 2004. Polar organic chemical integrative sampling and liquid chromatography–electrospray/ion-trap mass spectrometry for assessing selected prescription and illicit drugs in treated sewage effluents. *Arch. Environ. Contamination Toxicol.* **47**:427–439.

Kasprzyk-Hordern, B., R. M. Dinsdale, and A. J. Guwy. 2008. Multiresidue methods for the analysis of pharmaceuticals, personal care products and illicit drugs in surface water and wastewater by solid-phase extraction and ultra performance liquid chromatography-electrospray tandem mass spectrometry. *Anal. Bioanal. Chem.* **391**:1293–1308.

Kasprzyk-Hordern, B., R. M. Dinsdale, and A. J. Guwy. 2009a. Illicit drugs and pharmaceuticals in the environment - Forensic applications of environmental data, Part 2: Pharmaceuticals as chemical markers of faecal water contamination. *Environ. Pollut.* **157**:1778–1786.

Kasprzyk-Hordern, B., R. M. Dinsdale, and A. J. Guwy. 2009b. Illicit drugs and pharmaceuticals in the environment - Forensic applications of environmental data. Part 1: Estimation of the usage of drugs in local communities. *Environ. Pollut.* **157** :1773–1777.

Kasprzyk-Hordern, B., R. M. Dinsdale, and A. J. Guwy. 2009c. The removal of pharmaceuticals, personal care products, endocrine disruptors and illicit drugs during wastewater treatment and its impact on the quality of receiving waters. *Water Res.* **43**:363–380.

Loganathan, B., M. Phillips, H. Mowery, and T. L. Jones-Lepp. 2009. Contamination profiles and mass loadings of macrolide antibiotics and illicit drugs from a small urban wastewater treatment plant. *Chemosphere* **75**:70–77.

Mari, F., L. Politi, A. Biggeri, G. Accetta, C. Trignano, M. Di Padua, and E. Bertol. 2009. Cocaine and heroin in waste water plants: A 1-year study in the city of Florence, Italy. *Forensic Sci. Int.* **189**:88–92.

PHYSPROP (Physical properties) database, http://www.syrres.com/what-we-do/product. aspx?id=133, accessed on January 2010.

Postigo, C., M. J. López de Alda, and D. Barceló. 2008a. Analysis of drugs of abuse and their human metabolic byproducts in water by LC-MS/MS: a non-intrusive tool for drug abuse estimation at the community level. *Trends Anal. Chem.* **27**:1053–1069.

Postigo, C., M. J. López De Alda, and D. Barceló. 2008b. Fully automated determination in the low nanogram per liter level of different classes of drugs of abuse in sewage water by on-line solid-phase extraction-liquid chromatography-electrospray-tandem mass spectrometry. *Anal. Chem.* **80**:3123–3134.

Postigo, C., M. J. López De Alda, M. Viana, X. Querol, A. Alastuey, B. Artiñano, and D. Barceló. Determination of drugs of abuse in airborne particles by pressurized liquid extraction and liquid chromatography-electrospray-tandem mass spectrometry. *Anal. Chem.* **81**:4382–4388.

Postigo, C., M. J. López De Alda, and D. Barceló. 2010. Drugs of abuse and their metabolites in the Ebro River basin: occurrence in sewage and surface water, sewage treatment plants removal efficiency, and collective drug usage estimation. *Environ. Int.* **36**:75–84.

Richardson, S. D. 2008. Environmental mass spectrometry: Emerging contaminants and current issues. *Anal. Chem.* **80**:4373–4402.

United Nations Office on Drugs and Crime (UNODC). 2009. World Drug Report.

Van Nuijs, A. L. N., B. Pecceu, L. Theunis, N. Dubois, C. Charlier, P. G. Jorens, L. Bervoets, R. Blust, H. Meulemans, H. Neels, and Covaci A. 2009a. Can cocaine use be evaluated through analysis of wastewater? A nation-wide approach conducted in Belgium. *Addiction* **104**:734–741.

Van Nuijs, A. L. N., B. Pecceu, L. Theunis, N. Dubois, C. Charlier, P. G. Jorens, L. Bervoets, R. Blust, H. Neels, and A. Covaci. 2009b. Cocaine and metabolites in waste and surface water across Belgium. *Environ. Pollut.* **157**:123–129.

Van Nuijs, A. L. N., B. Pecceu, L. Theunis, N. Dubois, C. Charlier, P. G. Jorens, L. Bervoets, R. Blust, H. Neels, and A. Covaci. 2009c. Spatial and temporal variations in the occurrence of cocaine and benzoylecgonine in waste- and surface water from Belgium and removal during wastewater treatment. *Water Res.* **43**:1341–1349.

Zuccato, E., C. Chiabrando, S. Castiglioni, D. Calamari, R. Bagnati, S. Schiarea, and R. Fanelli. 2005. Cocaine in surface water: a new evidence-based tool to monitor community drug abuse. *Environ. Health* **4**:1–7.

Zuccato, E., C. Chiabrando, S. Castiglioni, R. Bagnati, R. Fanelli. 2008. Estimating Community drug abuse by wastewater analysis. *Environ. Health Perspect.* **116**:1027–1032.

CHAPTER 7

OCCURRENCE OF ILLICIT DRUGS IN WASTEWATER AND SURFACE WATER IN ITALY

SARA CASTIGLIONI and ETTORE ZUCCATO

7.1 INTRODUCTION

In the last few years, illicit drugs have been frequently detected in the aquatic environment, becoming a class of emerging contaminants. Some recent reviews (Postigo et al., 2008a; Zuccato and Castiglioni, 2009) suggest their widespread occurrence in wastewater, surface water (rivers and lakes), drinking water, and airborne particulates.

Analogously with pharmaceuticals, the main source of contamination is supposed to be the consumer itself. In fact, after ingestion, illicit drugs can be excreted unchanged or as main metabolites in consumers' urine and enter the urban sewage network. Once they reach wastewater treatment plants (WWTPs) through wastewater, they can be only partially removed during conventional water treatment processes, being often still detectable in treated water. WWTPs are designed specifically to mostly remove organic matter, nitrates, and phosphates from wastewater, which is ineffective in completely removing other substances, such as illicit drugs. Treated wastewater can, therefore, still bring along illicit drugs residues when discharged in surface water that is subsequently used to produce drinking water by specific advanced treatments.

Illicit drugs are a heterogeneous group of substances with different structures and physico-chemical properties, but they have all a polar chemical structure and similar behaviors in the environment. Because of their polarity, illicit drugs are mostly expected to be distributed in the aqueous phase, as recently suggested by several studies (Zuccato and Castiglioni, 2009; van Nuijs et al., 2010). However, some of these compounds have also moderate lipophilic properties and can, therefore, be

Illicit Drugs in the Environment: Occurrence, Analysis, and Fate Using Mass Spectrometry, Edited by
Sara Castiglioni, Ettore Zuccato, and Roberto Fanelli
Copyright © 2011 John Wiley & Sons, Inc.

adsorbed on the suspended solids of the water phase, or on the sludge in WWTPs (Kaleta et al., 2006). Despite their low volatility, illicit drugs have been also detected in airborne particles taken from the atmosphere of several cities (Postigo et al., 2009; Cecinato et al., 2009), probably because of their specific patterns of distribution and consumption. Illicit drugs are biologically active substances with pharmacological activity, and might also have potential actions and toxic effects in wildlife, as shown previously for therapeutic pharmaceuticals (Pomati et al., 2006). Thus, monitoring the presence of illicit drugs in the environment is useful to investigate their potential ecotoxicological implications related to their occurrence in complex mixtures in different environmental media.

The idea to search for illicit drugs in the environment derives from previous studies on therapeutic drugs, which were carried out since the end of the 1990s. During these investigations some authors observed that the amounts of pharmaceuticals measured in some environmental media roughly reflected the amounts prescribed to the population (Calamari et al., 2003). Since illicit drugs are consumed worldwide in amounts comparable to those of the therapeutic drugs (that is, 200 million individuals are current users of cocaine, heroin, amphetamine-like stimulants, marijuana, and other drugs; United Nations Office of Drugs and Crime, 2009), these investigators wondered whether illicit drugs too could be measured in the aquatic environment and their levels used to estimate the consumption of these substances by the population. The first investigation on this issue was conducted by our group in 2004 (Zuccato et al., 2005), and showed that cocaine and its main urinary metabolite (benzoylecgonine) were present either in urban wastewater and surface water in Italy. Benzoylecgonine was measured at concentrations between 420 and 750 ng/L and cocaine at lower concentrations (42–120 ng/L) in several WWTPs in Italy. The main Italian river (Po River) was also investigated and benzoylecgonine and cocaine concentrations were, respectively, 25 and 1.2 ng/L. Meanwhile, methamphetamine and methylendioxymethamphetamine (MDMA, or ecstasy) were measured in treated wastewaters in the United States at concentrations between 0.5 and 2 ng/L (Jones-Lepp et al., 2004). Our study in Italy was later extended to other metabolites of cocaine and to other common drugs of abuse, namely, opioids, amphetamines, cannabis, and to some related opioid pharmaceuticals, such as codeine and methadone. Specific analytical methods have been developed and used to measure these substances in wastewater (Castiglioni et al., 2006) and surface water (Zuccato et al., 2008a) in Italy, Switzerland, and the United Kingdom. The occurrence, behavior, and fate of illicit drugs in waste, surface, and even drinking water were subsequently investigated in several European countries and the United States (Postigo et al., 2008a; Zuccato and Castiglioni, 2009; Mari et al., 2009; van Nuijs et al., 2010), where these substances were found at concentrations up to the microgram per liter range.

This chapter reports the main outcomes from several investigations on the occurrence of illicit drugs in waste- and surface water in Italy, summarizing some results already published, and including some new findings obtained in recent investigations. This review will also analyze the behavior and fate of these substances in WWTPs in Italy and will discuss their impact in surface water.

7.2 ANALYTICAL METHODOLOGY

The multiresidue analytical methods that have been recently developed to measure illicit drugs in waste- and surface water, are mainly based on high-pressure liquid chromatography—tandem mass spectrometry (LC—MS/MS) (Castiglioni et al., 2008; Postigo et al., 2008a). In fact, liquid chromatography is the best technique for the analysis of polar substances and mass spectrometry is the most powerful technique to simultaneously detect several different compounds in complex matrices, such as wastewater. Nevertheless, Mari et al. (2009) have used gas chromatography—mass spectrometry (GC—MS) to measure illicit drugs in wastewater samples.

The first comprehensive analytical method for the determination of cocaine, amphetamines, morphine, and cannabinoid derivatives, methadone, and/or some of their metabolites was set up by our group in 2006 (Castiglioni et al., 2006) and was validated both in untreated and treated wastewater (Table 7.1). The original method was based on solid-phase extraction (SPE) and high-performance liquid chromatography—tandem mass spectrometry (HPLC—MS/MS) and it was later modified to be also applied to surface water samples and was used to detect these substances in several Italian rivers and lakes (Zuccato et al., 2008a). For these analyses, 24-h composite samples were obtained from WWTPs influents and effluents by pooling water collected at fixed intervals by automatic sampling devices. Surface water samples were obtained by pooling river or lake water aliquots collected every 20 min during a period of 2 h by a portable automatic sampler (Sigma 900 Standard, Hach Company, USA). After collection, water samples (500 mL for wastewater and 2 L for surface water) were frozen and stored in the dark at $-20°$C. Before extraction, samples were first filtered on a glass microfiber filter GF/A 1.6 μm (Whatman, Kent, U.K.) and then on a mixed 0.45-μm cellulose membrane filter (Whatman, Kent, U.K.). Samples (50—1000 mL) were then spiked with deuterated internal standards and the pH adjusted to 2.0 with 37% HCl before solid-phase extraction, which was performed by mixed reversed-phase, cation-exchange cartridges (Oasis-MCX). The cartridges were conditioned before use by washing with 6 mL methanol, 3 mL MilliQ water, and 3 mL water acidified to pH 2. Samples were then passed through the cartridges under vacuum, at a flow rate of 10 mL/min. Cartridges were vacuum-dried for 5 min and eluted with 3 mL of methanol and 3 mL of a 2% ammonia solution in methanol. The eluates were pooled and dried under a nitrogen stream. Dried samples were redissolved in 200 μL of MilliQ water, centrifuged, transferred into glass vials, and analyzed by high-pressure liquid chromatography—tandem mass spectrometry (HPLC—MS/MS) using an API 3000 triple quadrupole mass spectrometer, equipped with a turbo ion spray source (Applied Biosystems - Sciex, Thornhill, Ontario, Canada) and interfaced to LC Series 200 pumps and auto sampler (Perkin-Elmer, Norwalk, CT). Drugs were analyzed using an XTerra MS C18, 100×2.1 mm, 3.5-μm column (Waters Corp., Milford, MA), at a flow rate of 200 μL/min. Quantitative analyses were done in the selected-reaction monitoring (SRM) mode, measuring the fragmentation products of the protonated or deprotonated pseudo-molecular ions of each compound and deuterated analog (Castiglioni et al., 2006, 2008). Quantitation was made by isotope dilution, using the deuterated analog as an internal standard for each drug residue.

TABLE 7.1 Method Validation for the Analysis of Illicit Drugs and Their Metabolites in Untreated and Treated Wastewater and Surface Water[a]

Illicit Drugs	Recovery + SD (%) (Untreated WW[b])	Recovery + SD (%) (Treated WW[b])	Recovery + SD (%) (Surface Water)	LOQ[c] (Untreated WW[b])	LOQ[c] (Surface Water)
Cocaine and metabolites					
Benzoylecgonine	107 ± 8.9	107 ± 2.3	96 ± 6.7	1.98	0.10
Norbenzoylecgonine	85 ± 4.7	104 ± 2.5	99 ± 4.9	0.94	0.15
Cocaine	96 ± 4.8	94 ± 3.6	105 ± 1.9	1.4	0.13
Norcocaine	112 ± 6.9	119 ± 2.0	102 ± 4.4	1.92	0.15
Cocaethylene	109 ± 4	98 ± 2.3	105 ± 0.2	0.95	0.07
Opioids					
Morphine	88 ± 6.6	75 ± 3.4	85 ± 12	3.95	0.55
6-Acetylmorphine	106 ± 4.5	93 ± 7.6	87 ± 3.5	5.3	0.93
Morphine-3β-D-glucuronide	90 ± 10.5	67 ± 0.1	46 ± 1.5	0.63	0.14
Codeine	88 ± 3.2	102 ± 5.9	107 ± 8.6	3.8	0.62
6-Acetylcodeine	76 ± 6.5	105 ± 5.0	114 ± 8.3	2.6	0.31
Methadone	105 ± 3.1	90 ± 2.0	104 ± 2.8	1.14	0.07
EDDP	88 ± 3.2	85 ± 4.3	84 ± 10	1.64	0.10
Amphetamines					
Amphetamine	110 ± 4.5	103 ± 4.2	101 ± 4.5	5.4	0.65
Methamphetamine	112 ± 6.5	97 ± 3.4	108 ± 6.9	3.7	0.41
MDA	102 ± 3.3	98 ± 3.1	97 ± 6.4	8.7	1.18
MDMA	104 ± 2.3	97 ± 3.2	98 ± 2.3	6.3	0.35
MDEA	107 ± 3.7	89 ± 3.3	96 ± 4.3	4.19	0.38
Cannabinoids					
11-Nor-9-carboxy-Δ^9-THC	51 ± 1.3	61 ± 4.0	69 ± 1.1	1.75	0.48

[a]From Castiglioni et al., 2006; Zuccato et al., 2008a.
[b]WW, wastewater; [c]LOQ, limit of quantification.

The performance of the analytical method was good both in waste and surface water (Table 7.1). Recoveries were generally higher than 80%, except for THC-COOH (51%−69%) and morphine 3β-glucuronide (46%−67% in treated and surface water), and the overall variability of the method was lower than 10%. Limits of quantifications were in the low nanogram per liter range for untreated and treated wastewater and were lower than 1 ng/L for surface water. The instrumental repeatability and precision, the instrumental limits of detection and quantification, and the linearity of the analytical response have been tested and are reported in detail elsewhere (Castiglioni et al., 2006; Zuccato et al., 2008a).

The method developed by Mari et al. (2009) consisted of solid-phase extraction by Bond Elut Certify LRC cartridges (Varian, Harbor City, CA), and derivatization of samples with BSTFA. A gas chromatography−mass spectrometer (GC−MS, Thermoquest Trace GC/Finnigan Polaris Q MS) was employed for the analysis of cocaine, benzoylecgonine, and morphine. Recoveries were higher than 87% for all the compounds and the limits of quantification were lower than 25 ng/L.

7.3 ILLICIT DRUGS IN WASTEWATER

After the first study carried out in Italy in 2004 to investigate the presence of illicit drugs in wastewater (Zuccato et al., 2005), our group has conducted several other campaigns analyzing samples from different WWTPs in Italy and in Switzerland (Lugano), close to the Italian border. The plants we selected for these investigations were located in Milan, the largest city in the north of Italy, Como (a smaller city in the same area), and in four cities in Sardinia, an island in the south of Italy. The WWTP in Milan (Nosedo) processes wastewater from the central and eastern areas of Milan by means of two sewers and serves a population of 1,250,000 people with a mean flow rate of 370,000 m^3/day. The Como plant collects wastewater from the whole city, serving 100,000 people with a mean flow rate of 63,000 m^3/day. The four plants investigated in Sardinia serve between 20,000 and 300,000 people with mean flow rates between 7,000 and 80,000 m^3/day. The Lugano plant collects wastewater from the entire city area serving 120,000 people with a mean flow rate of 60,000 m^3/day. All the plants are equipped with pretreatment, primary, and secondary treatment facilities, that is, primary settling and activated sludge processes. The Nosedo plant is equipped with a further disinfection step, consisting of a treatment with peracetic acid. Mari et al. (2009) collected wastewater from the whole city of Florence at two collection points located, respectively, on the right and the left bank of the Arno River.

Concentrations of illicit drugs in untreated wastewater entering the WWTPs are reported in Table 7.2. Cocaine and its major metabolite benzoylecgonine (BE) were the most abundant compounds with concentrations up to the microgram per liter range. The highest levels of BE were detected in Milan (1.5 μg/L) and in Lugano (0.5 μg/L), and the lowest levels were found in Florence (0.1 μg/L). Cocaine was measured at lower concentrations than BE (50−465 ng/L) and other minor metabolites of cocaine such as norbenzoylecgonine, norcocaine, and cocaethylene were also detected at

TABLE 7.2 Concentrations (nanogram per liter) of Illicit Drugs in Untreated Wastewater in Some Cities in Italy and Switzerland (Lugano)

Illicit Drugs	Lugano (Mean ± SD) 4-week Monitoring (2006) (Castiglioni et al., 2006)	Milan (Mean ± SD) 5-week Monitoring (2008)	Como (Mean ± SD) 4-week Monitoring (2008)	Sardinia (4 Cities) (Mean ± SD) 1-week Monitoring (2009)	Florence (Mean ± SD) (12 Samples in 1 Year) (2006-2007) (Mari et al., 2009)
Cocaine and metabolites					
Benzoylecgonine	547 ± 169	1468 ± 211	462 ± 115	306 ± 220	127 ± 12
Norbenzoylecgonine	19 ± 6	42 ± 5	13 ± 4	10 ± 7	–
Cocaine	218 ± 58	465 ± 90	146 ± 53	87 ± 68	50 ± 9
Norcocaine	4.4 ± 0.9	8 ± 1	1.1 ± 1.2	1.6 ± 1.2	–
Cocaethylene	6.0 ± 2.6	12 ± 2	3.3 ± 1.1	2.4 ± 2.6	–
Opioids					
Morphine	204 ± 50	112 ± 22	1261 ± 1523	89 ± 56	12.3 ± 1.6
6-Acetylmorphine	10 ± 5	14 ± 14	2.0 ± 2.4	4.1 ± 4.8	–
Morphine-3β-D-glucuronide	18 ± 3	<LOQ[a]	12 ± 3	2.6 ± 4.6	–
Morphine-6β-D-glucuronide	–	<LOQ[a]	4.3 ± 2.9	7.7 ± 7.4	–
Codeine	186 ± 53	96 ± 12	8191 ± 9845	42 ± 7.0	–
6-Acetylcodeine	<LOQ[a]	<LOQ[a]	1.5 ± 2.7	<LOQ[a]	–
Methadone	50 ± 10	13 ± 2	15 ± 3	27 ± 19	–
EDDP	91 ± 19	24 ± 3	33 ± 7	38 ± 33	–
Amphetamines					
Amphetamine	<LOQ[a]	2 ± 3.4	<LOQ[a]	<LOQ[a]	–
Methamphetamine	<LOQ[a]	40 ± 17	12 ± 10	<LOQ[a]	–
MDA	<LOQ[a]	<LOQ[a]	<LOQ[a]	<LOQ[a]	–
MDMA	14 ± 13	28 ± 10	5.7 ± 4.3	0.9 ± 1.7	–
MDEA	<LOQ[a]	<LOQ[a]	<LOQ[a]	<LOQ[a]	–
Cannabinoids					
11-Nor-9-carboxy-Δ^9-THC	91 ± 25	60 ± 12	53 ± 10	70 ± 13	–

[a]LOQ, limit of quantification.

mean concentrations usually lower than 40 ng/L. Similar concentrations of cocaine and BE were found in Spain (Postigo et al., 2008b; Huerta-Fontela et al., 2008; Bijlsma et al., 2009), UK (Kasprzyk-Hordern et al., 2008), and the United States (Chiaia et al., 2008). Lower concentrations were observed in Belgium (van Nuijs et al., 2009a), Germany (Hummel et al., 2006), and Ireland (Bones et al., 2007), which was the only country where cocaine was measured in untreated wastewater at higher concentrations than BE.

Among the opioids, morphine and codeine were the most abundant compounds in untreated wastewater, and were usually found at concentrations in the hundreds nanogram per liter range, except in Florence (12 ng/L). Morphine concentrations in Lugano were higher than in Milan, probably reflecting higher consumptions of therapeutic morphine in this city. Other minor heroin metabolites (6-acetylmorphine, morphine-3β-glucuronide, and morphine-6β-glucuronide) were found at concentrations lower than 20 ng/L. Finally, methadone and its major metabolite 2-ethylidene-1,5-dimethyl-3,3-diphenylpyrrolidine (EDDP), were measured at concentrations between 10 and 90 ng/L; their levels were higher in Switzerland (Lugano) than in Italy. Morphine was also detected at higher concentrations in UK (Zuccato et al., 2008b) and Germany (Hummel et al., 2006), and at comparable concentrations in Spain (Boleda et al., 2007). Codeine, methadone, and EDDP were detected also in Spain (Boleda et al., 2007; Postigo et al., 2008b), Belgium (van Nuijs et al., 2009b), and Germany (codeine only) (Hummel et al., 2006) at concentrations lower than 200 ng/L.

Stimulatory drugs, such as amphetamines were measured in Italy at concentrations lower than 40 ng/L, and the highest levels were found in the city of Milan. Amphetamine was found in traces in Italy, but it was measured up to the microgram per liter range in the United States (Chiaia et al., 2008) and Spain (Bijlsma et al., 2009). Other amphetaminelike compounds, such as methamphetamine, 3,4-methylenedioxymethamphetamine (MDMA or ecstasy), 3,4-methylenedioxyamphetamine (MDA), and 3,4-methylenedioxyethamphetamine (MDEA) were found at similar levels in Europe (<200 ng/L) (Zuccato and Castiglioni, 2009), but, again, at higher concentrations (up to 2 μg/L) in the United States (Chiaia et al., 2008) and UK (Kasprzyk-Hordern et al., 2008).

The principal metabolite of cannabis, 11-nor-9-carboxy-Δ^9-THC (THC-COOH), was measured in all the WWTPs at concentrations between 50 and 90 ng/L, and its levels were higher in Switzerland (Lugano) than in Italy. THC-COOH was measured also in Spain (Boleda et al., 2007; Postigo et al., 2008b) at similar levels, and in UK (Zuccato et al., 2008b) at higher concentrations.

Illicit drug have been, therefore, commonly detected in untreated wastewater at high concentrations (up to the microgram per liter range) in several cities in Italy, other European countries, and the United States, confirming that residues of these substances derived from human excretion and are direclty conveyed to urban WWTPs through wastewater.

7.4   REMOVAL OF ILLICIT DRUGS IN WWTPS

The rate of removal of illicit drugs in WWTPs is potentially affected by several factors, such as molecular structure, treatment process, age of the activated sludge, environmental conditions, such as temperature and light intensity, and characteristics of the influents. Removal of illicit drugs during wastewater treatment was investigated in several WWTPs in Italy; the results are reported in Table 7.3. Cocaine and its metabolites, 6-acetylmorphine, morphine-glucuronides and THC-COOH were extensively

TABLE 7.3   Removal of Illicit Drugs in WWTPs in Italy (Milan and Como) and Switzerland (Lugano)

	Removal (%)		
	Milan	Lugano	Como
Cocaine and metabolites	100	57–96	71–100
Morphine	100	70	32
6-Acethylmorphine-morphine glucuronides	100	100	100
Codeine	100	19	19
Amphetamines	59–78	56	10–75
Methadone + EDDP	0.5–31	13–19	7–19
THC-COOH	100	90	100
Total removal rate	*97*	*68*	*89[a] (29)*

[a]Total removal rate calculated not considering morphine and codeine. See text for further explanation.

removed in WWTPs with rates higher than 80%. Removals of morphine and codeine were almost complete in Milan, but were lower in Lugano and Como. In the last plant, the low removal observed (29%) was because of the high concentrations of morphine and codeine (Table 7.2) and were, therefore, only partially removed. Amphetamines were not completely removed during wastewater treatments and had variable removal rates from 50% to 80% in the different WWTPs, probably because of the specific characteristics of the treatment plant. Finally, methadone and EDDP were resistant to degradation in WWTPs, with low removal rates (0.5−30%) in all cases. Similar removal rates were observed also in other countries such as Spain, Belgium, and UK (Huerta-Fontela et al., 2008; van Nuijs et al., 2009a; Zuccato and Castiglioni, 2009; Kasprzyk-Hordern et al., 2009; Postigo et al., 2010). Some substances (codeine, norcodeine, THC-COOH, nor-LSD, methamphetamine, and ecstasy) were found at higher concentrations in effluents than in influents (Boleda et al., 2007; Postigo et al., 2010). This because some of them are excreted mainly as glucuronide metabolites and are then deconjugated by the β-glucuronidase enzymes of the fecal bacteria in wastewater. The hydrolysis of morphine-3β-D-glucuronide to free morphine was also observed during a 3-day stability test in wastewater (Castiglioni et al., 2006).

Total removal rates in Milan, Lugano, and Como WWTPs were, respectively, 97%, 68%, and 89%, indicating a good, but incomplete, removal of illicit drugs in these plants. In the Como plant, the removal was only 29%, if we also take into account morphine and codeine. However, this low removal rate is a particular case, because of the very high concentrations occurring in WWTP influents.

In view of their partial removal during wastewater treatment, illicit drugs are persisting in treated wastewater, which is generally discharged into surface water (rivers, lakes, sea) or undergoes further treatment to produce drinking water. Therefore, substantial amounts of illicit drugs have the potential to directly enter surface or drinking water.

7.5 ILLICIT DRUGS IN SURFACE WATER

In Italy, illicit drugs have also been measured in some of the main rivers and lakes. For these investigations, river sampling sites were chosen with the aim of collecting waters downstream of largely populated areas in northern (Po, Lambro, and Olona Rivers) and central Italy (Arno River). The Po River, the largest Italian river (flow rate $600-1000$ m^3/s), was sampled at four sampling sites downstream of the inlets of the main influents and of the major cities. The Lambro River (flow rate 5 m^3/s) was sampled near its entrance into the Po River. The Olona River (flow rate 0.5 m^3/s) was sampled downstream of a WWTP discharge site. The Arno River (flow rate $4-10$ m^3/s) was sampled at four sampling sites upstream and downstream of some major tributaries and cities such as Florence and Pisa. Different lakes in northern Italy (Varese, Lugano, and Maggiore Lakes) were sampled with the aim of also assessing the presence of illicit drugs in large water basins (Fig. 7.1). Lake Maggiore is the second largest lake in northern Italy, with an area of 212 km^2, a total water volume of 37.5 km^3, and a renewal time of 4 years. Lake Varese has a surface area of 15 km^2, a water volume of 0.16 km^3, and a renewal time of 1.8 years. Lake Lugano has a surface area of 49 km^2, a water volume of 6.5 km^3, and a renewal time of 8.2 years. Samples were collected on the river banks close to major influents and effluents or where the lake was deepest.

Olona and Lambro Rivers are small and heavily polluted rivers flowing in the most inhabited and industrialized area in the north of Italy. Illicit drugs were detected at concentrations up to 183 ng/L (Table 7.4). BE and cocaine were the most abundant compounds and reached, respectively, 183 and 44 ng/L in the Olona and 50 and 15 ng/L in the Lambro. The other metabolites of cocaine were all detectable, even if at lower concentrations ($0.2-8$ ng/L). Among opioids, concentrations of morphine, codeine, methadone, and EDDP ranged from 3 to 40 ng/L, while 6-acetylmorphine (metabolite of heroin), and 6-acetylcodeine (metabolite of codeine) were undetectable. Methamphetamine and MDMA (ecstasy) were the only stimulatory drugs detectable in the Olona and Lambro Rivers; THC-COOH was detected only in the Lambro. Concentrations of drug residues were generally lower in the Po River, the major Italian river in northern Italy, because of a higher dilution factor. Cocaine and its metabolites BE and norbenzoylecgonine, codeine, methadone, and EDDP were in the low ng/L range ($0.5-4$ ng/L), while the other substances were in the high picogram per liter range, or undetectable (Table 7.4). In the Arno River, a major river in central Italy, concentrations of drug residues were generally higher than in the Po, ranging from 0.5 to 22 ng/L (Table 7.4).

Illicit drugs have also been measured in rivers in Spain (Llobregat and Ebro Rivers), UK (Thames, Taff and Ely Rivers), Belgium (3 rivers), Germany (11 rivers), Ireland (2 rivers), and the United States (4 rivers) (Zuccato and Castiglioni, 2009; Bartelt-Hunt et al., 2009; Postigo et al., 2010). Median levels were up to the tenths of nanogram per liter for cocaine and BE, a few nanograms per liter for amphetamines, morphine, codeine, methadone, and EDDP; trace amounts or below the limit of quantification (LOQ) were found for the other substances and metabolites.

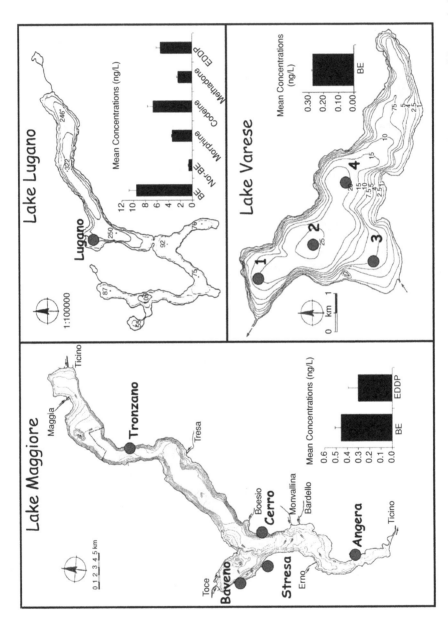

FIGURE 7.1 Sampling sites and mean concentrations (nanogram per liter) of illicit drugs in Lakes Maggiore, Varese, and Lugano in Italy and Switzerland (Zuccato et al., 2008a, with permission).

TABLE 7.4 Concentrations of Illicit Drugs (nanograms per liter) in Rivers in Italy[a]

Illicit Drugs	Po River (Mean ± SD) (Four Sampling Sites)	Arno River (Mean ± SD) (Four Sampling Sites)	Olona River (One Sampling Site)	Lambro River (One Sampling Site)
Cocaine and metabolites				
Benzoylecgonine	3.7 ± 1.2	21.8 ± 11.9	183	50
Norbenzoylecgonine	0.3 ± 0.1	1.6 ± 0.7	8.4	3.2
Cocaine	0.5 ± 0.2	1.7 ± 1.2	44	15
Norcocaine	<LOQ[b]	0.1 ± 0.1	3.6	0.4
Cocaethylene	<LOQ[b]	0.1 ± 0.0	1.3	0.2
Opioids				
Morphine	<LOQ[b]	3.0 ± 1.4	38	3.5
6-Acetylmorphine	<LOQ[b]	<LOQ[b]	<LOQ[b]	<LOQ[b]
Codeine	1.8 ± 0.6	6.2 ± 1.8	51	12
6-Acetylcodeine	<LOQ[b]	<LOQ[b]	<LOQ[b]	<LOQ[b]
Methadone	0.5 ± 0.2	4.8 ± 4.0	8.6	3.4
EDDP	1.0 ± 0.5	4.3 ± 2.1	18	9.9
Amphetamines				
Amphetamine	<LOQ[b]	<LOQ[b]	<LOQ[b]	<LOQ[b]
Methamphetamine	<LOQ[b]	<LOQ[b]	1.7	2.1
MDA	<LOQ[b]	1.2 ± 0.4	<LOQ[b]	<LOQ[b]
MDMA	0.2 ± 0.1	1.0 ± 0.4	1.7	1.1
Cannabinoids				
11-Nor-9-carboxy-Δ^9-THC	0.3 ± 0.2	0.5 ± 0.4	<LOQ[b]	3.7

[a]Data from Zuccato et al., 2008a.
[b]LOQ, limit of quantification.

Illicit drugs concentrations in lakes were lower than in rivers (Fig. 7.1). Only BE was measurable in Lake Varese and only picograms per liter levels of BE and EDDP were found in Lake Maggiore, the other substances being undetectable. In Lake Lugano, concentrations were substantially higher, with BE, norbenzoylecgonine, morphine, codeine, methadone, and EDDP in the nanogram per liter range, but the other substances were undetectable. The presence of trace amounts of illicit drugs also in large water basins, highlighted their "environmental persistence," which is mostly because of the continuous discharge of these substances from urban WWTPs or from the direct inlet of untreated wastewater in the environment. Illicit drugs were detected at higher concentrations in the Lake Lugano than in the Lake Maggiore, which might be ascribable to several factors, such as different renewal time (4 and 8.2 years, respectively, for the Maggiore and Lugano Lakes) and the different water volume and circulation of the two lakes.

The loads of illicit drugs flowing through the main Italian rivers have been cal-culated multiplying the concentrations (nanogram per liter) by the flow rates (cubic

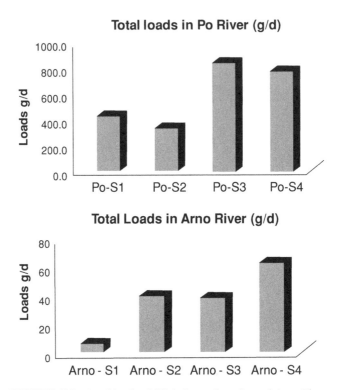

FIGURE 7.2 Total loads of illicit drugs along Po and Arno Rivers.

meters per second) and obtaining total amounts per day (grams per day) (Fig. 7.2). The loads were from 318 to 840 g/d in the Po River and from 6 to 65 g/d in the Arno River. In both rivers, loads increased substantially from source to mouth, in agreement with the increase of the population size discharging along the waterway. Huge amounts of illicit drugs, which are mostly biologically active substances, are, therefore, carried every day through rivers to the sea, with potential effects for ecosystems.

7.6 CONCLUSION

Illicit drugs have been detected in waste and surface waters in Italy indicating that these substances are common contaminants of the aquatic environment of populated areas. All the investigated substances were measured up to the microgram per liter range in untreated wastewater and benzoylecgonine, cocaine, morphine, codeine, and THC-COOH were the most abundant compounds. The removal of illicit drugs was investigated in several WWTPs in Italy and Switzerland and the total removal rates were generally higher than 60%. Methadone and its main metabolite EDDP were the most resistant to degradation, while amphetamines, codeine, and morphine were variably, but more efficiently, removed in the different WWTPs. Despite the good but incomplete removal during wastewater treatment processes, illicit drugs have been found in rivers and lakes at concentrations up to tenths of nanogram per liter. Their occurrence in surface water is probably ascribable to their continuous discharge in the environment more than to their persistence. Environmental concentrations are low, but risks for human health and the environment cannot be excluded. Morphine, cocaine, methamphetamine, MDA, and ecstasy have potent pharmacological activities, and their presence as complex mixtures in surface waters-together with those of the residues of many therapeutic drugs-may lead to unforeseen pharmacological interactions, possibly causing toxic effects to aquatic organisms. Biological effects from drug residues might, in fact, occur even at low environmental concentrations as discussed later in Chapter 14, Section V.

Moreover, the amount of residues of illicit drugs detected in untreated wastewater might roughly reflect the amounts consumed by the population producing the waste itself and our group recently proposed a novel "sewage epidemiology" approach, based on wastewater analysis for estimating drug consumption in communities (Zuccato et al., 2008b; see later, Chapter 16, Section V). The method is based on measurement of the residues of the illicit drugs excreted in consumers' urine into urban wastewater, and back-calculation of the local drug consumption from the measured levels. This method can provide evidence-based estimates of drug use in a defined area with the unique ability to monitor local consumption in real time.

REFERENCES

Bartelt-Hunt, S. L., D. D. Snow, T. Damon, J. Shockley, and K. Hoagland. 2009. The occurrence of illicit and therapeutic pharmaceuticals in wastewater effluent and surface waters in Nebraska. *Environ. Pollut.* **157**:786–791.

Bijlsma, L., J. V. Sancho, E. Pitarch, M. Ibáñez, and F. Hernández. 2009. Simultaneous ultra-high-pressure liquid chromatography-tandem mass spectrometry determination of amphetamine and amphetamine-like stimulants, cocaine and its metabolites, and a cannabis metabolite in surface water and urban wastewater. *J. Chromatogr. A* **1216**:3078–3089.

Boleda, M. R., M. T. Galceran, and F. Ventura 2007. Trace determination of cannabinoids and opiates in wastewater and surface waters by ultra-performance liquid chromatography-tandem mass spectrometry. *J. Chromatogr. A* **1175**:38–48.

Bones, J., K. V. Thomas, and B. Paull. 2007. Using environmental analytical data to estimate levels of community consumption of illicit drugs and abused pharmaceuticals. *J. Environ. Monit.* **9**:701–707.

Calamari, D., E. Zuccato, S. Castiglioni, R. Bagnati, and R. Fanelli. 2003. A strategic survey of therapeutic drugs in the rivers Po and Lambro in northen Italy. *Environ. Sci. Technol.* **37**:1241–1248.

Castiglioni, S., E. Zuccato, E. Crisci, C. Chiabrando, R. Fanelli, and R. Bagnati. 2006. Identification and measurement of illicit drugs and their metabolites in urban wastewater by liquid chromatography-tandem mass spectrometry. *Anal. Chem.* **78**:8421–8429.

Castiglioni, S., E. Zuccato, C. Chiabrando, R. Fanelli, and R. Bagnati. 2008. Mass spectrometric analysis of illicit drugs in wastewater and surface water. *Mass Spectrom. Rev.* **27**:378–394.

Cecinato, A., C. Balducci, and G. Nervegna. 2009. Occurrence of cocaine in the air of the World's cities: An emerging problem? A new tool to investigate the social incidence of drugs? *Sci. Total Environ.* **407**:1683–1690.

Chiaia, A. C., C. Banta-Green, and J. Field. 2008. Eliminating solid phase extraction with large-volume injection LC/MS/MS: analysis of illicit and legal drugs and human urine indicators in U.S. wastewaters. *Environ. Sci. Technol.* **42**:8841–8848.

Huerta-Fontela, M., M. T. Galceran, J. Martin-Alonso, and F. Ventura. 2008. Occurrence of psychoactive stimulatory drugs in wastewaters in north-eastern Spain. *Sci. Total Environ.* **397**:31–40.

Hummel, D., D. Löffler, G. Fink, and T. A. Ternes. 2006. Simultaneous determination of psychoactive drugs and their metabolites in aqueous matrices by liquid chromatography mass spectrometry. *Environ. Sci. Technol.* **40**:7321–7328.

Jones-Lepp, T. L., D. A. Alvarez, J. D. Petty, and J. N. Huckins. 2004. Polar organic chemical integrative sampling and liquid chromatography–electrospray/ion-trap mass spectrometry for assessing selected prescription and illicit drugs in treated sewage effluents. *Arch. Environ. Contamination Toxicol.* **47**:427–439.

Kaleta, A., M. Ferdig, and W. Buchberger. 2006. Semiquantitative determination of residues of amphetamine in sewage sludge samples. *J. Sep. Sci.* **29**:1662–1666.

Kasprzyk-Hordern, B., R. M. Dinsdale, and A. J. Guwy. 2008. Multiresidue methods for the analysis of pharmaceuticals, personal care products and illicit drugs in surface water and wastewater by solid-phase extraction and ultra performance liquid chromatography-electrospray tandem mass spectrometry. *Anal. Bioanal. Chem.* **391**:1293–1308.

Kasprzyk-Hordern, B., R. M. Dinsdale, and A. J. Guwy. 2009. The removal of pharmaceuticals, personal care products, endocrine disruptors and illicit drugs during wastewater treatment and its impact on the quality of receiving waters. Water Res. **43**:363–80.

Mari, F., L. Politi, A. Biggeri, G. Accetta, C. Trignano, M. Di Padua, and E. Bertol. 2009. Cocaine and heroin in waste water plants: a 1-year study in the city of Florence, Italy. *Forensic Sci. Int.* **189**:88–92.

Pomati, F., S. Castiglioni, E. Zuccato,R. Fanelli, D. Vigetti, C. Rossetti, and D. Calamari. 2006. Effects of a complex mixture of therapeutic drugs at environmental levels on human embryonic cells. *Environ. Sci. Technol.* **40**:2442–2447.

Postigo, C., M. J. Lopez de Alda, and D. Barceló. 2008a. Analysis of drugs of abuse and their human metabolites in water by LC-MS2: A non intrusive tool for drug abuse estimation at the community level. *TrAC-Trends Anal. Chem.* **27**:1053–1069.

Postigo, C., M. J. de Alda, and D. Barceló. 2008b. Fully automated determination in the low nanogram per liter level of different classes of drugs of abuse in sewage water by on-line solid-phase extraction-liquid chromatography-electrospray-tandem mass spectrometry. *Anal. Chem.* **80**:3123–3134.

Postigo, C., M. J. Lopez de Alda, M. Viana, X. Querol, A. Alastuey, B. Artiñano, and D. Barceló. 2009. Determination of drugs of abuse in airborne particles by pressurized liquid extraction and liquid chromatography-electrospray-tandem mass spectrometry. *Anal. Chem.* **81**:4382–4388.

Postigo, C., M. J. López de Alda, and D. Barceló. 2010. Drugs of abuse and their metabolites in the Ebro River basin: occurrence in sewage and surface water, sewage treatment plants removal efficiency, and collective drug usage estimation. *Environ. Int.* **36**:75–84.

United Nations Office of Drugs and Crime (UNODC). 2009. World Drug Report. http://www.unodc.org/documents/wdr/WDR_2009/WDR2009_eng_web.pdf.

van Nuijs, A. L., B. Pecceu, L. Theunis, N. Dubois, C. Charlier, P. G. Jorens, L. Bervoets, R. Blust, H. Neels, and A. Covaci. 2009a. Spatial and temporal variations in the occurrence of cocaine and benzoylecgonine in waste- and surface water from Belgium and removal during wastewater treatment. *Water Res.* **43**:1341−1349.

van Nuijs, A. L., I. Tarcomnicu, L. Bervoets, R. Blust, P. G. Jorens, H. Neels, and A. Covaci. 2009b. Analysis of drugs of abuse in wastewater by hydrophilic interaction liquid chromatography-tandem mass spectrometry. *Anal. Bioanal. Chem.* **395**:819−828.

van Nuijs, A. L., S. Castiglioni, I. Tarcomnicu, C. Postigo, M. J. Lopez de Alda, H. Neels, E. Zuccato, D. Barcelò, and A. Covaci. Illicit drug consumption estimations derived from wastewater analysis: A critical review. *Sci. Total Environ.* 10.1016/j.scitotenv.2010.05.030.

Zuccato, E., and S. Castiglioni. 2009. Illicit drugs in the environment. *Philos. Trans. A Math. Phys. Eng. Sci.* **367**:3965−3978.

Zuccato, E., C. Chiabrando, S. Castiglioni, D. Calamari, R. Bagnati, S. Schiarea, and R. Fanelli. 2005. Cocaine in surface waters: a new evidence-based tool to monitor community drug abuse. *Environ. Health* **4**:14.

Zuccato, E., S. Castiglioni, R. Bagnati, C. Chiabrando, P. Grassi, and R. Fanelli. 2008a. Illicit drugs, a novel group of environmental contaminants. *Water Res.* **42**:961−968.

Zuccato, E., C. Chiabrando, S. Castiglioni, R. Bagnati, and R. Fanelli. 2008b. Estimating community drug abuse by wastewater analysis. *Environ. Health Perspect.* **116**:1027−1032.

CHAPTER 8

OCCURRENCE OF ILLICIT DRUGS IN SURFACE WATER AND WASTEWATER IN THE UK

BARBARA KASPRZYK-HORDERN

8.1 INTRODUCTION

Illicit drugs, including the plant-derived substances (e.g., cannabis, cocaine, and heroin) and synthetic drugs (e.g., amphetamine, methamphetamine, and related designer drugs) are regarded as emerging environmental contaminants. They belong to the group of pharmacologically active compounds that also includes legal pharmaceuticals (prescription and over-the-counter pharmaceuticals) and veterinary medicines. Several reports have been published on the presence and fate of pharmacologically active compounds, such as veterinary medicines and pharmaceuticals (Capleton et al., 2006; Sarmah et al., 2006; Aga, 2007; Petrovic and Barcelo, 2007; Crane et al., 2008; Jjemba, 2008; Joss and Ternes, 2008; Kümmerer, 2008). However, illicit drugs have hardly been studied in the environment and only a few reports have been published on the occurrence of these compounds in surface water and/or wastewater. Investigations have taken place in the following countries: Italy (Zuccato et al., 2005, 2008a, b), Spain (Huerta-Fontela et al., 2008a, b; Postigo et al., 2008; Bijlsma et al., 2009), Ireland (Bones et al., 2007), United Kingdom (Zuccato et al., 2008a; Kasprzyk-Hordern et al., 2008a, 2009a), Belgium (Gheorghe et al., 2008; Van Nujis et al., 2009a, b, c), Switzerland (Castiglioni et al., 2006, Zuccato et al., 2008b), and the United States (Jones-Lepp et al., 2004; Chiaia et al., 2008; Bartelt-Hunt et al., 2009). Because of the limited extent of research undertaken in this field, there is limited data and minimal understanding of the environmental occurrence, transport, fate, and exposure for these compounds and their (very often) active metabolites. There is also no information available on the ecotoxicity of illicit drugs and their metabolites.

Illicit drugs enter the aquatic environment mainly through treated (or raw) sewage from domestic households. They can be present in the environment in the form

Illicit Drugs in the Environment: Occurrence, Analysis, and Fate Using Mass Spectrometry, Edited by
Sara Castiglioni, Ettore Zuccato, and Roberto Fanelli
Copyright © 2011 John Wiley & Sons, Inc.

of parent unaltered compounds or metabolites, or might undergo transformation during wastewater treatment to produce compounds of possible concern to humans and wildlife. Some of these compounds are ubiquitous and might be persistent in the environment. In addition, they are continuously introduced into the environment; therefore, even compounds of a low persistence might cause adverse effects. The other issue is the synergistic effect of different illicit drugs and other pharmacologically active compounds on organisms, through their combined parallel action. Illicit drugs are characterized by their polar and nonvolatile nature and, therefore, will not undergo volatilization from the aqueous environment, which extends the exposure of aquatic organisms to these compounds.

There are several illicit drugs that have been identified in the aquatic environment. Among them are: cocaine, amphetamines, opiates, cannabis, LSD (lysergic acid diethylamide), and their metabolites. Cocaine and its metabolites belong to the group of the most studied illicit drugs in the aquatic environment. Cocaine, a central nervous system stimulant, is produced as hydrochloride salt (which has limited medical use as a topical anesthetic for surgical procedures involving eye, ear, nose, and throat) or free base (known as crack) (King, 2009). It is usually quantified in surface water at concentrations not exceeding 50 ng/L (Fig. 8.1) (Huerta-Fontela et al., 2007, 2008b; Kasprzyk-Hordern et al., 2008a; Zuccato et al., 2008a; Van Nujis et al., 2009a, b). Benzoylecgonine, its major metabolite, is found in surface water at much higher levels, reaching a few hundred nanograms per liter (Zuccato et al., 2005, 2008a; Huerta-Fontela et al., 2007; Kasprzyk-Hordern et al., 2008a; Van Nujis et al., 2009a, b). Other metabolites of cocaine quantified in surface waters include: norbenzoylecgonine, norcocaine, and cocaethylene (Fig. 8.1) (Zuccato et al., 2008a). Measurable levels of cocaine and its metabolites in surface waters are linked to insufficient communal wastewater treatment, as both cocaine and its metabolites are present in raw and treated wastewater at high concentrations reaching, in the case of benzoylecgonine, 10 and 3, μg/L in wastewater influent and effluent, respectively (Fig. 8.1) (Castiglioni et al., 2006; Bones et al., 2007; Huerta-Fontela et al., 2007, 2008b; Chiaia et al., 2008; Kasprzyk-Hordern et al., 2008a; Postigo et al., 2008; Bijlsma et al., 2009; Mari et al., 2009; Van Nujis et al., 2009a, b, c).

Amphetamines, central nervous system stimulants, constitute the second group of the most studied illicit drugs. Among them are: amphetamine, methamphetamine, 3,4-methylenedioxymethamphetamine (MDMA), 3,4-methylenedioxyamphetamine (MDA) and 3,4-methylenedioxy-N-ethylamphetamine (MDEA). Amphetamine and methamphetamine have some limited therapeutic use in narcolepsy and attention deficit hyperactivity disorder, but most are manufactured in clandestine laboratories in Europe (King, 2009). Amphetamine is also formed as a metabolite of methamphetamine and several prescription drugs, such as selegiline (Kasprzyk-Hordern et al., 2009b). Amphetamines are frequently found in rivers across Europe at levels reaching 50 ng/L (Fig. 8.1) (Huerta-Fontela et al., 2007, 2008b; Kasprzyk-Hordern et al., 2008a; Zuccato et al., 2008a; Bartelt-Hunt et al., 2009). Amphetamine is the most abundant drug within the group of amphetamines and is found at the highest levels in surface water and wastewater. Concentrations of amphetamines in wastewater were found to vary between a few nanograms per liter and <5 μg/L in different

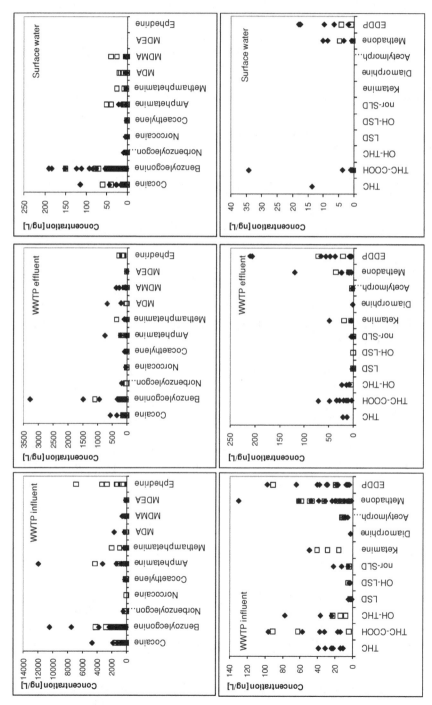

FIGURE 8.1 Maximum (●) and mean (□) concentrations of illicit drugs in surface water and wastewater across Europe and the United States (Castiglioni et al., 2006; Hummel et al., 2006; Boleda et al., 2007; Bones et al., 2007; Huerta-Fontela et al., 2007, 2008a, b; Chiaia et al., 2008; Gheorghe et al., 2008; Kasprzyk-Hordern et al., 2008a, 2009b; Postigo et al., 2008; Zuccato et al., 2008a; Bartelt-Hunt et al., 2009; Bijlsma et al., 2009; Loganathan et al., 2009; Mari et al., 2009; Van Nuijs et al., 2009a, c).

wastewater treatment plants (WWTPs) and different countries and are a reflection of local drug abuse trends (Fig. 8.1) (Castiglioni et al., 2006; Huerta-Fontela et al., 2007, 2008b; Chiaia et al., 2008; Kasprzyk-Hordern et al., 2008a; Postigo et al., 2008; Bartelt-Hunt et al., 2009; Bijlsma et al., 2009; van Nujis et al., 2009c). The levels of amphetamine and MDMA significantly exceeded the recorded average in wastewater influent only in the case of two samples and denoted 12 μg/L of amphetamine in the United Kingdom (Kasprzyk-Hordern et al., 2009a) and 27.5 μg/L of MDMA in Spain (value not included in Fig. 8.1) (Bijlsma et al., 2009). MDMA was also quantified at a very high level (21.2 μg/L; value not included in Fig. 8.1) in one treated wastewater sample in Spain (Bijlsma et al., 2009).

The presence of several opioids in surface water and wastewater was also reported (Castiglioni et al., 2006; Hummel et al., 2006; Bones et al., 2007; Postigo et al., 2008; Zuccato et al., 2008a; Mari et al., 2009; Wick et al., 2009). Among them are: illicit heroin (diamorphine) and several abused prescription drugs, such as codeine, dihydrocodeine, morphine, and hydrocodone. Diamorphine, a semisynthetic product of acetylation of morphine (a natural product in opium), is a narcotic analgesic used in the treatment of severe pain (King, 2009). Heroin and its specific metabolite 6-acetylmorphine were quantified in wastewater at low levels reaching 13 ng/L in the case of 6-acetylmorphine (Fig. 8.1) (Castiglioni et al., 2006; Postigo et al., 2008). Methadone, a prescription pain killer and also a drug used in the treatment of heroin addiction, and its metabolite EDDP have also been a subject of investigation (Castiglioni et al., 2006).

The most abused Δ^9-tetrahydrocannabinol (THC), an active constituent of cannabis, its major metabolite, 11-nor-9-carboxy-Δ^9-tetrahydrocannabinol (THC-COOH) and 11-hydroxy-THC (OH-THC), were also quantified in rivers and/or wastewater at low nanogram per liter levels (Fig. 8.1) (Castiglioni et al., 2006; Postigo et al., 2008; Zuccato et al., 2008a; Bijlsma et al., 2009).

Other studied illicit drugs include: LSD, a semisynthetic halucinogen, and its metabolites: 2-oxo-3-hydroxy lysergic acid diethylamide (OH-LSD) and nor-LSD. This potent hallucinogen was quantified in wastewater at single nanogram per liter levels reaching 3 ng/L (Fig. 8.1) (Postigo et al., 2008).

8.2 ABUSE OF ILLICIT DRUGS IN THE UNITED KINGDOM

The most serious drug abuse involves illicit products falling into the following categories: central nervous system stimulants, narcotic analgesics, hallucinogens, and hypnotics. The most prevalent of these are the plant-derived substances (e.g., cannabis, cocaine, and heroin), although it is predicted that the synthetic drugs (e.g., amphetamine, MDMA, and related designer drugs) are likely to pose a more significant social problem in the future. There is also an increasing recognition of the problems caused by misuse of medicinal products, primary benzodiazepine tranquilizers (King, 2009).

The 2007/08 British Crime Survey (BCS) estimated that within the group of 16- to 59-year-olds, around one in three people (35.8%) had ever used illicit drugs in

England and Wales, one in ten had used illicit drugs in the last year (9.3%), and one in twenty used these drugs in the last month (5.3%) (Hoare and Flatley, 2008). Findings from the 2006 Scottish Crime and Victimisation Survey (SCVS) show that: 36.6% of 16- to 59- year-olds have ever used drugs; 12.6% have used drugs in the last year; and 8% have used drugs in the last month. According to the Northern Ireland Crime Survey, prevalence is generally lower than in England and Wales, but shows similar patterns and trends (Eaton et al., 2007).

Cannabis is the world's most commonly used illicit drug (European Monitoring Centre for Drugs and Drug Addiction, 2007). The 2007−2008 BCS estimated that 7.4% of 16- to 59-year-olds in England and Wales used cannabis in the last year (Hoare and Flatley, 2008). The European Monitoring Centre for Drugs and Drug Addiction (EMCDDA) report on the state of the drugs problem in Europe placed the United Kingdom, alongside Italy, Spain, and the Czech Republic, among European countries with the highest cannabis abuse (EMCDDA, 2007).

Cocaine is the next most commonly used drug in the EU with 2.3% reporting use of any form of it (either cocaine powder or crack cocaine) in 2007−2008 in England and Wales in the last year (Hoare and Flatley, 2008). The UK is reported next to Spain, as the EU country with the highest (and increasing) cocaine prevalence levels (EMCDDA, 2007). Increases in prevalence of cocaine use in the last year among the 15–34-age-group have been registered in all countries reporting recent survey data (EMCDDA, 2007).

Use of ecstasy and amphetamine in 2007−2008 in England and Wales in the past year was estimated at 1.5% and 1.0%, respectively. The use of hallucinogens (LSD and magic mushrooms) in the last year was reported by 0.6% of 16- to 59-year-olds. Use of other drugs is slightly less prevalent with 0.5% reporting use of tranquilizers, 0.4% reporting using ketamine, and 0.1% using glues or anabolic steroids in the previous year. Other potent drugs are used in England and Wales on a smaller scale: opiate (heroin and methadone) use was reported by 0.2% of 16- to 59-year-olds (Hoare and Flatley, 2008).

Baseline estimates of market size for the United Kingdom (calculated with reference to seizures, purity, and survey-based estimates of usage) indicate that aggregate pure quantity of cocaine usage was 19.0 tons in the United Kingdom in 2003−2004 (8.9 tons of powder cocaine and 10.1 tons of crack cocaine). Aggregate pure quantity of cannabis accounted for 412.4 tons in 2003−2004. Amphetamines (amphetamine and methamphetamine) and ecstasy were estimated at 4.0 and 15.5 tons, respectively. Aggregate pure quantity of heroin accounted for 8.0 tons in the United Kingdom in 2003−2004. The baseline estimates for Wales and England only indicate the following pure aggregate quantities: 360.3, 3.6, 13.7, 16.8, and 7.0 tons of cannabis, amphetamines, ecstasy, cocaine, and heroin, respectively (Singleton et al., 2006).

The above discussion indicates that because of the high usage of illicit drugs in the United Kingdom, the presence of illicit drugs as well as their metabolites in the environment is to be expected.

8.3 OCCURRENCE OF ILLICIT DRUGS IN WASTEWATER AND SURFACE WATER IN THE UNITED KINGDOM

Very limited data exists on the presence and fate of drugs of abuse in the United Kingdom environment. Monitoring of illicit drugs in the United Kingdom has been undertaken in only one river in England (the River Thames) and two rivers in Wales (the River Taff and Ely).

8.3.1 Illicit Drugs in the River Thames

Zuccato et al. (2008a,b) studied the presence of several illicit drugs such as cocaine and its main metabolite benzoylecgonine, amphetamine, methamphetamine, MDA, MDMA, morphine, and THC-COOH in the River Thames and two WWTPs (Mogden WWTP and Beckton WWTP) during one sampling regime that took place in October 2005. Samples were collected from the upper rural course of the River Thames in Oxfordshire (New Bridge and Shillingford Bridge) and in its lower course, in the London area (near Chiswick Bridge, at the Houses of Parliament and near Tilbury). The results are presented in Table 8.1. Amphetamine and methamphetamine were not quantified in any analyzed sample. THC-COOH was quantified in only one sample in London, Chiswick Bridge at very low levels accounting for 1 ng/L. Cocaine was quantified in samples collected form the London area at levels reaching 6 ng/L near the Houses of Parliament. Benzoylecgonine, MDA, and MDMA were present in all samples collected from the River Thames. The highest levels of these drugs were quantified in the area of London, Chiswick Bridge and the Houses of Parliament and accounted for 17, 3, and 4 ng/L in the case of benzoylecgonine, MDA, and MDMA, respectively. It has to be emphasized here, however, that the River Thames in the London area is significantly affected by tidal movement contributing to possible dilution of the river water with sea water (Zuccato et al., 2008a).

8.3.2 Illicit Drugs in the Rivers Taff and Ely and the Impact of Wastewater Treatment on the Quality of River Water

Kasprzyk-Hordern et al. (2007, 2008a, b, c, 2009a, b) carried out a 10-month long monitoring program aimed at the verification of the presence and fate of two illicit drugs (amphetamine, cocaine and its metabolite, benzoylecgonine) in two different rivers in South Wales: Rivers Taff and Ely and two different WWTPs (WWTP Cilfynydd discharging treated effluent into the River Taff and WWTP Coslech discharging treated effluent into the River Ely) (Fig. 8.2). The River Taff is the tenth largest river in the United Kingdom. It has as its source the Brecon Beacons National Park which then flows through the urbanized areas of Merthyl Tydfil, Pontypridd, and Cardiff. The River Ely is a small and shallow river that flows through rural areas and finally reaches Cardiff, where it flows together with the River Taff into the Severn Estuary and the Bristol Channel. The two studied WWTP plants also differ significantly. WWTP Cilfynydd serves a population of 111,000. It treats mainly communal sewage and the technology utilized there is trickling filter beds. WWTP

TABLE 8.1 Concentrations of Illicit Drugs in the River Thames (October 2005)[a]

Sampling Points	Concentration [ng/L]							
	Amphetamine	Methamphetamine	MDA	MDMA	Cocaine	Benzoylecgonine	THC-COOH	Morphine
New Bridge	<0.65	<0.41	4	2	<0.13	4	<0.48	<0.55
Shillingford Bridge	<0.65	<0.41	2	2	<0.13	6	<0.48	5
London, Chiswick Bridge	<0.65	<0.41	3 ± 0.3	4 ± 0.4	4 ± 0.6	17 ± 0.8	1 ± 0.6	42 ± 4.7
London, Houses of Parliament	<0.65	<0.41	3	4	6	16	<0.48	9
London, Tilbury	<0.65	<0.41	<0.18	6 ± 0.3	4 ± 0.1	13 ± 0.6	<0.48	7 ± 0.7

[a]Zuccato et al., 2008a.

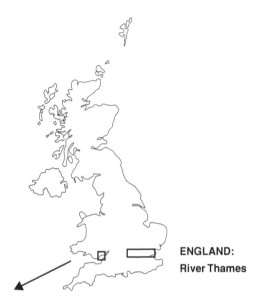

ENGLAND:
River Thames

WALES: River Taff and Ely

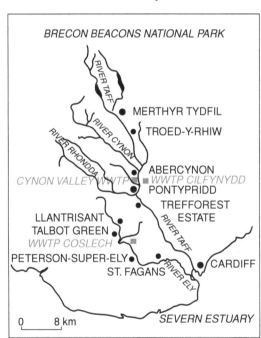

FIGURE 8.2 Location of sampling sites in the United Kingdom (modified from Kasprzyk-Hordern et al., 2008a, 2009a).

Coslech, on the other hand, serves approximately 30,000 inhabitants and treats both communal and industrial wastewater utilizing the most common biological treatment process, which is activated sludge. Both wastewater plants contribute significantly to the rivers' flows through a high volume of wastewater discharged into the two rivers. On average, a 23-fold dilution of wastewater effluent from WWTP Cilfynydd was observed in the River Taff during the period of the study. Only a 13-fold dilution was noted in the case of treated wastewater from WWTP Coslech reaching the River Ely (Kasprzyk-Hordern et al., 2008a, 2009a, b).

Both amphetamine and cocaine were found in the River Taff at low single nanogram per liter levels (Table 8.2). Their presence in the River Taff was found to be strongly associated with the discharge of insufficiently treated wastewater effluent. Benzoylecgonine was often found in the River Taff at levels 10 times higher than the parent compound reaching 100 ng/L. No studied illicit drugs were quantified in any sample collected in the Brecon Beacons National Park (at the source of the River Taff). In contrast, both amphetamine and benzoylecgonine were found with 100% frequency in receiving waters 2 km downstream of WWTP, which indicates a discharge of insufficiently treated wastewater effluent, but also it is a sign of high and steady consumption of illicit drugs in the region (Kasprzyk-Hordern et al., 2008a). In addition, all three compounds revealed high persistence in the River Taff as only a limited decrease of their concentration was observed even 27 km downstream of a discharge point (Table 8.2).

TABLE 8.2 Concentrations of Illicit Drugs in Samples Collected in the River Taff for a 10-Month Monitoring Program (November 2006–August 2007)

		Concentration (ng/L)		
Sampling Points		Amphetamine	Benzoylecgonine	Cocaine
Brecon Beacons National	L-H[a](M)	<1	<1	<0.3
Park, BB (the source of the River Taff)	%	0	0	0
Merthyl Tydfil, MT	L-H (M)	<1–3 (1)	<1	<0.3
(23.5 km downstream of BB)	%	20	0	0
Abercynon, AC (12 km	L-H (M)	<1–11 (3)	<1–1 (<1)	<0.3
downstream of MT; 1 km upstream of WWTP)	%	73	0	0
Pontypridd, PD (2 km	L-H (M)	2–13 (7)	<1–92 (46)	<0.3–4 (2)
downstream of WWTP)	%	100	100	45
Trefforest Estate, TE (7 km	L-H (M)	4–14 (7)	1–100 (36)	<0.3–7 (2)
downstream of PD)	%	100	78	33
Cardiff, CD (18 km	L-H (M)	<1–5 (3)	<1–123 (40)	<0.3–1 (<0.3)
downstream of TE)	%	87	78	11

[a]L, lowest recorded concentration; H, highest recorded concentration; M, mean concentration; %, frequency of detection.

TABLE 8.3 Concentrations of Illicit Drugs in Samples Collected in the River Ely for a 10-Month Monitoring Program (November 2006–August 2007)

Sampling Points		Concentration (ng/L)		
		Amphetamine	Benzoylecgonine	Cocaine
Llantrisant Forest, LF	L-H[a] (M)	<1–21 (4)	<1–41 (7)	<0.3
(10.5 km upstream of WWTP)	%	33	17	17
Talbot Green, TG (7 km	L-H (M)	<1–19 (2)	<1–53 (7)	<0.3
upstream of WWTP)	%	13	25	0
Peterson-super-Ely, PSE	L-H (M)	<1–3 (1)	<1–54 (16)	<0.3
(3.5 km downstream of WWTP)	%	25	75	0
Cardiff, CD (12 km	L-H (M)	<1–4 (2)	1–84 (23)	<0.3
downstream of PSE)	%	83	100	0

[a]L, lowest recorded concentration; H, highest recorded concentration; M, mean concentration; %, frequency of detection.

In contrast to the situation observed in the River Taff, no cocaine was quantified in the River Ely (Table 8.3). Amphetamine was quantified at very low concentrations and its presence varied at different sampling points (from 13% to 83% frequency). Benzoylecgonine was again quantified at the highest concentrations reaching 50 ng/L. Lower concentrations of illicit drugs in the River Ely when compared to the River Taff are strongly associated with lower concentrations of illicit drugs in final WWTP effluent resulting from better efficiency of treatment in WWTP Coslech than WWTP Cilfynydd (Tables 8.4 and 8.5). As can be observed from Table 8.5, activated sludge

TABLE 8.4 Concentrations of Illicit Drugs in Samples Collected for a 5-Month Monitoring Program (April 2007–August 2007) in Raw and Treated Wastewater

Sampling Points		Concentration (ng/L)		
		Amphetamine	Benzoylecgonine	Cocaine
WWTP	L-H[a] (M)	292–12020 (4310)	126–2114 (992)	21–1837 (521)
Cilfynydd–influent	%	100	100	100
WWTO	L-H (M)	19–739 (201)	202–3275 (1091)	48–324 (128)
Cilfynydd–effluent	%	100	100	100
WWTP Coslech–influent	L-H (M)	255–3225 (1196)	187–3715 (1082)	54–471 (207)
	%	100	100	100
WWTP Coslech–effluent	L-H (M)	<3–11 (2)	<1–29 (13)	<1
	%	14	86	0

[a]L, lowest recorded concentration; H, highest recorded concentration; M, mean concentration; %, frequency of detection.

TABLE 8.5   Efficiency of Illicit Drugs Removal during Wastewater Treatment Calculated for Samples Collected for a 5-Month Monitoring Program (April 2007–August 2007) in Raw and Treated Wastewater

	Removal Efficiency (%)		
Sampling point	Amphetamine	Benzoylecgonine	Cocaine
WWTP Cilfynydd (trickling filters)	95 ± 4	-25 ± 50	25 ± 80
WWTP Coslech (activated sludge)	100 ± 1	96 ± 7	100 ± 0

treatment utilized in WWTP Coslech gives almost 100% removal of amphetamine, cocaine, and benzoylecgonine, while trickling filter beds utilized in WWTP Cilfynydd resulted in 95% removal of amphetamine, only 25% removal of cocaine, and no removal of benzoylecgonine. This indicates the poor efficiency of trickling filter beds technology when compared to activated sludge. As a result of effective activated sludge treatment in WWTP Coslech, no cocaine and low benzoylecgonine and amphetamine levels accounting for an average of 13 and 2 ng/L, respectively, were released with wastewater effluent into receiving waters (Table 8.4). In contrast, ineffective trickling filters treatment in WWTP Cilfynydd resulted in the release into receiving waters of much higher quantities of the studied drugs, accounting for, on average, 128, 1091, and 201 ng/L in treated wastewater effluent in the case of cocaine, benzoylecgonine, and amphetamine, respectively (Table 8.4). The highest recorded levels over a few months study denoted: 324, 3275, and 739 ng/L of cocaine, benzoylecgonine, and amphetamine, respectively (Kasprzyk-Hordern et al., 2009a).

There is limited data on the efficiency of the removal of illicit drugs during wastewater treatment. Castiglioni et al. (2006) reported 88%–100% degradation of benzoylecgonine, 95%–100% degradation of cocaine, and 100% degradation of amphetamine in WWTPs with activated sludge treatment. Bones et al. (2007) also reported high removal of benzoylecgonine (92%), but lower removal of cocaine (72%) during activated sludge treatment. Huerta-Fontela et al. (2008b) studied 42 different WWTPs utilizing mainly activated sludge treatment and observed varying, but high, efficiencies of amphetamine removal ranging from 52% to over 99% in different wastewater treatment plants. The removal of cocaine and benzoylecgonine was also very effective and denoted >88% (Huerta-Fontela et al., 2008). High removal of amphetamine and cocaine (>99%) during wastewater treatment was also reported by Bijlsma et al. (2009), although it was observed that in the case of high loads of illicit drugs occurring during weekends, the efficiency of the removal of illicit drugs decreased to 35% and 85% in the case of benzoylecgonine and amphetamine, respectively. Postigo et al. (2008) reported 49%−99% removal of amphetamine, 69%−99% removal of benzoylecgonine, and 67%−99% removal of cocaine. The removal rates were found to be wastewater treatment-plant dependent (Postigo et al., 2008). The above discussion indicates that illicit drugs such as cocaine and amphetamine are effectively removed during wastewater treatment, especially when activated sludge treatment is implemented. It has to be, however, emphasized that high removal of illicit drugs during activated sludge treatment might be a result of not only biodegradation, but also other processes, such as partitioning of these compounds to sludge.

TABLE 8.6 Loads of Illicit Drugs in Rivers and WWTPs in Wales for a 10-Month Sampling Period[a]

		Load (g/day)					
		Amphetamine		Benzoylecgonine		Cocaine	
Sampling point		Mean	Max	Mean	Max	Mean	Max
River Taff	MT	0.0	0	0.0	0	0.0	0
	PD	8.0	29.3	39.0	104.9	1.2	5.4
River Ely	TG	1.6	12.8	4.6	53.2	0.0	0
	PSE	0.1	0.5	2.6	8.1	0.0	0
WWTP Cilfynydd	Influent	109.8	256.5	27.9	76.2	15.3	66.4
	Effluent	6.3	15.8	30.4	69.9	3.7	6.9
WWTP Coslech	Influent	19.3	41.8	16.4	48.1	3.5	6.1
	Effluent	0	0.2	0.3	0.6	0	0

[a]Mean and maximum values calculated for the results obtained.

Unfortunately, despite the high efficiency of treatment processes, the removal of illicit drugs is not sufficient to entirely eliminate the possibility of illicit drugs being discharged into receiving waters. The calculated mean and the highest recorded environmental loads of illicit drugs monitored in the Rivers Taff and Ely over the period of 10 months at two sampling points located above and below WWTP discharge are presented in Table 8.6. Mean daily loads recorded in the River Taff were as follows: 0 g/day for all illicit drugs at Merthyl Tydfill sampling point located approximately 13 km upstream of WWTP discharge, and 8, 1.2, and 39 g/day for amphetamine, cocaine, and benzoylecgonine, respectively, at Pontypridd sampling point located approximately 2 km downstream of WWTP discharge. Lower, although still noticeable, average loads of illicit drugs were also observed in the case of the River Ely. At the Talbot Green sampling point (approximately 7 km upstream of WWTP discharge) average daily loads denoted 1.6, 4.6, and 0 g/day of amphetamine, benzoylecgonine, and cocaine, respectively. At Peterson-super-Ely sampling point, located approximately 3.5 km downstream from the discharge point, average daily loads were found to be lower and denoted: 0.1, 2.6, and 0 g/day of amphetamine, benzoylecgonine, and cocaine, respectively. This results, as discussed above, from both the lower capacity of WWTP Coslech and also the better efficiency of treatment when compared with WWTP Cilfynydd (Table 8.6) (Kasprzyk-Hordern et al., 2008a, 2009a).

The comparison of the results obtained for the three rivers (the Rivers Thames, Taff, and Ely) (Tables 8.1, 8.2, and 8.3) in the United Kingdom indicates that the presence of illicit drugs in the aqueous environment is of regional character and reflects drug abuse trends in the studied catchment area. There are also other vital parameters that affect concentration levels of illicit drugs in the studied rivers. Among them are: efficiency of wastewater treatment, dilution of wastewater in receiving waters, rainfall, and regional climatic variations. Similar cocaine levels were observed in the case of the Rivers Taff and Thames. Benzoylecgonine was, in general, quantified at slightly higher concentrations in the Welsh rivers. Surprisingly, no amphetamine was

TABLE 8.7   Seasonal Variations of Loads of Illicit Drugs in the River Taff[a]

	Load (g/day)				
	Amphetamine	Benzoylecgonine	Cocaine	Flow (m^3/s)	Temp (°C)
January	6.8	39.2	0.0	13.8	7
February	9.9	60.3	5.3	8.8	7
March	6.0	14.6	0.0	14.2	10
April	3.3	27.6	1.4	5.3	13
May	2.4	22.6	0.0	7.6	15
June	2.1	40.4	0.9	5.1	17
July	16.1	104.9	5.4	14.3	13
August	6.6	36.5	0.0	9.1	14

[a]PD sampling point.

quantified in the River Thames. This is in contrast with findings in both the Rivers Taff and Ely.

8.4   EFFECT OF SEASONAL VARIATIONS AND VARIABLE FLOW CONDITIONS ON ILLICIT DRUG CONCENTRATIONS IN WELSH RIVERS

The results obtained for the 10-month sampling period in the Rivers Taff and Ely re-vealed that concentrations of illicit drugs vary. Seasonal variations of loads of chosen illicit drugs in the River Taff are presented in Table 8.7. The results do not indicate any particular pattern concerning the levels of illicit drugs loads, which might be sea-sonally dependent. The main factors affecting the levels of illicit drugs concentrations include weather conditions (mainly rainfall and sunlight), proximity to wastewater plant, and the efficiency of wastewater treatment. Because of the particularly mild climate in South Wales resulting from the proximity to the sea, the temperature of river water did not change significantly over the year and varied from 7 to 17°C. In general, concentrations of illicit drugs were found to be considerably higher during dry weather conditions and significantly decreased during wet weather conditions be-cause of a significant dilution of the river water with rain water (Kasprzyk-Hordern et al., 2008a). River flows in the Rivers Taff and Ely vary substantially over the year. During prolonged dry periods, river flows are significantly reduced and many of the numerous small springs tend to dry up. This is when the river is largely fed by treated wastewater effluent. Following heavy rainfall, river flows can increase dramatically, which, when combined with the mountainous and steep topography of the upper catchment, contributes to high and rapid flows (Environment Agency, 2000 a, b).

8.5   ILLICIT DRUGS IN THE UNITED KINGDOM AQUEOUS ENVIRONMENT IN THE CONTEXT OF EUROPEAN FINDINGS

Concentration levels of illicit drugs in the United Kingdom aqueous environment represents, similarly to other countries, the usage trends of illicit drugs in local communities. As the United Kingdom is considered to be among the EU countries with the highest prevalence of illicit drugs, high environmental loads of these drugs are to be expected. As mentioned above, the United Kingdom is reported next to Spain as the EU country with the highest and steadiest increasing cocaine prevalence levels. It is also known to be among the EU countries with the highest prevalence of the usage of amphetamines and ecstasy (EMCDDA, 2007). As estimated by the sewage epidemiology approach, amphetamine consumption in London, Milan, and Lugano clearly indicates a much higher prevalence of this drug in the United Kingdom when compared to Italy or Switzerland (Zuccato et al., 2008b). The results obtained from the two studied WWTPs in Wales indicate that consumption of cocaine, also calculated with the utilization of the sewage epidemiology approach, accounts for on average 900 mg/day/1000 people and was found to vary, depending on the day of collection, from 100 to 3700 mg/day/1000 people (Kasprzyk-Hordern et al., 2009b). Estimated by Zuccato et al. (2008b), cocaine consumption in Milan and Lugano accounted for approximately 1000 and 600 mg/day/1000 people, respectively. Van Nuijs et al. (2009a) reported the following consumption levels in Belgium: 98–1829 mg/day/1000 people. However, it has to be remembered that these calculations have a regional character only and might be associated with a high error if extrapolated on a national scale.

The comparison of cocaine, benzoylecgonine, and amphetamines concentration levels in the United Kingdom rivers with European findings suggests similar patterns of the occurrence of illicit drugs in the aquatic environment. Cocaine, benzoylecgonine, and amphetamine were found in the Rivers Taff and Ely at concentrations reaching 7, 123, and 21 ng/L, respectively (Kasprzyk-Hordern et al., 2008a). Zuccato et al. (2008a) quantified cocaine and benzoylecgonine in rivers Olona, Lambro, and Po in Italy at levels reaching 44 and 183 ng/L, respectively. The levels of cocaine and benzoylecgonine in Belgium also exceeded 20 and 100 ng/L, respectively, in some rivers (Van Nuijs et al., 2009a). In Spanish Llobregat River, cocaine and benzoylecgonine were quantified in surface water at maximum 10 and 111 ng/L, respectively (Huerta-Fontela et al., 2007, 2008a). Methamphetamine was not quantified in the River Thames. However, it was reported at up to 2.1 ng/L levels in rivers Olona and Lambro in Italy (Zuccato et al., 2008a) and in Llobregat River in Spain (Huerta-Fontela et al., 2008a). MDA, which was not present in Italian rivers was reported at single nanogram per liter in the River Thames (Zuccato et al., 2008a) and reached 15 ng/L in the Llobregat River (Huerta-Fontela et al., 2008a). On the other hand, MDMA was quantified in all studied rivers: Thames, Olona, Lambro (Zuccato et al., 2008a), and Llobregat (Huerta-Fontela et al., 2007, 2008a) at low nanogram per liter levels, reaching the highest levels of 40 ng/L in the Llobregat River.

8.6 CONCLUSIONS AND FUTURE RESEARCH DIRECTION

Because of very limited data available on the occurrence and fate of illicit drugs in the UK, it is very difficult to draw any explicit conclusions on the possible environmental risk associated with the presence of these compounds in the environment. However, the research undertaken clearly indicates that illicit drugs as emerging contaminants are ubiquitous and persistent in the environment. Although they are present in the aquatic environment at low parts per trillion levels, their possible effect on living organisms should not be underestimated. This is because illicit drugs reveal very high pharmacological potency in humans, even when administered at very low levels. For example, LSD is among the most potent drugs known, being active in humans at doses from about 20 μg (King, 2009). Its possible potency and toxicity in aquatic organisms is not known. In addition, illicit drugs usually simultaneously occur in the environment with other pharmacologically active compounds and as a result synergistic action of several active chemicals is to be expected. Communal wastewater and its insufficient treatment are considered to be the main source of environmental contamination.

Therefore, more research is needed in order to understand the fate of illicit drugs during wastewater treatment and in the environment (both aquatic and terrestrial). In particular, the susceptibility of illicit drugs to biological, chemical, and physical processes occurring in the environment, such as microbial degradation, photodegradation, sorption to sludge particles, and soil sediments needs to be extensively studied. As several metabolites of illicit drugs are known to be pharmacologically active, studies on their occurrence and fate in the environment are of equal importance. Studies on the possible acute and chronic toxic effects of illicit drugs on aquatic organisms are nonexistent and this topic needs urgent attention.

The chirality of illicit drugs has to be also considered as it is a major parameter determining the potency and toxicity of drugs (Kasprzyk-Hordern, 2010a). Most illicit drugs (e.g., amphetamines and opiates) are chiral compounds. A chiral molecule has at least one chiral center (e.g., asymmetric carbon) as a result of which it shows optical activity. It exists in the form of two enantiomers, being the nonsuperimposable mirror images of each other. Enantiomers of the same drug have similar physicochemical properties, but differ in their biological properties, such as distribution, metabolism, and excretion, as these processes (because of stereospecific interactions of enantiomers with biological systems) usually favor one enantiomer over the other. In addition, because of different pharmacological activity, chiral drugs can differ in toxicity. Amphetamine is a great example as it exists as a pair of enantiomers, which differ in pharmacological activity and metabolism. The S-enantiomer of amphetamine is known to be much more potent than R-enantiomer and, as a result, it might reveal much higher toxicity toward certain organisms. Furthermore, degradation of chiral drugs during wastewater treatment and in the environment can be stereospecific and can lead to chiral products of varied toxicity. Distribution of different enantiomers of the same chiral drug in the aquatic environment and biota can also be stereospecific. Biological processes can lead to stereoselective enrichment or depletion of the enantiomeric composition of chiral drugs. As a result the very same drug might reveal different activity and toxicity and this will depend on its origin and exposure to several factors governing its fate in the environment. The preliminary research undertaken by Kasprzyk-Hordern et al. (2010b) aimed at enantioselective analysis of drugs of abuse: amphetamines (amphetamine, methamphetamine, MDEA, MDMA and MDA), ephedrines (ephedrine, pseudoephedrine, norephedrine), and venlafaxine during wastewater treatment indicated their nonracemic composition. It was, for example, observed that in the case of methamphetamine, only the more potent S(+)-enantiomer was detected in all treated wastewater samples. The reverse situation was observed in the case of amphetamine, where less potent $R(-)$-enantiomer was present in both raw and treated wastewater at slightly higher concentrations than $S(+)$-enantiomer. The study of enantiomeric composition of MDMA during wastewater treatment indicated its enrichment in one enantiomer only. A similar situation was observed in the case of initially racemic venlafaxine. This might suggest enantioselective processes occurring during treatment, although more comprehensive research has to be undertaken to support such a hypothesis.

REFERENCES

Aga, D. S. editor. 2007. *Fate of Pharmaceuticals in the Environment and in Water Treatment Systems*. Boca Raton, FL: CRC Press.

Bartelt-Hunt, S. L., D. D. Snow, T. Damon, J. Shockley, and K. Hoagland. (2009). The occurrence of illicit and therapeutic pharmaceuticals in wastewater effluent and surface waters in Nebraska. *Environ. Pollut.* **157**:786–791.

Bjilsma, L., J. V. Sancho, E. Pitarch, M. Ibáñez, and F. Hernández. (2009). Simultaneous ultra-high-pressure liquid chromatography-tandem mass spectrometry determination of amphetamine and amphetamine-like stimulants, cocaine and its metabolites, and a cannabis metabolite in surface water and urban wastewater. *J. Chromatogr. A* **1216**:3078–3089.

Boleda, M. R., M. T. Galceran, and F. Ventura. (2007). Trace determination of cannabinoids and opiates in wastewater and surface waters by ultra-performance liquid chromatography-tandem mass spectrometry. *J. Chromatogr. A* **1175**:38–48.

Bones, J., K. V. Thomas, and B. Paull. (2007). Using environmental analytical data to estimate levels of community consumption of illicit drugs and abused pharmaceuticals. *J. Environ. Monit.* **9**:701–707.

Capleton, A. C., C. Courage, P. Rumsby, P. Holmes, E. Stutt, A. B. A. Boxall, and L. S. Levy. (2006). Prioritising veterinary medicines according to their potential indirect human exposure and toxicity profile. *Toxicol. Lett.* **163**:213–223.

Castiglioni, S., E. Zuccato, E. Crisci, Ch. Chiabrando, R. Fanelli, and R. Bagnati. (2006). Identification and measurement of illicit drugs and their metabolites in urban wastewater by liquid chromatography-tandem mass spectrometry. *Anal. Chem.* **78**:8421–8429.

Chiaia, A. C., C. Banta-Green, and J. Field. (2008). Eliminating solid phase extraction with large-volume injection LC/MS/MS: analysis of illicit and legal drugs and human urine indicators in U.S. wastewaters. *Environ. Sci. Technol.* **42**:8841–8848.

Crane, M., A. B. A. Boxall, and K. Barrett, editors. 2008. *Veterinary Medicine in the Environment*. Boca Raton, Fl: CRC Press.

Eaton, G., C. Davies, L. English, A. Lodwick, M. A. Bellis, and J. McVeigh. 2007. United Kingdom drug situation: Annual report to the European Monitoring Centre for Drugs and Drug Addiction. European Monitoring Centre for Drugs and Drug Addiction.

Environment Agency (2000a). Local Environment Agency Plan, Taff Area, Environmental Overview. www.environmentagency.gov.uk/.

Environment Agency (2000b). Local Environment Agency Plan, Ely and Vale of Glamorgan Area, Environmental Overview. www.environment-agency.gov.uk/.

European Monitoring Centre for Drugs and Drug Addiction (EMCDDA). The state of the drugs problem in Europe, Annual Report. 2007. European Monitoring Centre for Drugs and Drug Addiction.

Gheorghe, A., A. van Nujis, B. Pecceu, L. Bervoets, P. G. Jorens, R. Blust, H. Neels, and A. Covaci A. (2008). Analysis of cocaine and its principal metabolites in waste and surface water using solid phase extraction and liquid chromatography—ion trap tandem mass spectrometry. *Anal. Bioanal. Chem.* **391**:1309–1319.

Hoare, J., and J. Flatley. 2008. Drug Misuse Declared: Findings from the 2007/08, British Crime Survey, England and Wales, Home Office Statistical Bulletin, October.

Huerta-Fontela, M., M. T. Galceran, and F. Ventura. (2007). Ultraperformance liquid chromatography-tandem mass spectrometry analysis of stimulatory drugs of abuse in wastewater and surface waters. *Anal. Chem.* **79**:3821–3829.

Huerta-Fontela, M., M. T. Glaceran, and F. Ventura. (2008a). Stimulatory drugs of abuse in surface waters and their removal in a conventional drinking water treatment plant. *Environ. Sci. Technol.* **42**:6809–6816.

Huerta-Fontela, M., M. T. Glaceran, J. Martin-Alonso, and F. Ventura. (2008b). Occurrence of psychoactive stimulatory drugs in wastewaters in north-eastern Spain. *Sci. Total Environ.* **397**:31–40.

Hummel, D., D. Löffler, G. Fink, and T. A. Ternes. (2006). Simultaneous determination of psychoactive drugs and their metabolites in aqueous matrices by liquid chromatography mass spectrometry. *Environ. Sci. Technol.* **40**:7321–7328.

Jemba, P. K. 2008. Pharma-ecology: The Occurrence and Fate of Pharmaceuticals and Personal Care Products in the Environment. Hoboken, New Jersey: Wiley (Blackwell).

Jones-Lepp, T. L., D. A. Alvarez, J. D. Petty, and J. N. Huckins. (2004). Polar organic chemical integrative sampling and liquid chromatography–electrospray/ion-trap mass spectrometry for assessing selected prescription and illicit drugs in treated sewage effluents. *Arch. Environ. Contamination Toxicol.* **47**:427–439.

Joss, A., and T. A. Ternes. (editors.). 2008. Human pharmaceuticals, hormones and fragrances: the challenge of micropollutants in urban water management. London: IWA Publ.

Kasprzyk-Hordern, B., R. M. Dinsdale, and A. J. Guwy. (2007). Multi-residue method for the determination of basic/neutral pharmaceuticals and illicit drugs in surface water by solid-phase extraction and ultra performance liquid chromatography – positive electrospray ionisation tandem mass spectrometry. *J. Chromatogr. A* **1161**:132–145.

Kasprzyk-Hordern, B., R. M. Dinsdale, A. J. Guwy. (2008a). The occurrence of pharmaceuticals, personal care products, endocrine disruptors and illicit drugs in surface water in South Wales, UK. *Water Res.* **42**:3498–3518.

Kasprzyk-Hordern, B., R. M. Dinsdale, and A. J. Guwy. (2008b). The effect of signal suppression and mobile phase composition on the simultaneous analysis of multiple classes of acidic/neutral pharmaceuticals and personal care products in surface water by solid-phase extraction and ultra performance liquid chromatography – negative electrospray tandem mass spectrometry. *Talanta* **74**:1299–1312.

Kasprzyk-Hordern, B., R. M. Dinsdale, and A. J. Guwy. (2008c). Multiresidue methods for the analysis of pharmaceuticals, personal care products and illicit drugs in surface water and wastewater by solid-phase extraction and ultra performance liquid chromatography-electrospray tandem mass spectrometry. *Anal. Bioanal. Chem.* **391**:1293–1308.

Kasprzyk-Hordern, B., R. M. Dinsdale, A. J. Guwy. (2009a). The removal of pharmaceuticals, personal care products, endocrine disruptors and illicit drugs during wastewater treatment and its impact on the quality of receiving waters. *Water Res.* **43**:363–380.

Kasprzyk-Hordern, B., R. M. Dinsdale, and A. J. Guwy. (2009b). Illicit drugs and pharmaceuticals in the environment – forensic applications of environmental data. Part 1:

Estimation of the usage of drugs in local communities. *Environ. Pollut.* **157**:1773–1777.

Kasprzyk-Hordern, B. (2010a). Pharmacologically active compounds in the environment and their chirality. *Chem. Soc. Rev.* **39**:4466–4503.

Kasprzyk-Hordern, B., V. V. R. Kondakal, and D. R. Baker. (2010b). Enantiomeric analysis of drugs of abuse in wastewater by chiral liquid chromatography coupled with tandem mass spectrometry. *J. Chromatogr. A.* **1217**:4575–4586.

King, L. A. 2009. Forensic Chemistry of Substance Misuse. London: RSC Publishing.

Kümmerer, K. (editor.). 2008. Pharmaceuticals in the Environment: Sources, Fate, Effects and Risks. New York: Springer.

Loganathan, B., M. Philips, H. Mowery, T. L. Jones-Lepp. (2009). Contamination profiles and mass loadings of macrolide antibiotics and illicit drugs from a small urban wastewater treatment plant. Chemosphere **75**:70–77.

Mari, F., L. Politi, A. Biggeri, G. Accetta, C. Trignano, M. D. Padua, and E, Bertol. (2009). Cocaine and heroin in waste water plants: a 1-year study in the city of Florence, Italy. *Forensic Sci Int.* **189**:88–92.

Petrovic, M., and D. Barcelo. (editor.). 2007. Analysis fate and removal of pharmaceuticals in the water cycle (comprehensive analytical chemistry). Amsterdam: Elsevier.

Postigo, C., M. J. Lopez de Alda, D. Barceló. 2008. Fully automated determination in the low nanogram per liter level of different classes of drugs of abuse in sewage water by on-line solid-phase extraction-liquid chromatography-electrospray-tandem mass spectrometry. *Anal. Chem.* **80**:3123–3134.

Sarmah, A. K., M. T. Meyer, and A. B. A. Boxall. (2006). A global perspective on the use, sales, exposure pathways, occurrence, fate and effects of veterinary antibiotics (VAs) in the environment. *Chemosphere* **65**:725–759.

Singleton, N., R. Murray, and L. Tinsley. (editors.) 2006. Measuring different aspects of problem drug use: methodological developments. Home Office Online Rep 16/06; www.homeoffice.gov.uk/rds/pdfs06/rdsolr1606.pdf.

Van Nuijs, A. L. N., B. Pecceu, L. Theunis, N. Dubois, C. Charlier, P. G. Jorens, L. Bervoets, R. Blust, H. Neels, and A. Covaci. (2009a). Cocaine and metabolites in waste and surface water across Belgium. *Environ. Pollut.* **157**:123–129.

Van Nuijs, A. L. N., B. Pecceu, L. Theunis, N. Dubois, C. Charlier, P. G. Jorens, L. Bervoets, R. Blust, H. Neels, and A. Covaci A. (2009b). Spatial and temporal variations in the occurrence of cocaine and benzoylecgonine in waste- and surface water from Belgium and removal during wastewater treatment. *Water Res.* **43**:1341–1349.

Van Nuijs, A. L. N., I. Tarcomnicu, L. Bervoets, R. Blust, P. G. Jorens, H. Neels, and A. Covaci. (2009c). Analysis of drugs of abuse in wastewater by hydrophilic interaction liquid chromatography-tandem mass spectrometry. *Anal. Bioanal. Chem.* **395**:819–828.

Wick, A., G. Fink, A. Joss, H. Siegrist, and T. A. Ternes. (2009). *Water Res.* **43**:1060–1074.

Zuccato, E., S. Castiglioni, and R. Fanelli (2005). Identification of the pharmaceuticals for human use contaminating Italian aquatic environment. *J. Hazard. Mater.* **122**:205–209.

Zuccato, E., S. Castiglioni, R. Bagnati, C. Chiabrando. P. Grassi, R. Fanelli. (2008a). Illicit drugs, a novel group of environmental contaminants. *Water Res.* **42**:961–968.

Zuccato, E., C. Chiabrando, S.Castiglioni, C. Bagnati, and R. Fanelli. (2008b). Estimating community drug abuse by wastewater analysis. *Environ. Health Perspect.* **116**:1027–1032.

CHAPTER 9

ON THE FRONTIER: ANALYTICAL CHEMISTRY AND THE OCCURRENCE OF ILLICIT DRUGS IN SURFACE WATERS IN THE UNITED STATES

TAMMY JONES-LEPP, DAVID ALVAREZ, and BOMMANNA LOGANATHAN

9.1 INTRODUCTION

While environmental scientists focused on industrial and agricultural pollutants (e.g., PCBs, volatile organics, dioxins, benzene, DDT) in the 1970s and 1980s, the subtle connection between personal human activities, such as drug consumption, and the subsequent release of anthropogenic drugs and drug metabolites into the natural environment was overlooked. There was evidence of this possible connection nearly 30 years ago when Garrison et al. (1976) reported the detection of clofibric acid (the bioactive metabolite from a series of serum triglyceride-lowering drugs) in a groundwater reservoir that had been recharged with treated wastewater (Garrison et al., 1976). A year later, Hignite and Azarnoff (1977) reported finding aspirin, caffeine, and nicotine in wastewater effluent, and then Watts et al. (1984) reported the presence of three pharmaceuticals (erythromycin, tetracycline, and theophylline), bisphenol A, and other suspected endocrine-disrupting compounds (EDCs) in a river water sample (Hignite and Azarnoff, 1977; Watts et al., 1984). Following those three journal articles, nothing was published for nearly a decade regarding the drug−human−environmental connection. Renewed interest in the subject was reported by Daughton and Ternes's seminal and authoritative work published in 1999 (Daughton and Ternes, 1999). Since the 1999 publication of Daughton and Ternes, the number of publications from the scientific community regarding the human drug consumption and environmental interaction have increased from two

Illicit Drugs in the Environment: Occurrence, Analysis, and Fate Using Mass Spectrometry, Edited by Sara Castiglioni, Ettore Zuccato, and Roberto Fanelli
Copyright © 2011 John Wiley & Sons, Inc.

publications in the1980s to currently over 300 scientific publications per year. Most of these publications report methods for the detection of common pharmaceuticals and over-the-counter (OTCs) drugs. However, very few publications have dealt with the occurrence, transport, and fate of illicit drugs in the environment.

In the United States, Snyder et al. (2001) reported the presence of hydrocodone, codeine, and diazepam (valium), in a stream entering into Lake Mead, Nevada (Snyder et al., 2001). While these drugs are not considered illicit substances, they are considered controlled substances, compounds that the Drug and Enforcement Agency (DEA) lists as schedule III and IV drugs, as substances for potential abuse. (DEA, http://www.usdoj.gov/dea/pubs/ abuse/1-csa.htm). Then, for the first time, the presence of an illicit substance, methamphetamine, was reported by Khan and Ongerth, in wastewater effluent from a large US city in California and announced publicly at the 2003 National Ground Water conference (Khan and Ongerth, 2003). Jones-Lepp et al. (2004) reported for the first time in the peer-reviewed literature the detection of two illicit drugs, methamphetamine and methylenedioxymethamphetamine (MDMA, ecstasy), collected from wastewater treatment plant (WWTP) effluent streams in Nevada and South Carolina (Jones-Lepp et al., 2004).

In the United States, there are multiple possible sources of release of illicit drugs into waterways. The largest possible contributor of illicit drugs would be from consumer consumption and subsequent excretion into the municipal sewer systems and transport through the WWTP process into streams, lakes, rivers, or wetlands (Jones-Lepp et al., 2004; Chiaia et al., 2008; Loganathan et al., 2009; Bartelt-Hunt et al., 2009). A smaller contribution could be from consumer consumption and subsequent excretion into septic systems, or other nonseweraged systems (e.g., boat privies, outhouses), and then leakage from the septics into surrounding source waters, creeks, bays, and wetlands (Jones-Lepp, 2006). Another possible source of illicit drugs can be from runoff from biosolids that have been applied as soil amendments to crops, municipal parkways, or during forest restoration (Kaleta et al., 2006; Kinney et al., 2006; Jones-Lepp and Stevens, 2007; Edwards et al., 2009). A likely source of illicit drugs could be from clandestine drug laboratories. For example, during the illegal manufacturing of methamphetamine, well over 50 hazardous chemicals are either used, or produced, as methamphetamine by-products (US EPA, 2008). All of these hazardous compounds, including methamphetamine, have the potential to enter the environment through improper disposal into the city sewer or individual septic systems, or via shallow drainage ditches directly onto surrounding soils (commonly used in remote methamphetamine operations), and through burn or burial pits (US EPA, 2008).

Another aspect of environmental monitoring of illicit drugs is socioeconomic. Daughton in 2001 was the first researcher to comment on developing an environmental monitoring program for the use of illicit substances (Daughton, 2001). Daughton proposed using sewerage monitoring to provide data on the daily influxes of drugs from a community and applying this data to obtain a realistic perspective on the overall magnitude and extent of community illicit drug use. Using Daughton's premise, two epidemiology studies have been completed in Europe (Italy and Spain) (Zuccato et al., 2005; Postigo et al., 2008). Recently, in 2009, the first epidemiologic study, using

Daughton's premise, was completed and published in the United States (Banta-Green et al. 2009).

Besides environmental monitoring data and, as important, is the lack of data regarding the ecotoxicity of the pharmaceuticals and illicit drugs. The missing eco-toxicity data makes estimations of predicted no-effect concentrations (PNEC), and hazard and risk assessments almost impossible, or at worse, a "best guess" scenario. Some researchers try to derive risk assessment data from the use of models that use quantitative structure-activity relationships (QSARs) and other measurements.

In the absence of empirical environmental data, one might be tempted to use such models as EPA's Ecological Structure Activity Relationships (ECOSAR) program, which is insufficiently accurate to actually predict ecotoxicity (Fent el al., 2006). For example, the collapse of the vulture populations in India because of exposure to minimal (subtherapeutic doses) amounts of diclofenac would never have been predicted with modeling (Oaks et al., 2004). Even more critical is generating risk assessments for those organisms that live in the aquatic environment. Even though acute toxicity may not be a high risk, chronic exposure to sublethal doses may alter an aquatic organisms feeding and mating behaviors. Brown et al. (2007) demonstrated the deficiencies of trying to model bioconcentration factors (BCFs) versus actual field measurements in fish plasma (Brown et al., 2007). There were extreme differences for some of the compounds measured and Brown points out the importance of using real-life exposures to test theoretical models at an early stage in model development (Brown et al., 2007).

Ecotoxicological consequences of illicit drugs being deposited into environmental matrices, particularly water, have not been closely examined. Therefore, it can only be surmised that these substances may have the potential to adversely affect biota that are continuously exposed to them, even at very low levels. The potential for chronic effects on human health is also unknown and of increasing concern because of the multiuse (continuously recycled in a closed loop) character of water, as in densely populated arid areas. The focus of this chapter will be on the state-of-the-art in sampling, extraction, and analysis of illicit drugs in the waterways of the United States. However, since much of the work with illicit drugs has been performed outside the United States, some of that data will also be given as examples. Better characterization of illicit drugs in the environment forms the foundation of improved risk assessments and sound science-based environmental policy.

9.2 PHYSICOCHEMICAL PROPERTIES OF ILLICIT DRUGS

The persistence of illicit drugs or any chemical in an aquatic ecosystem depends on its physicochemical and ecosystem-specific properties. Among these are concentration of dissolved/suspended organic matter, solubility, microbial population, etc. (Baughman et al., 1978; Loganathan and Kannan, 1994). Persistence of methamphetamine, MDMA, and related compounds in aquatic systems are a function of physical (e.g., volatilization from, and adsorption to, suspended solids and sediment), chemical (hydrolysis and photolysis), and biological removal (microbial degradation and uptake) mechanisms in addition to flow and other water characteristics (Loganathan et al., 2001). Considering the chemical makeup of illicit drugs, the volatilization of these compounds from natural water and sediment mixture is minimal, because of adsorption onto suspended solids or sediment (Loganathan et al. 2009). Very limited information is available on the half-lives of illicit drugs in water, sediment, and biota. For example, cocaine hydrochloride's water solubility is 0.17 g/100 mL, whereas its solubility in ether is 28.6 mg/100 mL, and the boiling point is about 188°C. These characteristics indicate that it is compatible with organic matter and will adsorb onto solid materials (Claustre and Bresch-Rieu, 1999). Photolysis of small molecules, such as methamphetamine and MDMA, may be possible in clear surface waters; however, photolysis rates for these chemicals are not available.

The pK_a, along with log D_{OW} (the pH-dependent n-octanol-water distribution ratio), can provide strong evidence of whether compounds will be in an ionized state and their hydrophobicity (Wells, 2006). These two physicochemical properties can help determine whether they will be retained in water, biosolids, sediment, and/or biological medium. For example, the pK_a's and log D_{OW} of methamphetamine, MDMA, cocaine, all weak bases, were 9.9 pK_a/−0.23 log D_{OW}, 10.38 pK_a/−1.11 log D_{OW}, and 8.6 pK_a/1.83 log D_{OW}, respectively (pK_a: methamphetamine, Logan, 2002; MDMA, Tsujikawa et al. 2009; and cocaine, Domènech et al. 2009; log D_{OW} was calculated using SPARC program, at pH 7, http://ibmlc2.chem.uga.edu/sparc/index.cfm). Although all three compounds have been detected in the water column, the log D_{OW}'s would suggest that only methamphetamine and MDMA will make it through the WWTP process and into the water column, while cocaine may be more likely to partition to the solids (Logan, 2002; Garrett et al. 1991; Jones, 1998). Structures and select physicochemical properties of a few common illicit drugs are given in Fig. 9.1 and Table 9.1.

There are four efficiency studies available that look at the removal of illicit drugs from WWTPs (van Nuijs et al., 2009; Huerta-Fontela et al., 2008a; Castiglioni et al., 2006a; Loganathan et al., 2009). However, we can use the data from van Nuijs et al. (2009) and Loganathan et al. (2009) to illustrate the importance of using log D_{OW}, in conjunction with pK_a, to predict removal and partitioning. If we consider the log D_{OW} of cocaine and methamphetamine, 8.6 pK_a/1.83 log D_{OW} and 9.9 pK_a/−0.23 log D_{OW}, respectively, one would predict that cocaine (log D_{OW} > 1) would be removed from wastewater more efficiently than methamphetamine (log D_{OW} < 1). Indeed, van Nuijs et al. (2009) showed that cocaine is nearly 100% removed by those WWTPs using conventional activated sludge (CAS) treatment. Loganathan et al. (2009) calculated the removal efficiency of methamphetamine at 55% at another WWTP that also uses CAS (van Nuijs et al., 2009b; Loganathan et al., 2009).

9.3   SAMPLING OF ILLICIT DRUGS IN SURFACE WATERS

The techniques used for collecting samples of surface waters, or of any environmental matrix, for the detection of illicit drugs are no different than would be used for any other chemical class. Illicit drugs, like many OTC and prescription pharmaceuticals,

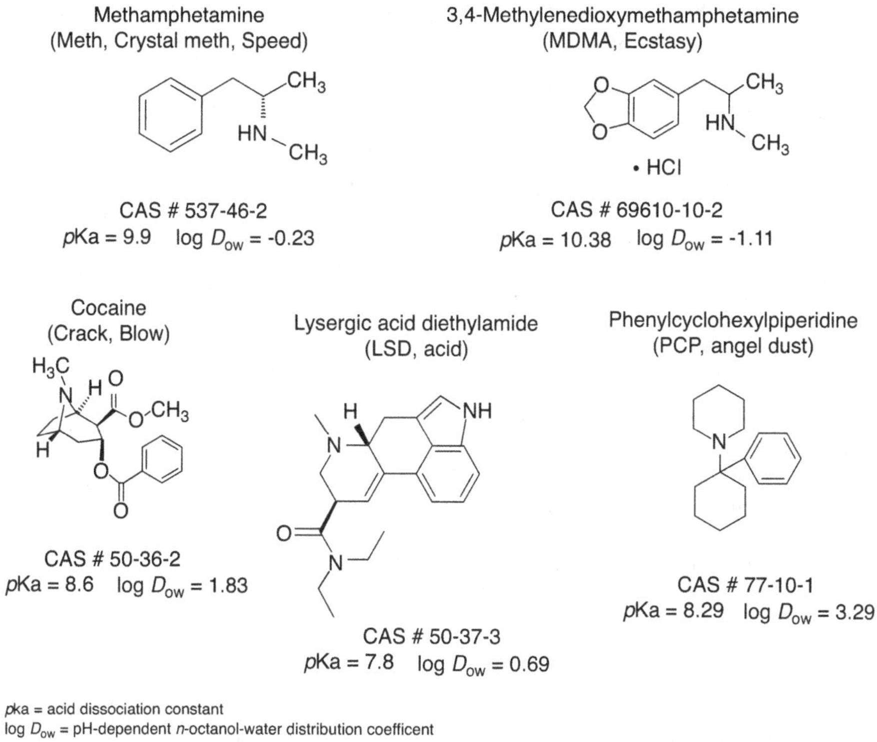

Methamphetamine
(Meth, Crystal meth, Speed)

CAS # 537-46-2
pKa = 9.9   log D_{ow} = -0.23

3,4-Methylenedioxymethamphetamine
(MDMA, Ecstasy)

• HCl

CAS # 69610-10-2
pKa = 10.38   log D_{ow} = -1.11

Cocaine
(Crack, Blow)

CAS # 50-36-2
pKa = 8.6   log D_{ow} = 1.83

Lysergic acid diethylamide
(LSD, acid)

CAS # 50-37-3
pKa = 7.8   log D_{ow} = 0.69

Phenylcyclohexylpiperidine
(PCP, angel dust)

CAS # 77-10-1
pKa = 8.29   log D_{ow} = 3.29

$p$ka = acid dissociation constant
log D_{ow} = pH-dependent n-octanol-water distribution coefficent

FIGURE 9.1   Chemical names, common names, structures, and select properties of common illicit drugs.

can have vast differences in their chemical structure resulting in a wide range of water solubility, photolytic stability, and other physicochemical parameters. The specific parameters, important in determining the storage and extraction conditions, have little to no impact on the selection of the sample collection method.

The decision on the sampling method to use is constrained by the type of information needed to answer a specific hypothesis and by the available resources (both logistical and financial). Instantaneous or time-integrated, whole water, or dissolved (filtered), one sample or replicates, and how much and what types of quality control measures will be used are all options that need to be considered as part of the sample collection plan (Alvarez and Jones-Lepp, 2010). The development of a sound sampling plan will help eliminate problems in the field and ensure a representative sample will be collected to meet the needs of the study.

9.3.1 Sampling Techniques

The collection of surface water samples generally falls into two classes of methods: active or passive. Active sampling techniques involve physically taking a sample

TABLE 9.1   Several Common Illicit Drugs and Their Precursor and Product Ions Formed by ESI−MS/MS

Illicit Drug Molecular Weight (CAS No.)	Precursor Ions	Product Ions	LODs	Reference
Methamphetamine 149.3 amu (537-46-2)	150.0 $(M+H)^+$	119 $(M+H-CH_3NH_2)^+$ 91 $(M+H-CH(CH_3)$ $NH(CH_3))^+$	1.5 ng 1.5 ng/L	Jones-Lepp et al., 2004 Chiaia et al., 2008
MDMA 193.1 amu (69610-10-2)	194.1 $(M+H)^+$	163 $(M+H-CH_3NH_2)^+$	1.0 ng 1.0 ng/L	Jones-Lepp et al., 2004 Chiaia et al., 2008
Cocaine 303.4 amu (50-36-2)	304.1 $(M+H)^+$	182.3 $(M+H-C_7H_5O_2)^+$	2.0 ng/L 20 fg	Chiaia et al., 2008
LSD 323.4 amu (50-37-3)	324.4 $(M+H)^+$	223.3 $(M+H-C_5H_{11}NO)^+$	0.5 ng/L	Chiaia et al., 2008
PCP [1-(1-phenylcyclohexyl) piperidine] 243.4 amu (77-10-1)	244.2 $(M+H)^+$	159.4 $(M+H-C_5H_{11}N)^+$	2.5 ng/L	Chiaia et al., 2008

either by manual or automatic means. Grab sampling methods are among the most common of active methods, which in the most simplistic form is filling a container with water at a specific location. This is performed by "hand-dipping" a container from the shore or boat or by lowering a container into the water from a structure, such as a bridge. If discrete samples are desired to be taken from a specific depth in the water column, a variety of systems such as the Kemmerer, Van Dorn, and double check-valve bailers can be used (Lane et al., 2003). Depth and width integrated samples can be collected using specialized samplers, which can be moved either vertically or horizontally across a body of water. Composite samples are often taken to achieve a representative sample of a larger body of water or to obtain an average water sample over time. Composite samples are generated by combining smaller volumes of water in a single container either manually or by using a automated sampler. Automated samplers are often used in remote locations or locations where water flow may be intermittent. They can be programmed to take samples at predetermined intervals or be started by an external sensor, such as a flow meter or depth gauge.

The majority of the published studies for illicit drugs use a simple grab sampling technique of collecting a 1-L water sample in a glass bottle (Buchberger and Zaborsky, 2007; Huerta-Fontela et al., 2008b; Loganathan et al., 2009). Other studies used automated sampling devices to take 24-h composite samples of 1 to 2 L of untreated WWTP influent (raw sewage) and treated effluent samples (Castiglioni et al., 2006b; Zuccato et al., 2008). Postigo et al. (2008) also collected 24-h composite samples of influent and effluent samples, but only needed a final sample size of 5 mL because of the use of an on-line solid-phase extraction system coupled to a liquid chromatography electrospray tandem mass spectrometer (Postigo et al., 2008).

Passive sampling techniques are those that require no manual or mechanical means for the sampling to occur. The samplers are placed in the water for a defined period of time and chemical uptake (sampling) occurs by diffusion or partitioning process. Passive samplers have advantages over active samplers in that they can be deployed for extended periods (months) in remote locations; episodic events such as runoff, spills, etc., are not missed; they allow detection of trace concentrations of chemicals that may not be possible with standard 1– to 2-L sample sizes; in the case of time-integrative samplers, they provide time-weighted average concentrations of chemicals, which are a fundamental part of ecological risk assessments (Alvarez and Jones-Lepp, 2010).

Time-integrative and equilibrium samplers make up the bulk of the passive sampling techniques. Among these, the semipermeable membrane device (SPMD), the polar organic chemical integrative sampler (POCIS), solid-phase microextraction (SPME), polymer sheets, polymers on glass (POGs), and the Chemcatcher are the most common (Alvarez et al., 2007; Mills et al., 2007). Jones-Lepp et al. (2004) were the first to demonstrate the utility of passive sampling devices in illicit drug-monitoring studies. Since then, three other publications describe the use of passive samplers to sample for illicit drugs (Alvarez et al., 2007; Mills et al., 2007; Bartelt-Hunt et al., 2009). In all of these cases, the POCIS was used as it has the ability to sample chemicals containing varied functional groups over a range of polarities common with illicit drugs. Although many of the other passive sampling devices would be capable of sampling certain drugs, they are much more limited in the range of chemical classes that could be sampled.

9.3.2 Handling and Storage Considerations

In general, the collection of environmental waters for the detection of illicit drugs should follow common handling and storage protocols. Samples are generally collected in amber glass containers and shipped chilled (<4–6°C) via overnight carrier to the laboratory. As with most emerging contaminants, the use of additives as sample preservatives is not required. Upon receipt at the laboratory, the samples should be stored chilled and extracted within 7–14 days. As with all laboratory procedures, storage and holding times for any new chemical should be evaluated prior to sample collection to ensure the integrity of the samples.

9.3.3 Quality Control

The types and amount of quality control used during the field component of a study can vary depending on the data requirements of the study. At a minimum, field blanks should be used to identify any contamination either through direct contact or airborne exposure of the sample. Other quality control samples to be considered include equipment blanks, if the same sampling equipment is repetitively used, trip blanks (contaminant-free water samples which accompany the field collected samples from the field to the laboratory but are not exposed to the air), and positive control samples (water samples fortified with the target analytes used to measure any loss or degradation of the analytes because of the handling and storage methods).

9.4   ANALYTICAL METHODS FOR ILLICIT DRUGS

While this chapter is devoted to detection of illicit drugs in water, we will also briefly mention the analytical methods for environmental media other than water. Many analytical challenges are offered to environmental chemists by the variety of environmental matrices, e.g., sediments, water, plants, biosolids/sludges, and soils, in their quest to tease out individual chemicals from these complex matrices. Additives and naturally occurring chemicals can cause substantial interferences during both extraction and detection methodologies. Since most illicit drugs usually occur in the environment at part-per-trillion (ppt) levels, the analytical methods can require intensive separation and cleanup procedures to isolate and concentrate the chemical from the matrix before analysis.

9.4.1   Extraction Techniques

Solid-phase extraction (SPE) is the most widely reported method for the extraction of pharmaceuticals and illicit drugs from aqueous matrices. In this section, we will look at SPE, as well as large-volume injection (LVI) and direct injection as extraction techniques (Jones-Lepp, 2006; Loganathan et al., 2009; Chiaia et al., 2008; Banta-Green et al., 2009).

9.4.1.1   *Solid-Phase Extractions (SPE)*   The SPE sorbents are chosen for their ability to retain the pharmaceuticals of interest based upon a variety of the physicochemical properties of the analytes of interest (e.g., pK_a, D_{ow}, polarity). The SPE sorbent most frequently reported for recovery of illicit drugs, is the hydrophobic lipophilic-balanced (HLB) sorbent-containing cartridges. Mixed-cation exchange (MCX) sorbents have also been used. Jones-Lepp (2006) and Loganathan et al. (2009) reported using the HLB [6-mL capacity, 0.2 g, 30-μm, obtained from Waters Corporation (Milford, MA)] sorbent for the extraction of pharmaceuticals and illicit drugs and, the USEPA's Office of Water Method 1694 recently published recommends the HLB sorbent cartridges/discs for aqueous extractions of pharmaceuticals (Jones-Lepp, 2006; Loganathan et al., 2009; USEPA, 2007). However, Boles and Wells (2010), in a review of analytical methods for amphetaminelike compounds, point to a number of analytical studies using both MCX and HLB sorbents (Boles and Wells, 2010). They conclude, along with van Nuijs (2009), that MCX and HLB are interchangeable as SPE sorbents (Boles and Wells, 2010; van Nuijs et al., 2009). The choice of one sorbent over another depends on the compounds of interest and what interferences would be removed (Boles and Wells, 2010; van Nuijs et al., 2009a).

9.4.1.2 Large-Volume Injection (LVI) Chiaia et al. (2008), report directly coupling a large-volume injector (1800 μL) to a tandem mass spectrometer (Chiaia et al., 2008). Their method allowed them to detect part-per-trillion to part-per-billion (ppb) levels of methamphetamine, amphetamine, ephedrine, cocaine, cocaine metabolites (e.g., benzoylecgonine, norcocaine, norbenzoylecgonine), hydrocodone, oxycodone, methadone, MDMA, MDMA metabolites (e.g., MDA, MDEA, MBDB), LSD, and PCP). Banta-Green et al. (2009) used the LVI technique, directly coupled to a liquid chromatography−mass spectrometry−mass spectrometry (LVI−LC/MS/MS), to determine the utility of community-wide drug testing (Banta-Green et al., 2009). They surveyed 96 WWTPs for the presence of the illicit drugs and their metabolites, and then back-calculated the target community's drug use (Banta-Green et al., 2009).

9.4.1.3 Pressurized Liquid Extraction (PLE) Very few papers have been written describing the extraction of illicit drugs from solid matrices. Stein et al. (2008) describe a PLE method for extracting psychoactive compounds from sediments and Jones-Lepp and Stevens (2007) also describe a PLE method for extracting methamphetamine and MDMA from biosolids (Stein et al., 2008; Jones-Lepp and Stevens, 2007). Because of the complexity and variable sizes of environmental solids, the samples usually need to be dried, pulverized, and homogenized before extraction. Briefly, small amounts of homogenized solid samples (usually < 2 g) are subsampled and extracted. Depending upon what matrix and what analytes are being extracted, the proper solvents, pressures, and temperatures are chosen (Stein et al., 2008; Jones-Lepp and Stevens, 2007).

9.4.2 Detection Techniques

9.4.2.1 Ion-Mobility Spectrometry (IMS) It is interesting to note that in 1976 Karasek and colleagues used ion-mobility spectrometry (IMS) to detect heroin and cocaine at atmospheric pressure (Karasek et al. 1976). In the 1980s Lawrence further developed IMS to detect other illicit drugs from solid surfaces and for atmospheric sampling (Lawrence, 1987, 1984). More recently, Hill's research group expanded the utilization of IMS to amphetamine, methamphetamine, PCP, morphine, THC, LSD, and heroin, coupling the IMS to a mass spectrometer for more specificity (Wu et al., 2000).

9.4.2.2 Mass Spectrometry (MS) The majority of detection techniques for pharmaceuticals and illicit drugs are liquid chromatography−mass spectrometry (LC-MS) based. To date, the only instruments reported in the United States for detecting illicit drugs in environmental matrices are mass spectrometers. The reality is that most environmental matrices are complex and only the mass accuracy and specificity given by mass spectrometry can overcome the large amounts of interferences found in real-world matrices. There are a variety of mass spectrometers now being used as detectors coupled to liquid chromatographs (LC). Available as mass detectors are ion-trap mass spectrometers (ITMS), quadrupole time-of-flight mass spectrometers (q-TOFMS), triple-quadrupole mass spectrometers (QqQ), magnetic-sector mass spectrometers, and, most recently, Orbitrap − mass spectrometers. A variety of mass spectrometers have been used and all US researchers have reported using the tandem mass-spectrometry (MS/MS) mode when detecting illicit drugs, as well as other emerging contaminants. The MS/MS mode is where a precursor ion [typically a $(M+H)^+$ in the positive mode, or $(M-H)^-$ ion in the negative mode] is formed in the LC/MS source. The ion formed is transported to an area of the MS where it is energized and collided (either in a QqQ, ITMS, q-TOFMS, or a magnetic-sector mass spectrometer) subsequently producing product ions. Product ions are typically the loss of various functional groups from the analytes, for example $(M+H-OH)^+$ or $(M+H-CH_3)^+$. Table 9.1 shows several illicit drugs, their precursor, and product ions as reported in the literature. In the United States, Jones-Lepp et al. (2004) used micro-liquid chromatography-electrospray/ion-trap mass spectrometry ($\mu-LC-ES/ITMS$) to assess and detect four prescription drugs (azithromycin, fluoxetine, omeprazole, and levothyroxine) and two illicit drugs (methamphetamine and MDMA) in wastewater effluent (Jones-Lepp et al., 2004). Chiaia et al. (2008) and Banta-Green et al. (2009) coupled LVI to a tandem mass spectrometer (triple-stage quadrupole) to accurately identify and quantify a variety of illicit and prescription drugs and their metabolites (Chiaia et al., 2008; Banta-Green et al., 2009). Bartelt-Hunt et al. (2009) used a QqQ to accurately identify and quantify a variety of prescription drugs, non-prescription drugs (e.g., DEET, caffeine), and the illicit drugs, methamphetamine, cocaine, MDMA, etc. (Bartelt-Hunt et al., 2009).

9.4.2.3 *Accurate Illicit Drug Identification*   When using LC−MS techniques for identifying known and unknown chemicals, it cannot be emphasized enough that the analyst must use a MS/MS technique in order to accurately identify analytes. For example, MDMA and caffeine while having different molecular weights have overlapping product ions (mass 163 *m/z*). However, they have different precursor to product pathways. MDMA with a molecular weight of 193 *m/z*, forms 194 *m/z*, $(M+H)^+$, forming the predominant product ion, 163.0 *m/z*, $(M-CH_3NH_2+H)^+$, using collision-induced dissociation (CID). While caffeine having a molecular weight of 194 *m/z* (one amu different from MDMA), forms 195 *m/z*, $(M+H)^+$, and under CID, forms predominantly the product ion 138 *m/z*, $(M-CH_3NCO)^+$, with mass 163 *m/z* also formed, but less abundantly. Therefore, if an analyst was monitoring the 163 *m/z* ion channel, and detected 163 *m/z*, near or at the same retention time as caffeine, they might misidentify that compound as MDMA, when, in fact, it is caffeine. Another example would be between methamphetamine and *N,N*'-dimethylphenethylamine (DMPEA, a widely used industrial chemical, used as a flavoring agent). These two chemicals are isobaric ions of each other, both have exactly the same molecular mass (149.0 *m/z*), but are slightly different in chemical structure. Fortunately, under CID LC−ESI MS/MS conditions, these two chemicals form unique predominant product ions, 119 *m/z* $(M+H-CH_3NH_2)^+$ and 105 *m/z* $(M+H-N(CH_3)^2)^+$. However, both compounds also form 91 *m/z* as a secondary product ion $(M+H-CH-N-(CH_3)2)^+$. If a researcher chose to monitor mass 91 *m/z*, instead of 119 *m/z*, for methamphetamine (and there are those who have reported doing so in the literature) then a false positive

for methamphetamine could occur. Therefore, it is important that the proper product and transition ions are chosen to ensure specificity and accuracy.

9.4.3   Occurrence of Illicit Drugs in US Waterways

Jones-Lepp et al. (2004) report detecting both methamphetamine and MDMA (ecstasy) in the low parts per trillion range from two sewage effluents, one in the southwest and the other in the southeast regions of the United States (Jones-Lepp et al., 2004). Jones-Lepp reported finding in 2006 methamphetamine at two sites, one from an urban creek in Las Vegas, Nevada and the other in the state of Maine. Methamphetamine was detected at 5 ng/L in the urban creek, which is surrounded by homes that were on septic tanks. Methamphetamine was also detected at 7 ng/L at the sewage effluent outfall of a large WWTP in Maine (Jones-Lepp, 2006). Chiaia et al. (2008), reported detecting methamphetamine at five of the seven WWTPs sampled from throughout the United States, with concentrations ranging from 10 to 2000 ng/L, and MDMA at five of the seven plants, with concentrations ranging from 3 to 70 ng/L (Chiaia et al., 2008). Chiaia et al. (2008) also reported finding cocaine at all seven of the WWTPs sampled (ranging from 10 to 860 ng/L), as well as the prescription opiates: hydrocodone, oxycodone, and methadone. Bartelt-Hunt et al. (2009) sampled eight sites across the state of Nebraska for a variety of pharmaceuticals and methamphetamine (Chiaia et al., 2008; Bartelt-Hunt et al., 2009). They detected methamphetamine at seven sites, except one upstream from the Lincoln WWTP, ranging from 2 to 350 ng/L (effluent from Omaha WWTP). The lower levels of methamphetamine were detected not only in WWTP effluents, but also in streams that were upstream from large city WWTPs (Bartelt-Hunt et al., 2009). This finding can possibly indicate the presence of clandestine drug labs, as well as input from septic tank leakages into these feeder streams. Banta-Green et al. (2009) sampled 96 WWTPs effluents from across the state of Oregon for methamphetamine, MDMA, and cocaine (Banta-Green et al., 2009). At all 96 WWTPs, methamphetamine was detected, while MDMA was detected at less than one-half of WWTPs, and benzoylecgonine (a cocaine metabolite) was primarily detected in the urban WWTPs effluents (Banta-Green et al., 2009).

A recent, extensive study [conducted by Jones-Lepp (EPA), Alvarez (USGS) and Sanchez (University of Arizona, Yuma Agricultural Center)] along the Colorado River shows the input of illicit drugs into the Colorado River from various sources. The Colorado River, is the main water source (e.g., drinking, agricultural, and industrial) for millions of people living in the southwestern part of the United States (e.g., Nevada, Arizona, California, Utah, and Colorado) and western Mexico. Samples were taken throughout the Colorado River basin, from the upper basin, starting at Glenwood Springs, Colorado, to the lower basin, ending in Somerton, Arizona (see Fig. 9.2). Using a modified version of the method (Oasis MCX, instead of Oasis HLB, SPE cartridges) established by Jones-Lepp (2006), methamphetamine, MDMA, and pseudoephedrine were detected in most of the effluents of the WWTPs sampled and at three different non-WWTP sites (Crystal Beach, AZ, New River, CA, and Cedar Pocket, AZ)(see Table 9.2).

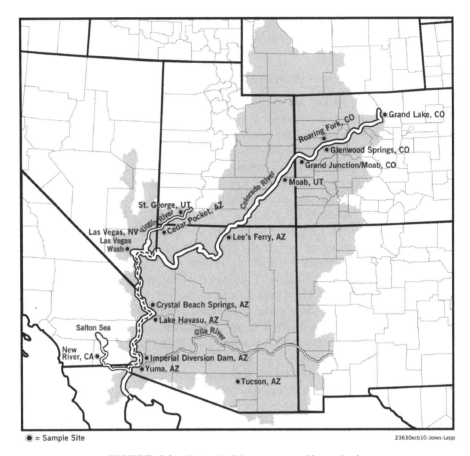

= Sample Site

FIGURE 9.2   Colorado River: upper and lower basin.

Pseudoephedrine (a similar in structure to methamphetamine and MDMA) was detected in the Virgin River (a tributary of the Colorado River) at Cedar Pocket, AZ. Cedar Pocket is located along the Virgin River and is approximately 18 km downstream from the St. George, UT, WWTP, which empties into the Virgin River. One possibility for detection at this site is the negative log $D_{OW} = -1.85$, at pH 7, as calculated for pseudoephedrine. This negative value indicates that this compound is more hydrophilic, and, therefore, more likely to stay in the water column, as compared to methamphetamine and MDMA.

Methamphetamine, at 220 ng/L, was detected in the New River, CA. The New River, is interesting, geographically speaking, as the New River flows out of Mexicali, Mexico, and back into Calexico, United States, to the Salton Sea sink in California. There are raw human waste sources and illegal methamphetamine manufacturing laboratories along the New River, starting in Mexico, and back to the Salton Sea, that could contribute to this drug entering into the waterway (personal communication with anonymous US Border Patrol officer).

TABLE 9.2   Concentrations of Methamphetamine, MDMA, and Pseudoephedrine from Colorado River Basin[a]

Sampling Site	Sample Type	Amount Detected (ng/L)		
		Methamphet.	MDMA	Pseudoephedrine
Grand Lake, CO (headwaters)	CR	ND	ND	ND
Glenwood Springs, CO	WWTP	253	74	ND
Glenwood Springs, CO	CR	ND	ND	ND
Roaring Fork, CO	CR	ND	ND	ND
Grand Junction/Fruita, CO	CR	ND	ND	ND
Moab, UT	WWTP	ND	ND	ND
Moab, UT	CR	ND	ND	ND
St. George, UT	WWTP	ND	ND	350
Cedar Pocket, AZ	VR	ND	ND	230
Lee's Ferry, AZ	CR	ND	ND	ND
Las Vegas Wash[b]	LVW	230	ND	ND
Crystal Beach, AZ[c]	CR	ND - 22	ND - 36	ND
Lake Havasu, AZ[d]	WWTP	103 (ND – 480)	4 (ND – 17)	330 (ND – 780)
Yuma, AZ	WWTP	650	ND	ND
Gila River, AZ	GR	ND	ND	ND
Tucson, AZ[e]	WWTP	245	ND	372
Imperial Diversion Dam, AZ	CR	ND	ND	ND
Somerton, AZ	WWTP	84	ND	ND
New River, CA	NR	221	ND	ND

[a]ND, not detected. Sample type: CR, Colorado River; GR, Gila River; LVW, Las Vegas Wash below convergence of three WWTPs effluents; NR, New River; VR, Virgin River; WWTP, wastewater treatment plant.
[b]Average from two sampling events.
[c]Range of concentrations of three sampling events (min – max).
[d]Average from three WWTPs (Northwest Regional, Mulberry, and Island) over one year, where $n = 7$ sampling events (min – max).
[e]Average of $n = 9$ sampling events from 02/08 to 07/08 .

The third non-WWTP site, was offshore, in the middle of the Colorado River, near Crystal Beach, AZ. This site was sampled three times, May, July, and November of 2007, and methamphetamine and MDMA were detected only once, at 22 and 36 ng/L, respectively, in the July 2007 sample.

9.5   CONCLUSIONS

We can see from this chapter, that there are several viable methods available, depending upon the analytical need, to separate, concentrate, quantify, and reliably detect these compounds. The caveat is that mass spectrometry is the only definitive detection method and must be used in the MS/MS mode to ensure accurate detection of not only the illicit compounds, but other emerging contaminants. Papers showing the detection of illicit drugs in the United States are still few in number (Table 9.3).

TABLE 9.3   Analytical Methods and Illicit Drugs Identified in US Waterways

Reference	Illicit Drugs Identified	Extraction Method	Environmental Media
Chiaia et al., 2008	Methamphetamine, MDMA, cocaine, cocaine metabolites	Large volume injection	Wastewater
Bartelt-Hunt et al., 2009	Methamphetamine	POCIS	Wastewater
Banta-Green et al., 2009	Cocaine, cocaine metabolites	Large volume injection	Sewerage
Jones-Lepp et al., 2004	Methamphetamine, MDMA	POCIS	Wastewater
Jones-Lepp et al., 2006	Methamphetamine	SPE	Source water, wastewater
Jones-Lepp and Stevens, 2007	Methamphetamine	PLE	Biosolids
Khan and Ongerth, 2003	Methamphetamine	Unknown	Wastewater
Loganathan et al., 2009	Methamphetamine, MDMA	SPE	Wastewater

However, we can discern from these few studies that illicit drugs, and their metabolites, are making their way into US waterways. There are potential ecotoxicological and sociological ramifications from these findings not yet addressed. Lacking are the ecotoxicological studies to determine whether the levels of illicit drugs detected are of significance to both ecological and human health, both for acute and chronic exposures. It is of socioeconomic significance that, using the methods outlined in this chapter, researchers have been able to demonstrate the utility of back-calculating from the amounts of illicit drugs found in sewerages, and WWTP effluents, to community usages (Banta-Green et al., 2009).

The methods and approaches presented in this chapter to detect illicit drugs will provide information needed for developing a framework for exposure and ecotoxicological studies to ensure accurate risk assessments for future regulatory efforts.

ACKNOWLEDGMENTS

We would like to thank Dr. Charles Sanchez (University of Arizona, Yuma Agricultural Center) and Dr. Doyle Wilson (City of Lake Havasu) for their intensive sampling efforts along the Colorado River and Lake Havasu.

The United States Environmental Protection Agency, through its Office of Research and Development, collaborated in the research described here with the USGS-Columbia Environmental Research Center and Murray State University. It has been subjected to Agency review and approved for publication.

REFERENCES

Alvarez, D., J. Petty, J. Huckins, T. Jones-Lepp, D. Getting, J. Goddard, and S. Manahan. 2004. Development of a passive, in situ, integrative sampler for hydrophilic organic contaminants in aquatic environments, *Environ. Toxicol. Chem.* **23**:1640–1648.

Alvarez, D. A., and T. L. Jones-Lepp. 2010. Sampling and analysis of emerging pollutants. In *Water Quality Concepts, Sampling, and Chemical Analysis*, edited by Y. Li, and K. Migliaccio, pp. 199–226. Boca Raton, FL: Taylor Francis/CRC Press.

Alvarez, D. A., J. N. Huckins, J. D. Petty, T. L. Jones-Lepp, F. Stuer-Lauridsen, D. T. Getting, J. P. Goddard, and A. Gravell. 2007. Tool for monitoring hydrophilic contaminants in water: polar organic chemical integrative sampler (POCIS). In *Passive Sampling Techniques. Comprehensive Analytical Chemistry*, edited by R. Greenwood, G. Mills, and B. Vrana. Vol **48**, pp. 171–197 Amsterdam: Elsevier.

Banta-Green, C. J., J. A. Field, A. C. Chiaia, D. L. Sudakin, L. Power, and L. de Montigny. 2009. The spatial epidemiology of cocaine, methamphetamine and 3,4-methylenedioxy methamphetamine (MDMA) use: a demonstration using a population measure of community drug load derived from municipal wastewater. *Addiction* **104**:1874–1880.

Bartelt-Hunt, S. L., D. D. Snow, T. Damon, J. Shockley, and K. Hoagland. 2009. Illicit and therapeutic pharmaceuticals in wastewater effluent and surface waters in Nebraska. *Environ. Pollut.* **157**:786–791.

Baughman, G. L., and R. R. Lassiter 1978. Prediction of environmental pollutant concentration. In Estimating the Hazard of Chemical Substances to Aquatic Life, edited by J. Cairns, Jr., D. L. Dickson, and A. W. Maki. Pp. 35–54. West Conshocken, PA: ASTM (American Society for Testing and Materials) STP 657.

Boles, T., and M. Wells. 2010. Analysis of amphetamine-type stimulants as emerging pollutants. *J. Chromatogr. A* **1217**:2561–2568.

Brown, J. N., N. Paxéus, L. Förlin, and D. G. J. Larsson. 2007. Variations in bioconcentration of human pharmaceuticals from sewage effluents into fish plasma. *Environ. Toxicol. Pharmacol.* **24**:267–274.

Buchberger, W. and P. Zaborsky. 2007. Sorptive extraction techniques for trace analysis of organic pollutants in the aquatic environment. *Acta Chim. Slov.* **54**:1–13.

Castiglioni, S., R. Bagnati, R. Fanelli, F. Pomati, D. Calamari, and E. Zuccato. 2006a. Removal of pharmaceuticals in sewage treatment plants in Italy. *Environ. Sci. Technol.* **40**: 357–363.

Castiglioni, S., E. Zuccato, E. Crisci, C. Chiabrando, R. Fanelli, R. Bagnati. 2006b. Identification and measurement of illicit drugs and their metabolites in urban wastewater by liquid chromatography-tandem mass spectrometry. *Anal. Chem.* **78**:8421–8429.

Chiaia, A., C. Banta-Green, and J. Field. 2008. Large volume injection HPLC/MS/MS: a demonstration for illicit drugs, prescription opioids and urine indicators in municipal wastewater across the USA. *Environ. Sci. Technol.* **42**:8841–8848.

Claustre, A., and I. Bresch-Rieu. 1999. Cocaine: international programme on chemical safety. *Poisons Infor. Monogr.* 139.

Daughton, C. 2001. Illicit drugs in municipal sewage. In *Pharmaceutical and Personal Care Products in the Environment—Scientific and Regulatory Issues*, edited by C. Daughton and T. Jones-Lepp, pp. 348–364. Washington, DC: American Chemical Society Symposium Series 791. p.

Daughton, C., and T. Ternes. 1999. Pharmaceuticals and personal care products in the environment: agents of subtle change? *Environ. Health Perspect.* **107**:907–938.

Domènech, X., J. Peral, I. Munoz. 2009. Predicted environmental concentrations of cocaine and benzoylecgonine in a model environmental system. *Water Res.* **43**:5236–5242.

Edwards, M., E. Topp, C. D. Metcalfe, H. Li, N. Gottschall, P. Bolton, W. Curnoe, M. Payne, A. Beck, S. Kleywegt, and D. R. Lapen. 2009. Pharmaceutical and personal care products in tile drainage following spreading and injection of dewatered municipal biosolids to an agricultural field. *Sci. Total Environ.* **407**:4220–4230.

Fent, K., A. A. Weston, and D. Caminada. 2006. Review: Ecotoxicology of human pharmaceuticals. *Aquat. Toxicol.* **76**:122–159.

Garrett, E. R., K. Seyda, P. Marroum 1991. High performance liquid chromatographic assays of the illicit designer drug "Ecstasy," a modified amphetamine, with applications to stability, partitioning and plasma protein binding. *Acta Pharmacol. Nord.* **3**:9–14.

Garrison, A.W., J. D. Pope, and F. R. Allen. 1976. GC/MS Analysis of organic compounds in domestic wastewaters. In *Identification and Analysis of Organic Pollutants in Water*, edited by C. H. Keith, pp. 517–556. Ann Arbor, MI: Ann Arbor Science Publ

Hignite, C., and D. Azarnoff. 1977. Drugs and drug metabolites as environmental contaminants: Chlorophenoxyisobutyrate and salicylic acid in sewage water effluent. *Life Sci.* **20**: 337–341.

Huerta-Fontela, M., M. T. Galceran, and F. Ventura. 2008a. Stimulatory drugs of abuse in surface waters and their removal in a conventional drinking water treatment plant. *Environ. Sci. Technol.* **42**:6809–6816.

Huerta-Fontela, M., M. T. Galceran, J. Martin-Alonso, and F. Ventura. 2008b. Occurrence of psychoactive stimulatory drugs in wastewaters in north-eastern Spain. *Sci. Total Environ.* **397**:31–40.

Jones, R. 1998. Pharmacokinetics of Cocaine: Considerations when assessing cocaine use by urinalysis. *Natl. Inst. Drug Abuse Monogr* **175**:221–234.

Jones-Lepp, T. 2006. Chemical markers of human waste contamination: Analysis of urobilin and pharmaceuticals in source waters. *J. Environ. Monit.* **8**:472–478.

Jones-Lepp, T., and R. Stevens. 2007. Pharmaceuticals and personal care products in biosolids/sewage sludge: the interface between analytical chemistry and regulation. *Anal. Bioanal. Chem.* **387**:1173–1183.

Jones-Lepp, T., D. Alvarez, J. Petty, and J. Huggins. 2004. Polar organic chemical integrative sampling (POCIS) and LC-ES/ITMS for assessing selected prescription and illicit drugs in treated sewage effluents. *Arch. Environ. Contaminants Toxicol.* **47**:427–439.

Jones-Lepp, T. L., D. A. Alvarez, B. Englert, and A. L. Batt. 2009. Pharmaceuticals and hormones in the environment. In: *Encyclopedia of Analytical Chemistry*. Hoboken New Jersey: John Wiley & Sons.

Kaleta, A., M. Ferdig, and W. Buchberger. 2006. Semiquantitative determination of amphetamine in sewage sludge samples. *J. Sep. Sci.* **29**:1662–1666.

Karasek, F. W., H. H. Hill, and S. H. Kim. 1976. Plasma chromatography of heroin and cocaine with mass identified mobility spectra. *J. Chromatogr.* **117**:327–336.

Khan, S. J., and J. E. Ongerth. 2003. Mass-balance analysis of pharmaceutical residues within a sewage treatment plant. In *Proceedings of the Symposium at the Meeting of the National Ground Water Association,* March 19–21, Minneapolis, MN.

Kinney, C. A., E. T. Furlong, S. D. Zaugg, M. R. Burkhardt, S. L. Werner, J. D. Cahill, and G. R. Jorgensen. 2006. Survey of organic wastewater contaminants in biosolids destined for land application. *Environ. Sci. Technol.* **40**:7207–7215.

Lane, S. L., S. Flanagan, and F. D. Wilde. 2003. Selection of equipment for water sampling (ver. 2.0); *U.S. Geological Survey Techniques of Water-Resources Investigations,* book 9, chap. A2. http://ppubs.water.usgs.gov/twri9A2/ (accessed May 2, 2008).

Lawrence, A. H. 1984. Ion mobility spectrometry/mass spectrometry of some prescription and illicit drugs. *J. Biochem. Biophys. Methods* **9**:277–306.

Lawrence, A. H. 1987. Detection of drug residues on the hands of subjects by surface sampling and ion mobility spectrometry. *Forensic Sci. Int.* **34**:73–83.

Logan, B. K. 2002. Methamphetamine – Effects on human performance and behavior. *Forensic Sci. Rev.* **14**:133–151.

Loganathan, B. G., M. Phillips, H. Mowery, and T. L. Jones-Lepp. 2009. Contamination profiles and mass loadings of macrolide antibiotics and illicit drugs from a small urban wastewater treatment plant. *Chemosphere* **75**:70–77.

Loganathan, B. G., and K. Kannan. 1994. Global organochlorine contamination: An overview. *AMBIO: A J. Human Environ.* **23**:187–191.

Loganathan, B. G., K. Kannan, K. S. Sajwan, and D. A. Owen. 2001. Butyltin compounds in freshwater ecosystems. In Persistent, bioaccumulative and toxic chemicals I: Fate and exposure, edited by R. L. Lipnick J. Hermens, K. Jones, and D. Muir, p. 308. Washington, DC: American Chemical Society (Symposium Series, 772).

Mills, G. A., B. Vrana, I. Allan, D. A. Alvarez, J. N. Huckins, and R. Greenwood. 2007. Trends in monitoring pharmaceuticals and personal-care products in the aquatic environment by use of passive sampling devices. *Anal. Bioanal. Chem.* **387**:1153–1157.

Oaks, J. L., M. Gilbert, M. Z. Virani, R. T. Watson, C. U. Meteyer, B. A. Rideout, H. L. Shivaprasad, S. Ahmed, M. J. I. Chaudhry, M. Arshad, S. Mahmood, A. Ali, and A. A. Khan. 2004. Diclofenac residues as the cause of vulture population decline in Pakistan. *Nature (London)* **427**:630–633.

Postigo, C., M. J. Lopez de Alda, and D. Barceló. 2008. Fully automated determination in the low nanogram per liter level of different classes of drugs of abuse in sewage water by on-line solid-phase extraction-liquid chromatography-electrospray-tandem mass spectrometry. *Anal. Chem.* **80**:3123–3134.

Snyder, S., K. Kelly, A. Grange, G. Sovocool, E. Snyder, and J. Giesy. 2001. Pharmaceuticals and personal care products in the waters of Lake Mead, Nevada. In *Pharmaceutical and Personal Care Products in the Environment—Scientific and Regulatory Issues*, edited by C. Daughtonand T. Jones-Lepp, pp. 116–139 Washington, DC: American Chemical Society, (Symp. Ser. 791).

Stein, K., M. Ramil, G. Fink, M. Sander, and T. Ternes. 2008. Analysis and sorption of psychoactive drugs onto sediment. *Environ. Sci. Technol.* **42**:6415–6423.

Tsujikawa, K., K. Kuwayama, H. Miyaguchi, T. Kanamori, Y. Iwata, and H. Inoue. 2009. Degradation of N-hydroxy-3,4-methylenedioxymethamphetamine in aqueous solution and its prevention. *Forensic Sci. Int.* **193**:106–111.

US EPA. 2008. RCRA hazardous waste identification of methamphetamine production process by-products, report to Congress under the USA PATRIOT Improvement and Reauthorization Act of 2005, Washington D.C. last accessed 2009 23, October; http://www.epa.gov/waste/hazard/wastetypes/wasteid/downloads/rtc-meth.pdf.

US EPA. 2007. Method 1694: Pharmaceuticals and personal care products in water, soil, sediment, and biosolids by HPLC/MS/MS, EPA-821-R-08-002, Washington DC: US EPA.

van Nuijs, A., I. Tarcomnicu, L. Bervoets, R. Blust, P. G, Jorens, H. Neels, and A. Covaci. 2009a. Analysis of drugs of abuse in wastewater by hydrophilic interaction liquid chromatography–tandem mass spectrometry. *Anal. Bioanal. Chem.* **395**:819–828.

van Nuijs, A., B. Pecceu, L. Theunis, N. Dubois, C. Charlier, P. Jorens, L. Bervoets, R. Blust, H. Neels, and A. Covaci A. 2009b. Spatial and temporal variations in the occurrence of cocaine and benzoylecgonine in waste- and surface water from Belgium and removal during wastewater treatment. *Water Res.* **43**:1341–1349.

Watts, C. D., B. Crathorn, M. Fielding, and C. P. Steel. 1984. Identification of non-volatile organics in water using field desorption mass spectrometry and high performance and high performance liquid chromatography. In *Analysis of Organic Micropollutants in Water: Proceedings of the Third European Symposium,* 1983 September 19−21: Oslo, Norway, edited by G. Angeletti, and A. Bjorseth, pp. 120–131. Oslo: Reidel Publ;

Wells, M. 2006. Log DOW: Key to understanding and regulating wastewater-derived contaminants. *Environ. Chem.* **3**:439–449.

Wu, C., W. Siems, and H. Hill. 2000. Secondary electrospray ionization ion mobility spectrometry/mass spectrometry of illicit drugs. *Anal. Chem.* **72**:396–403.

Zuccato, E., C. Chiabrando, S. Castiglioni, D. Calamari, R. Bagnati, S. Schiarea, and R. Fanelli. 2005. Cocaine in surface waters: A new evidence-based tool to monitor community drug abuse. *Environ. Health Perspect.* **4**:14.

Zuccato, E., S. Castiglioni, R. Bagnati, C. Chiabrando, P. Grassi, and R. Fanelli. 2008. Illicit drugs, a novel group of environmental contaminants. *Water Res.* **42**:961–968.

CHAPTER 10

MONITORING NONPRESCRIPTION DRUGS IN SURFACE WATER IN NEBRASKA (USA)

SHANNON L. BARTELT-HUNT and DANIEL D. SNOW

10.1 INTRODUCTION

Recent improvements in analytical detection methods for trace organic compounds have allowed for the quantification of numerous organic microcontaminants, including pharmaceuticals, hormones, and personal-care products in surface waters. Effluent from wastewater treatment plants (WWTPs) have been implicated as one of the primary sources of these compounds to surface water (Lee and Rasmussen, 2006; Glassmeyer et al., 2005; Miao et al., 2004). Although the presence of therapeutic pharmaceuticals in WWTP effluent has been well-established, there have been a limited number of studies investigating the occurrence of nonprescription pharmaceuticals, including illicit drugs in WWTP effluent. A small number of studies have reported the presence of illicit drugs in WWTP effluents at nanogram per liter levels, including methamphetamine, MDMA (ecstasy), tetrahydrocannabinol (THC), and cocaine and cocaine metabolites (Bartelt-Hunt et al., 2009; Zuccato et al., 2008; Bones et al., 2007; Zuccato et al., 2005; Jones-Lepp et al., 2004). There are also few published studies of the ecotoxicological impacts of chronic low-level exposures of therapeutic or illicit pharmaceuticals in aquatic systems (Pounds et al., 2008; Fent et al., 2006), although these chemicals may have effects at environmentally relevant concentrations (Raldua et al., 2008; Schreiber and Szewzyk, 2008). Because these compounds are biologically active, both ecotoxicological and human health impacts are of potential concern.

Although the occurrence and concentration of illicit and therapeutic pharmaceuticals are primarily documented using discrete sampling events, there are fewer data available indicating variability or time-weighted average (TWA) concentrations of these compounds in receiving waters downstream of WWTP outfalls. Traditional

Illicit Drugs in the Environment: Occurrence, Analysis, and Fate Using Mass Spectrometry, Edited by Sara Castiglioni, Ettore Zuccato, and Roberto Fanelli
Copyright © 2011 John Wiley & Sons, Inc.

water sampling approaches, such as grab and composite sampling, have been shown to be effective for documenting the occurrence of pharmaceuticals, but these sampling techniques only capture information at the time of sample collection, and may miss events, such as changes in the flow regime, chemical inputs, and/or the influence of precipitation (MacLeod et al., 2007). In addition, continuous on-line sampling methods may be prohibitively expensive. One device that has been developed for use in sampling trace organic compounds is the polar organic chemical integrative sampler (POCIS). This sampling device is designed to trap polar organic compounds from water. Its ease of use and apparent resistance to biofouling make it particularly attractive for determining TWA concentrations of organic compounds in water (Alvarez et al., 2004). POCIS samplers have been used previously for both qualitative and semiquantitative evaluation of pharmaceuticals, pesticides, and hormones in surface waters (Bartelt-Hunt et al., 2009; Sellin et al., 2009; Arditsoglou and Voutsa, 2008; Harman et al., 2008; Zhang et al., 2008; Alvarez et al., 2007; MacLeod et al., 2007; Jones-Lepp et al., 2004; Alvarez et al., 2004).

10.2 USE OF PASSIVE SAMPLERS FOR MONITORING OCCURRENCE AND FATE OF ILLICIT DRUGS IN SURFACE WATERS

Passive samplers have been used in environmental monitoring for over 30 years with an original focus on air monitoring applications. More recently, passive samplers have been developed to monitor pollutant concentrations in water, soil, and sediments. The need for using passive samplers for organic microcontaminant identification is primarily because of low levels of contamination present in environmental media and the high analyte recovery in passive samplers at low concentrations (Alvarez et al., 2004).

POCIS were designed to sequester hydrophilic compounds from water. It is comprised of a solid sequestration media inside a polyethersulfone (PES) membrane, which is held together by stainless steel compression rings. The sampler has three compartments: the water boundary layer, the diffusive membrane, and the receiving phase. The water boundary layer is the layer adjacent to the bulk water environment. The diffusive membrane allows specific contaminants from the water boundary layer to reach the receiving phase. The receiving phase is an infinite sink for the contaminants and maintains a concentration close to zero, which results in optimal mass transfer by diffusion. The only limitation for mass transfer is the actual surface area available for contaminant transfer. The commercially available divinylbenzene N-vinylpyrrolidone copolymer (Oasis HLBTM, Waters Corporation, Milford, MA) is the preferred sorbent media for pharmaceuticals because it typically provides analyte recovery rates exceeding 95% (Alvarez et al., 2004).

The process of compound accumulation on the sorbent media is a first-order reaction. First-order kinetic models include an integrative, curvilinear, and equilibrium partitioning phases. During the integrative phase, the sampler acts as an infinite sink for contaminants with linear uptake (Alvarez et al., 2004); the samplers have been demonstrated to remain in the integrative phase for exposures of at least 30 days (Alvarez et al., 2007).

In order to use the POCIS to estimate TWA concentrations of contaminants, an uptake rate (R_s) must be determined experimentally for compound of interest (Alvarez et al., 2004). The uptake rate can be determined as:

$$R_s = (D_w/L_w)A \qquad (10.1)$$

where D_w is the aqueous diffusion coefficient on the contaminant, L_w is the aqueous film layer thickness, and A is the available surface area. Once an uptake rate has been calculated, the TWA water concentration of the contaminant of interest can be calculated as:

$$C_w = C_s M_s / R_s t \qquad (10.2)$$

where C_w and C_s are the analyte concentration in water and sorbent, respectively; M_s is the mass of the sorbent, R_s is the uptake rate determined from equation above; and t is the exposure time.

POCIS have the advantage of being able to retain contaminants from the integrative phase, while still being able to acquire additional contaminants. They have the ability to handle large volumes of water over time with the addition of evaluating variations in contaminant concentration and flow rates (Alvarez et al., 2004). Although the POCIS has been used in numerous studies investigating the occurrence of organic wastewater contaminants (Bartelt-Hunt et al., 2009; Sellin et al., 2009; MacLeod et al., 2007; Togola and Budzinski, 2007; Alvarez et al., 2004; Jones-Lepp et al., 2004), it has a distinct limitation. Uptake rates have to be calculated for compounds of interest before POCIS can be used to estimate aqueous concentrations. Published sampling rates are available for only a small number of illicit or therapeutic pharmaceuticals (Bartelt-Hunt et al., 2009). In addition, calculated uptake rates have been demonstrated to be sensitive to a number of environmental factors including: membrane fouling, turbulence and its effect on contaminant mass-transfer across the filter membrane, salinity, and temperature (Togola and Budzinski, 2007).

10.3 DETERMINATION OF LABORATORY UPTAKE RATES FOR PHARMACEUTICALS

Uptake rates were measured in the laboratory by submerging a POCIS sampler containing Oasis HLB sorbent media in a 2-L beaker of ultrapure water spiked with the pharmaceuticals of interest (Table 10.1) at a concentration of 5000 ng/L under flowing conditions. The experiment was performed in triplicate. The water temperature was monitored at approximately 25°C. A negative control experiment was also performed in duplicate, which consisted of a beaker containing water and POCIS, but excluded any spiked compounds to assess the potential for contamination during the experiment. A positive control experiment was performed, consisting

TABLE 10.1 Compounds Evaluated and LC–MS Parameters

Compound	Use	CAS No.	Mol. Weight (g/mol)	Retention Time (min)	MRM	Collision Energy (eV)	Cone Voltage (V)	IDL (ng)
Nonprescription pharmaceuticals & personal care products								
Caffeine	Stimulant	58-08-2	194.19	11.94	195>138	18	32	0.33
1,7-Dimethylxanthine	caffeine metabolite	611-59-6	180.16	11.25	181>124	20	32	0.61
Cotinine	nicotine metabolite	486-56-6	176.22	10.30	177>78	20	35	0.28
D-Amphetamine	Stimulant	51-64-9	135.21	10.90	136>91	16	18	0.70
Methamphetamine	Stimulant	537-46-2	149.23	10.99	150>91	20	20	0.43
Internal standards								
Phenyl-^{13}C$_6$–sulfamethazine		57-68-1	284.1	11.95	285>124	25	30	
Δ-9-Methamphetamine		537-46-2	158.1	10.99	159>93	18	20	
^{13}C$_3$-caffeine		58-08-2	197.1	11.87	198>140	18	32	

of a 2-L beaker containing ultrapure water spiked with the same concentration of contaminants, but containing no POCIS. The purpose of the positive control was to monitor natural decay of the pharmaceuticals unrelated to POCIS uptake. The beakers were covered with foil and 100- mL water samples were obtained from each beaker at 0, 3, 7, 14, and 30 days. At the end of the 30-day exposure period, the POCIS was removed. All aqueous samples and the POCIS were stored at −20°C until analysis. The average flow rate in the beakers during the experiment was determined to be 5 cm/s.

10.3.1   Extraction Methodology

Handling and elution of POCIS followed procedures described previously (Bartelt-Hunt et al., 2009; Alvarez et al., 2004; Jones-Lepp et al., 2004). After the 30-day exposure, each individual POCIS device was removed from the beaker. The POCIS was disassembled and the two filter membranes carefully separated to expose the polymer sorbent. Approximately 20 mL of high-purity methanol was used to wash the sorbent from the membrane surface directly into silane-treated vials. Vials containing the methanol and sorbent slurry were held at −20°C until they could be processed for analysis. For the elution step, the methanol−sorbent slurry was poured into silane-treated glass gravity flow chromatography columns and target compounds eluted by adding an additional 50 mL of high-purity methanol through the resin column and drained by gravity into 120 mL evaporation tubes (RapidVAP, Labconco, Kansas City, MO). Approximately 1 ng of Δ-9-methamphetamine and $^{13}C_3$-caffeine internal standards were added to the eluate and used for quantification. Extracts were evaporated under nitrogen to approximately 1 mL and quantitatively transferred to autosampler vials for analysis by liquid chromatography−tandem mass spectrometry (LC−MS/MS). Standards and spiking solutions were prepared from stock solutions (5 μg/μL) in methanol. Calibration solutions (2, 5, 12.5, 25, and 50 pg/μL) for the POCIS extracts were prepared in 50:50 methanol:water. All standards and extracts were stored in amber vials at −20°C. Aqueous samples (25 mL) from the calibration experiments were spiked with internal standards, extracted, and analyzed using a Spark Holland Symbiosys Environ on-line extraction system with Oasis HLB cartridges. Calibration solutions for the water analysis were prepared in purified water, extracted, and analyzed by liquid chromatography−tandem mass spectrometry under identical conditions as water samples collected from the calibration experiments.

10.3.2   Liquid Chromatography−Tandem Mass Spectrometry

POCIS extracts and water samples were analyzed for the pharmaceuticals of interest as listed in Table 10.1. Standards and extracts were analyzed on a Quattro Micro triple quadrupole with a Waters 2695 high-pressure liquid chromatography (HPLC) and autosampler. Electrospray ionization in positive ion mode was used for detection of target compounds by multiple reaction monitoring (MRM) with argon collision gas. A Thermo (Bellefonte, PA) Betabasic-18 column (250 × 2.1 mm, 5 μm, 50°C) was used for separation at a flow rate of 0.2 ml/min with a gradient of methanol with 0.1% formic acid in water. Mass spectrometer operational parameters were optimized by infusing each compound separately (Table 10.1). The source conditions were: capillary 2.5 kV, extractor 2 V, RF lens 0.8 V, source temperature 90°C, desolvation temperature 400°C, cone gas flow at 30 L/h, and desolvation gas flow at 700 L/h. Compound retention times, ionization modes, and MRM transitions are listed in Table 10.1. A five-point internal standard calibration curve was used for quantification of each analyte. Δ-9-Methamphetamine was used as the internal standard for methamphetamine and phenyl-$^{13}C_6$-sulfamethazine and $^{13}C_3$-caffeine was used as the internal standard for all other target compounds. Eight replicate analysis of the lowest standard (2 pg/μL) was used to determine the signal-to-noise ratio and sensitivity of the instrumental method. Based on the standard deviation of the lowest standard (2 pg/μL), the estimated detection limits for most compounds are less than 1 pg/μL, corresponding to 1 ng recovered from the POCIS. Approximately 10 ng of each analyte were mixed with 20 mL of methanol and processed with the POCIS samples through empty liquid chromatography columns to check for loss during elution, evaporative concentration, and transfer to autosampler vials. Recovery of target compounds was calculated by dividing the measured amounts by the added quantity of each compound and averaged 123 ± 30% for all contaminants. Two laboratory reagent blanks were processed with the POCIS samples, with all compounds below instrument detection limits listed in Table 10.1.

10.3.3 Calculation of Uptake Rates

Uptake rates for each compound of interest were determined by applying Eq. (2) to data obtained in the laboratory calibration experiment. Uptake rates were calculated using the change in average concentration measured in replicate experiments between each set of sequential sampling points over the 30-day exposure period. Uptake rates calculated for each set of sampling points were averaged to produce an uptake rate descriptive of the overall exposure period (Table 10.2). Pharmaceutical concentrations in the control beakers at the end of the exposure period were typically greater than 85% of the initial concentration and, thus, nonspecific losses were considered negligible. Of the compounds of interest, previously published uptake rates are available only for methamphetamine. A value of 0.089 L/d was published by Jones-Lepp et al. (2004). Our measured value of 0.28 ± 0.10 L/d compares favorably considering that

TABLE 10.2 Experimentally Determined Uptake Rates

Compound	Experimental Uptake Rate Under Flowing Conditions (L/d)
Caffeine	0.20 ± 0.06
1,7-Dimethylxanthine	0.25 ± 0.11
Cotinine	0.12 ± 0.04
D-Amphetamine	0.16 ± 0.07
Methamphetamine	0.28 ± 0.10

variability of uptake rates within a factor of two to three is consistent with variability in contaminant concentrations observed in the field, based on continuous monitoring over an extended period (Togola and Budzinski, 2007).

10.4 OCCURRENCE OF ILLICIT DRUGS IN NEBRASKA SURFACE WATERS INFLUENCED BY WWTP EFFLUENT

In a previously published study (Bartelt-Hunt et al., 2009), we evaluated the occurrence and concentration of twenty pharmaceuticals (both illicit and therapeutic) and related metabolites in surface waters upstream and downstream of WWTP outfalls and in WWTP effluent in Nebraska. In this chapter, we discuss the results obtained for the five nonprescription pharmaceuticals evaluated (Table 10.1). The treatment plants monitored as part of this study serve populations ranging from 20,000 to approximately 420,000 and represent a variety of secondary treatment technologies, including trickling filter, activated sludge, trickling filters and activated sludge in parallel, and biological nutrients. The purpose of this study was (1) to evaluate instream TWA concentrations of pharmaceuticals in surface waters impacted by WWTP effluents and in WWTP effluent in Nebraska; (2) to evaluate the use of POCIS as a semiquantitative tool for assessing organic compound concentrations in surface waters; and (3) to document the presence of an illicit drug, methamphetamine, in WWTP effluents originating from rural municipalities.

10.4.1 Sampling Locations

Surface waters were sampled upstream and downstream of the WWTP discharge structure at Lincoln, Nebraska, Grand Island, Nebraska, and Columbus, Nebraska. In addition, samplers were installed downstream of the Hastings, Nebraska WWTP. Samples were also obtained from the effluent channel just prior to discharge in the Missouri River at the Omaha, Nebraska WWTP because of the inaccessibility of the receiving water body at this location. Additional information related to each WWTP including the community population, the average daily flow, and the secondary treatment technique employed at each facility may be found in Table 10.3.

10.4.2 POCIS and Sampling Methodology

POCIS, holders, and deployment canisters were obtained from Environmental Sampling Technologies (EST, Inc, St. Joseph, MO). Each stainless steel canister was fitted with three pharmaceutical POCIS filled with Oasis HLB sorbent (Waters Corporation, Milford, MA). Each POCIS had a surface area of 41 cm^2 and contained 200 mg of sorbent medium. At each sampling location, canisters were deployed for a 7-day exposure period. At the Grand Island site, the downstream POCIS was recovered 4 weeks after deployment, ostensibly because of vandalism. At sites where upstream and downstream samples were obtained, POCIS devices were simultaneously deployed at each location. All sites were sampled between August and November 2006,

TABLE 10.3   Wastewater Treatment Facilities Sampled in Nebraska[a]

Facility Location	Receiving Water Body	Sampling Locations	Community Population (Year 2006)	Secondary Treatment Technique	Average Daily Flow (mg/d)
Columbus, Nebraska	Loup River	Upstream and downstream of discharge	20,909	Activated sludge	3.4
Hastings, Nebraska	West Fork of the Big Blue River	Downstream of discharge	25,437	Trickling filter	3.6
Grand Island, Nebraska	Wood River	Upstream and downstream of discharge	44,546	Biological nutrients	12
Lincoln, Nebraska	Salt Creek	Upstream and downstream of discharge	241,167	Activated sludge and trickling filter	18
Omaha, Nebraska	Missouri River	Effluent	419,545	Trickling filter	27.2

[a]Bartelt-Hunt et al., 2009.

when base flow is a primary source of water in the receiving water bodies. The sampling period was also outside of the required seasonal disinfection period, so no disinfection processes were ongoing at the time of sampling. POCIS extraction and analysis were performed as described above and in Bartelt-Hunt et al. (2009).

Estimated aqueous concentrations of nonprescription pharmaceuticals (Table 10.1) recovered from POCIS samplers at each sampling location are presented in Table 10.4. Aqueous concentrations were calculated using Eq. (2) with experimentally determined uptake rates (Table 10.2).

Of the five nonprescription pharmaceuticals and metabolites, four were consistently detected at the sampling sites. D-Amphetamine was not detected in any of the POCIS residues. In general, the concentration of a given compound in the downstream sample was higher than in the upstream sample for each location where POCIS were deployed simultaneously upstream and downstream of the WWTP discharge structure. This result indicates that WWTP effluent is a significant source of pharmaceutical loading to the receiving waters, a finding in agreement with other studies, which have also demonstrated WWTP effluent to be a significant source of pharmaceutical loading to the environment (Batt et al., 2006; Kim et al., 2007; Lee and Rasmussen, 2006; Roberts and Thomas, 2006). The estimated concentration of these compounds ranged from 1.1 ng/L to approximately 295 ng/L in the upstream samples and 1.8 ng/L to approximately 1350 ng/L in the downstream samples. Higher concentrations of a particular compound in the upstream samples suggests that it may be persistent in the environment or that surface run-off sources are more significant than treated WWTP effluent during the sampling period. For example, caffeine was detected upstream of the Lincoln WWTP at a concentration of 295 ng/L, but the

TABLE 10.4 Pharmaceutical Mass Recovered in POCIS and Estimated Aqueous Concentrations of Pharmaceuticals Detected in Passive Samplers Upstream and Downstream of WWTP Discharges and in WWTP Effluent[a]

| | Grand Island, NE | | | | Columbus, NE | | | |
| | Upstream | | Downstream | | Upstream | | Downstream | |
	ng/POCIS	ng/L	ng/POCIS	ng/L	ng/POCIS	ng/L	ng/POCIS	ng/L
Caffeine	55 ± 14.4	39 ± 10	2455.8 ± 775.0	438 ± 138	36.0 ± 3.1	25.7 ± 2.2	31.1 ± 0.8	22.2 ± 0.6
1,7-Dimethylxanthine	4.3 ± 0.3	2.5 ± 0.2	213 ± 70.3	30.4 ± 10.0	4.4 ± 0.5	2.5 ± 0.3	5.2 ± 0.5	3.0 ± 0.3
D-Amphetamine	ND[a]	ND	ND	ND	ND	ND	ND	ND
Cotinine	7.5 ± 0.4	8.9 ± 0.5	19.2 ± 2.9	5.7 ± 0.9	8.1 ± 0.3	9.6 ± 0.1	9.3 ± 0.9	11.0 ± 0.3
Methamphetamine	2.2 ± 0.1	1.1 ± 0.1	40.7 ± 7.2	5.2 ± 0.9	2.1 ± 0.1	1.1 ± 0.1	3.6 ± 0.7	1.8 ± 0.4

| | Lincoln, NE | | | | Hastings, NE | | Omaha, NE | |
| | Upstream | | Downstream | | Downstream | | Effluent | |
	ng/POCIS	ng/L	ng/POCIS	ng/L	ng/POCIS	ng/L	ng/POCIS	ng/L
Caffeine	413.4 ± 562.0	295 ± 401	35.2 ± 4.8	25.1 ± 3.4	1872.9 ± 314.8	1337 ± 224.8	1023.0 ± 185.8	730 ± 132
1,7-Dimethylxanthine	6.6 ± 1.9	3.8 ± 1.1	39.2 ± 17.7	22.4 ± 10.1	409.1 ± 41.5	233.7 ± 23.7	254.9 ± 43.0	145.7 ± 24.6
D-Amphetamine	ND	ND	ND	ND	ND	ND	ND	ND
Cotinine	17.7 ± 8.0	21.1 ± 2.4	54.4 ± 26.8	64.8 ± 15.3	349.0 ± 109.9	415.4 ± 32.7	244.5 ± 61.4	291 ± 18
Methamphetamine	ND	ND	39.0 ± 7.3	19.9 ± 3.7	96.9 ± 20.2	49.4 ± 10.3	541.7 ± 121.2	276.4 ± 61.8

[a]Modified from Bartelt-Hunt et al., 2009.

[b]ND, not detected.   Au: b is not cited in table.

downstream concentration was 25.1 ng/L. Similarly, cotinine was detected at a concentration of 8.9 ng/L upstream of the Grand Island, NE WWTP, but the downstream concentration was 5.7 ng/L. It should be noted, however, that at other locations caffeine and cotinine were detected at higher concentrations in the downstream samples.

Trace levels of the illicit drug methamphetamine were detected at most sampling locations. The methamphetamine metabolite, D-amphetamine, was not detected in any of the POCIS. These data represent only the second report of methamphetamine in WWTP effluent and in streams impacted by WWTP effluent. Previously, Jones-Lepp et al. (2004) reported estimated methamphetamine concentrations of 1.3 and 0.8 ng/L from POCIS in WWTP effluent and in-stream samples, respectively. These samples were obtained at a WWTP located in Las Vegas, Nevada, a large metropolitan area. In this study, estimated methamphetamine concentrations in about one-half of the samples were comparable to those reported from Las Vegas, while estimated levels in the WWTP effluent and downstream samples from three sites were 276 ng/L (effluent) and 1.1 to 49.4 ng/L (downstream). The effluent sample was obtained from Omaha, NE, which has a population of 491,000, while the downstream samples were obtained from municipalities with populations ranging from 20,000 to 240,000. The differences between data obtained from these locations and previously reported results may reflect differences in drug use, wastewater treatment practices, and timing of sample collection. Trace concentrations of methamphetamine, an illicit pharmaceutical, was present in all but one sampling location at higher concentrations than have been previously reported for this compound.

10.5   FATE OF ILLICIT DRUGS IN SURFACE WATERS DOWNSTREAM OF WWTP DISCHARGES

To further investigate the behavior of nonprescription pharmaceuticals and personal care products in the environment, a second study was performed at two field sites: Salt Creek, downstream from the WWTP in Lincoln, Nebraska and the West Fork of the Big Blue River, downstream from the Hastings, Nebraska WWTP (Table 10.3). At each site, POCIS were placed at four locations: in the effluent prior to discharge, in the stream within the effluent mixing zone, at locations approximately 500, 1000, and 1500 m downstream from the effluent discharge. POCIS were housed in triplicate in a steel canister and the canisters were secured to avoid displacement during the exposure period. POCIS were deployed on May 5, 2009 and retrieved on May 28, 2009, for a 23-day exposure period. POCIS extraction and analysis were performed as described above and in Bartelt-Hunt et al. (2009).

Figure 10.1 shows the change in estimated concentrations of 1,7- dimethylxanthine, caffeine, and methamphetamine detected in receiving waters as a function of distance from the WWTP outfall. Cotinine and D-amphetamine were not detected in POCIS samplers at either site. Estimated concentrations of caffeine and 1,7- dimethylxanthine were higher in the receiving water downstream from the Hastings, Nebraska WWTP relative to the receiving water downstream of the Lincoln, Nebraska WWTP. In-stream estimated concentrations of methamphetamine were

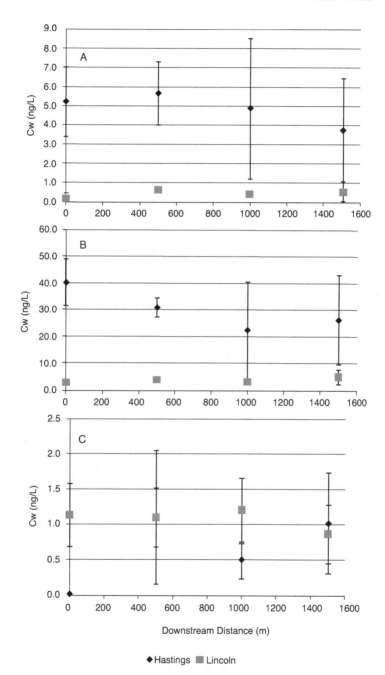

FIGURE 10.1 Estimated in-stream concentrations of 1,7-dimethylxanthine (A), caffeine (B), and methamphetamine (C) observed in POCIS samplers in receiving waters versus distance downstream from WWTP effluent at two locations in Nebraska.

approximately 1 ng/L. Data obtained during this experiment indicate that all three compounds were not significantly degraded over distances up to 1500 m from the WWTP outfall. Although to our knowledge, there has been no previous data investigating the fate or persistence of methamphetamine in receiving waters, our data for methamphetamine is consistent with previous studies indicating that methamphetamine is recalcitrant and is not removed during wastewater treatment (Loganathan et al., 2009).

REFERENCES

Alvarez, D. A., J. N. Huckins, J. D. Petty, T. Jones-Lepp, F. Stuer-Laridsen, D. T. Getting, J. P. Goddard, and A. Gravell, 2007. Tool for monitoring hydrophilic contaminants in water: polar organic chemical integrative sampler (POCIS). In Comprehensive Analytical Chemistry Edited by R. Greenwood, G. Mills, and B. Vrana, pp. 171–197. Amsterdam: Elsevier.

Alvarez, D. A., J. D. Petty, J. N. Huckins, T. L. Jones-Lepp, D. T. Getting, J. P. Goddard, and S. E. Manaham, 2004. Development of a passive, in situ, integrative sampler for hydrophilic organic contaminants in aquatic environments. *Environ. Toxicol. Chem.* **23**:1640–1648.

Arditsoglou, A., and D. Voutsa, 2008. Passive sampling of selected endocrine disrupting compounds using polar organic chemical integrative samplers. *Environ. Pollut.* **156**:316–324.

Batt, A. L., I. B. Bruce, D. S. Aga, 2006. Evaluating the vulnerability of surface waters to antibiotic contamination from varying wastewater treatment plant discharges. *Environ. Pollut.* **142**:295–302.

Bartelt-Hunt, S. L., D. D. Snow, T. Damon, J. Shockley, and K. Hoagland, 2009. The occurrence of illicit and therapeutic pharmaceuticals in wastewater effluent and surface waters in Nebraska. *Environ. Pollut.* **157**:786–791.

Bones, J., K. V. Thomas, and B. Paull, 2007. Using environmental analytical data to estimate levels of community consumption of illicit drugs and abused pharmaceuticals. *J. Environ. Monit.* **9**:701–707.

Fent, K., A. A. Weston, and D. Caminada, 2006. Ecotoxicology of human pharmaceuticals. *Aquatic Toxicol.* **76**:122–159.

Glassmeyer, S., E. T. Furlong, D. W. Kolpin, J. D. Cahill, S. D. Zaugg, S. L. Werner, M. T. Meyer, and D. D. Kryak, 2005. Transport of chemical and microbial contaminants from known wastewater discharges: potential for use as indicators of human fecal contamination. *Environ. Sci. Technol.* **39**:5157–5169.

Harman, C., O. Boyum, K. N. Tollefsen, K. Thomas, and M. Grung, 2008. Uptake of some selected aquatic pollutants in semipermeable membrane devices (SPMDs) and the polar organic chemical integrative sampler (POCIS). *J. Environ. Monit.* **10**:239–247.

Jones-Lepp, T. L., D. A. Alvarez, J. D. Petty, and J. N. Huckins, 2004. Polar organic chemical integrative sampling and liquid chromatography-electrospray/ion-trap mass spectrometry for assessing selected prescription and illicit drugs in treated sewage effluents. *Arch. Environ. Contamination Toxicol.* **47**:427–439.

Kim, S. D., J. Cho, I. S. Kim, B. J. Vanderford, and S. A. Snyder, 2007. Occurrence and removal of pharmaceuticals and endocrine disruptors in South Korean surface, drinking, and waste waters. *Water Res.* **41**:1013–1021.

Lee, C. J., and T. J. Rasmussen, 2006. Occurrence of organic wastewater compounds in effluent-dominated streams in northeastern Kansas. *Sci. Total Environ.* **371**:258–269.

Loganathan, B., M. Phillips, H. Mowery, and T. L. Jones-Lepp, 2009. Contamination profiles and mass loadings of macrolide antibiotics and illicit drugs from a small urban wastewater treatment plant. *Chemosphere* **75**:70–77.

MacLeod, S. L., E. L. McClure, and C. S. Wong, 2007. Laboratory calibration and field deployment of the polar organic chemical integrative sampler for pharmaceuticals and personal care products in wastewater and surface water. *Environ. Toxicol. Chem.* **26**:2517–2529.

Miao, X., F. Bishay, M. Chen, and C. D. Metcalfe, 2004. Occurrence of antimicrobials in the final effluents of wastewater treatment plants in Canada. *Environ. Sci. Technol.* **38**:3533–3541.

Pounds, N., S. Maclean, M. L. Webley, D. Pascoe, and T. Hutchinson, 2008. Acute and chronic effects of ibuprofen in the mollusc Planorbis carinatus (Gastropoda: Planorbidae). *Ecotoxicol. Environ. Safety* **70**:47–52.

Raldua, D., M. Andre, and P. Babin, 2008. Clofibrate and gemfibrozil induce an embryonic malabsorption syndrome in zebrafish. *Toxicol. Appl. Pharmacol.* **228**:301–14.

Roberts, P. H., and K. V. Thomas, 2006. The occurrence of selected pharmaceuticals in wastewater effluent and surface waters of the lower Tyne catchment. *Sci. Total Environ.* **356**:143–153.

Schreiber, F., and U. Szewzyk, 2008. Environmentally-relevant concentrations of pharmaceuticals influence the initial adhesion of bacteria. *Aquatic Toxicol.* **87**:227–233.

Sellin, M.K., D. D. Snow, D. L. Akerly, and A. S. Kolok. 2009. Estrogenic compounds downstream from three small cities in eastern Nebraska: Occurrence and biological effect. *J. Am. Water Res. Assoc.* **45**:1–8.

Togola, A., and H. Budzinski, 2007. development of polar organic integrative samplers for analysis of pharmaceuticals in aquatic systems. *Anal. Chem.* **79**:6734–6741.

Zhang, Z., A. Hibberd, and J. L. Zhao, 2008. Analysis of emerging contaminants in sewage effluent and river water: comparison between spot and passive sampling. *Anal. Chim. Acta* **607**:37–44.

Zuccato, E., S. Castiglioni, R. Bagnati, C. Chiabrando, P. Grassi, R. Fanelli, 2008. Illicit drugs, a novel group of environmental contaminants. *Water Res.* **42**:961–968.

Zuccato, E., C. Chiabrando, S. Castiglioni, D. Calamari, R. Bagnati, S. Schiarea, and R. Fanelli, 2005. Cocaine in surface waters: a new evidence-based tool to monitor community drug use. *Environ. Health* **4**:14–21.

MASS SPECTROMETRIC ANALYSIS OF ILLICIT DRUGS IN THE ENVIRONMENT

Illicit Drugs in Drinking Water

CHAPTER 11

PRESENCE AND REMOVAL OF ILLICIT DRUGS IN CONVENTIONAL DRINKING WATER TREATMENT PLANTS

MARIA HUERTA-FONTELA, MARIA TERESA GALCERAN, and FRANCESC VENTURA

11.1 ILLICIT DRUGS IN DRINKING WATER

The presence of chemical pollutants released into the aquatic environment as a consequence of anthropogenic activities has been recognized as an important issue in environmental chemistry. Water analysis provides exhaustive and reliable information about human habits and activities. Recently, illicit drugs were identified as a new unexpected group of water contaminants, thus confirming this "water squealer" ability. To date, these compounds have been found in several water matrices around the world and the concern regarding their presence in these resources is exponentially growing, because of their potent psychoactive properties and their unknown effects on the aquatic environment. The presence and detection of illicit drugs in water resources is related to the combination of two main effects: their excretion as unchanged or slightly transformed metabolites allows easy identification of these compounds; and the high consumption rates reported for them contribute to relatively high concentration measurable levels in aquifers.

These compounds directly enter the water system by means of sewage waters, as was demonstrated in the first study reporting the presence of cocaine and its major metabolite, benzoylecgonine, in Italian wastewaters (Zuccato et al., 2005). Since then, several studies dealing with the analysis of illicit drugs in wastewaters showed that some of them survive common wastewater treatments. Compounds such as cocaine, amphetamine, ecstasy, or cannabinoids are found, at concentrations in

Illicit Drugs in the Environment: Occurrence, Analysis, and Fate Using Mass Spectrometry, Edited by Sara Castiglioni, Ettore Zuccato, and Roberto Fanelli
Copyright © 2011 John Wiley & Sons, Inc.

the nanogram per liter range in wastewater treatment plant (WWTP) effluents in Italy, Germany, Spain, Belgium, Ireland, United Kingdom, and the United States (Jones-Lepp et al., 2004; Castiglioni et al., 2006; Hummel et al., 2006; Huerta-Fontela et al., 2007, 2008a, b; Bones et al., 2007; Boleda et al, 2007; Kasprzyk-Hordern et al., 2008a, 2009; Gheorghe et al., 2008; Biljsma et al., 2009; van Nuijs et al., 2009; Zuccato and Castiglioni, 2009; Petrovic et al., 2009; Postigo et al., 2010). For instance, Castiglioni et al. (2006) measured cocaine and benzoylecgonine at 11 and 29 ng/L, respectively, in effluent samples, and Hummel et al. (2006) found benzoylecgonine (45 ng/L) in treated waters in Germany. Cocaine was also found in treated waters in Spain at concentrations ranging from 900 (Postigo et al., 2010) to 1400 ng/L (Huerta-Fontela et al., 2008a). The poor removal of illicit drugs in WWTPs has a direct effect on the rivers, lakes, or seas where WWTPs discharge their effluents. Several studies have also reported relatively high levels of several different illicit drugs, such as cocaine, amphetamines, ecstasy, LSD, or cannabinoids, in surface waters. For instance, the main carboxylic metabolite of tetrahydrocannabinol (THC-COOH) was detected at concentrations ranging from 0.5 to 7 ng/L in Italian rivers (Zuccato et al., 2008), at 1 ng/L in one United Kingdom river (Bones et al., 2007), and at 15 ng/L in Spain (Boleda et al., 2009). However, the "water trip" does, not end at this stage, because surface waters are often used for drinking water production. The removal efficiencies in drinking water treatment plants (DWTPs) and the production of new and additional disinfection by-products (DBP) during treatments affect the quality of tap water directly, as they are responsible for both the complete elimination of illicit drugs from the water cycle and the introduction of new by-products that can reach human beings.

Up to now, few studies on the presence of these compounds in final treated waters (Hummel et al., 2006) have been conducted and still fewer studies dealing with the removal or behavior of these compounds through treatments have been found (Huerta-Fontela et al., 2008b; Zuccato et al., 2008; Boleda et al., 2009). In 2006, Hummel et al. studied the presence of several psychoactive drugs in wastewater, river water, and final treated tap water in Germany. Benzoylecgonine, the main metabolite of cocaine, and opiates (codeine and morphine) were among the compounds investigated and were not found at levels above their limit of quantification (LOQ) (2−10 ng/L) in the two tap water samples analyzed. In the same study, some other compounds, such as carbamazepine or primidone, were detected in the final drinking waters. The presence and behavior of stimulatory drugs, such as MDMA (ecstasy), cocaine, and LSD in a DWTP, was evaluated by Huerta-Fontela in another study in 2008. The survey, conducted over six consecutive months, included the evaluation of the effectiveness of both the conventional and the advanced treatments used in the facility investigated. Benzoylecgonine was found to be the most persistent compound, found in 22 out of 24 treated water samples at concentrations ranging from 3 to 130 ng/L. At the same time, Boleda et al. (2009) published a similar report evaluating the removal of depressor drugs, such as opiates and cannabinoids, in the same drinking water facility. The results showed the presence of methadone and its main metabolite (EDDP) in the final treated water at concentrations of 20 and 60 ng/L, respectively. To the best of our knowledge, only these two surveys evaluating the presence of illicit drugs in DWTPs exist. More research is needed in order to improve our knowledge of the behavior and presence of these psychoactive drugs in tap waters.

11.2   ILLICIT DRUGS THROUGH DRINKING WATER TREATMENT

The treatments used to keep tap water safe for human consumption and to improve its quality by removing undesired contaminants are similar around the world. The most common processes are physical treatments, such as filtration, flocculation, and sedimentation. There are other processes involving ion exchange (that is, reverse osmosis), adsorption (activated carbon), and chemical disinfection, such as chlorination or ozonation. Facilities select the treatments or the combinations of treatment depending on several different factors, such as the characteristics of the raw water, the specific problems affecting the water to be treated and, obviously, the costs of each treatment. Generally, the effectiveness of drinking water treatments in eliminating or degrading contaminants depends on the physicochemical properties of the compounds, the quality of the source water, and the treatment processes applied.

Table 11.1 summarizes the physicochemical properties of illicit drugs found most frequently in water. Log K_{ow} is a valuable parameter used to predict the partition behavior of the selected compounds. This parameter together with water solubility can give us a fair idea of the presence and distribution of illicit drugs in the aquatic environment. For instance, fentanyl with a log K_{ow} of 4 and a water solubility of 200 mg/L is more prone to be found in the sludge phase. However, the information given by log K_{ow} should be carefully evaluated, since the removal efficiency of drinking water treatments can be affected by additional parameters, such as interactions with raw water materials, hydrogen bonding, or cation exchange that may modify predicted behavior.

Generally, the physicochemical properties of the compounds, together with the structural information, suggest their behavior during conventional and well-known drinking water treatments. The knowledge of the chemical properties of these substances and the experience obtained studying the behavior of other contaminants can help us predict the behavior of illicit drugs during DWT. Table 11.2 correlates certain reactive groups present in the structure of illicit drugs with their expected behavior in drinking water treatments. However, it must be considered that external effects, such as raw water composition (namely, presence of bromide, ammonia, temperature) or treatment operational status (that is, carbon filter aging, ozone generation efficiency) will play an important and, in some cases, determinant role.

11.2.1   Filtration, Coagulation, and Sedimentation

The water treatment in most frequent use since the early twentieth century is a combination of the following steps: coagulation, flocculation—sedimentation, and filtration. Filtration is used by many water treatment facilities to remove particles from the water. It clarifies water and enhances the effectiveness of disinfection. Coagulation and flocculation are also common DWTP unit operations that may decrease the concentrations of potential drinking water contaminants in the source water.

TABLE 11.1 Structures of the Main Illicit Drugs Investigated and Their Physicochemical Properties

Classification	Compound		CAS number	MW	pKa	Log K_{ow}	Solubility (mg/L; water)
Amphetamine–type stimulants	Amphetamine	R1=H	30-62-9	135	10.1	1.76	28,000[b]
	METH	R1=CH3	537-46-2	179	9.7	1.64	22,400[b]
	MDA	R1=H	4764-17-4	149	9.9	2.07	13,300[b]
		R2=O-CH2-O-					
	MDMA	R1=CH3	42542-10-9	193	9.9[b]	2.28[b]	2700[b]
		R2=O-CH2-O-					
	MDEA	R1=CH2-CH3	82801-81-8	207	10.3[b]	2.77[b]	6100[b]
		R2=O-CH2-O-					
Alkaloids	Cocaine	R1=CH3	50-36-2	303	8.6	2.30	1800
		R2=CH3					
	Norcocaine	R1=H	18717-72-1	289	9.4	1.72	4900
		R2=CH3					
	Cocaethylene	R1=CH3	529-38-4	317	8.8	2.66	—
		R2=CH2-CH3					
	BE	R1=CH3	519-09-5	289	—	<0	88,000[b]
		R2=H					
	Norbenzoylecgonine	R1=H	519-09-5	289	4.3	<0	—
		R2=H					
Cannabinoids	THC	R1=CH3	1972-08-3	314	10.6	7.6	2800
	THC-COOH	R1=COOH	56354-06-4	445	4.1	6.57	—
	THC-OH	R1=CH2-OH	36557-05-8	330	3.7	5.33	—
Opiates	Heroin	R1=CH3 R2=COCH3 R3= COCH3	561-27-3	369	7.95	1.58	600
	Morphine	R1=CH3 R2=H R3=H	57-27-2	285	8.21	0.89	149
	Morphine-3β-gluc.[a]	R1=CH3 R2=Gluc R3=H	—	461	8.2/2.9	<0	—
	Normorphine	R1=H R2=H R3=H	466-97-7	271	9.8	0.5	256,000
	Codeine	R1=CH3 R2=CH3 R3=H	76-57-3	299	8.2	1.19	9000
	Norcodeine	R1=CH3 R2=CH3 R3=H	467-15-02	285	9.23	0.69	3,920,000
	6-AM	R1=CH3 R2=H R3=COCH3	2784-73-8	327	9.6	1.72	—
	Methadone		76-99-3	309	8.94	3.93	48.5

Name	Structure	CAS	MW			
EDDP		66729-78-0	278	—	2.97	—
Fentanyl		437-38-7	336	7.3	4.1	200
Ketamine		6740-88-1	237	7.5	3.12[b]	2800
LSD		50-37-3	323	7.8	2.95	370[b]
PCP		77-10-01	243	8.7	1.90	11,000

Recreational drugs

[a]Gluc.:Glucuronide.
[b]Estimated values.

TABLE 11.2 Predicted Reactivity of Illicit Drugs through Conventional Drinking Water Treatments

Compound	Treatments			
	Chlorination	Chlorine dioxide	Ozonation	Carbon adsorption
Amphetamine	—NH$_2$	—	—	—
METH	—NHR	—	—	—
MDA	—NH$_2$	—	—	—
MDEA	—NHR	—	—	—
MDMA	—NHR	—	—	—
Cocaine	—[a]	NR$_3$	—	—
Norcocaine	—NHR	—	—	—
Cocaethylene	—	NR$_3$	—	—
BE	—	NR$_3$	—	—
Norbenzoylecgonine	—NHR	—	—	—
Ketamine	—NHR	—	—	Log $K_{ow} > 3$
LSD	—NHR	NR$_3$	—	—
PCP	—	NR$_3$	—	—
Heroin	Ph-R(donor)	NR$_3$	C=C	—
Morphine	—OH	NR$_3$	C=C	—
Normorphine	—NHR>—OH	—	C=C	—
Codeine	—	NR$_3$	C=C	—
Norcodeine	—NHR>—OH	NR$_3$	C=C	—
6-AM	—OH	NR$_3$	C=C	—
Methadone	—	NR$_3$	—	Log $K_{ow} > 3$
EDDP	—	NR$_3$	C=C	Log $K_{ow} > 3$
Fentanyl	—	NR$_3$	—	Log $K_{ow} > 3$
THC	—OH	—	C=C	Log $K_{ow} > 3$
THC-COOH	—OH	—	C=C	Log $K_{ow} > 3$
THC-OH	—OH	—	C=C	Log $K_{ow} > 3$

[a]—, No reactive sites.

No studies solely examining the behavior of illicit drugs during these treatments have been published. However, studies performed with pharmaceutical compounds report low or negligible removal rates during the processes of clarification and sand filtration. For instance, Nakada et al. (2007) reported that compounds with log $K_{ow} < 3$ showed removal percentages lower than 50% after sand filtration, suggesting that hydrophobicity can be a limiting factor during the filtration process. In addition, Stackelberg et al. (2007) reported reductions in the concentration of compounds with log K_{ow} (<3) lower than 15% during clarification, confirming that this is not a primary route for degradation and removal of organic compounds.

Given the physicochemical properties of illicit drugs, it can be extrapolated that filtration, coagulation, or clarification are not the appropriate steps for removing these compounds. An exception can be found for cannabinoids, fentanyl, LSD, ketamine, and methadone, which may show higher removals because of their relatively high hydrophobicity.

11.2.2 Oxidation Process

Because of their suitability for disinfection and oxidation (e.g., taste and odor control, micropollutant removal), chemical oxidants are commonly used in water treatment processes. In drinking water treatment systems, the oxidants commonly used are chlorine, chlorine dioxide, chloramines, and ozone.

11.2.2.1 Chlorination   For nearly a century, chlorine has been the most widely used oxidant for the disinfection of drinking water. Specifically, 98% of drinking water facilities use chlorine-based disinfectants among their treatments. Chlorine is often used at one or two points during treatment as a pretreatment for a primary disinfection and as a post-treatment to ensure a residual concentration of chlorine through the distribution system, thus avoiding microbial regrowth.

Oxidations, additions to unsaturated bonds, and electrophilic substitutions at nucleophilic sites are the most usual reaction pathways of chlorine. These reactions generally introduce small modifications in the reacting compound, which may cause one of the major drawbacks of chlorination treatments: the formation of chlorinated organic compounds as disinfection by-products. The different structures and moieties of illicit drugs suggest different behavior with chlorine treatment. Most of the drugs studied and found in water systems contain amino groups. It is known that chlorine has high reactivity with amino moieties by means of a water-assisted mechanism.

Generally, the rate constants of amines decrease as follows: $-NH_2 > -NHR > -NR_2$, with the rate constant of tertiary amines at least two orders of magnitude lower than that of primary or secondary amines. As suggested by Abia et al. (1998), the reaction occurs between the amine moiety, which undergoes a proton transfer to water, and the HOCl molecule, leading to the formation of chloramines (Reaction 11.1). Therefore, compounds such as amphetamine-type stimulants (ATS), ketamine, LSD (and its metabolites) or some drug metabolites like normorphine, norcodeine, norcocaine, or norbenzoylecgonine should show high reactivity and removal efficiency with chlorine treatment.

Reaction 11.1   Suggested chlorination for amphetamine according to mechanism suggested by Abbia et al. (1998).

For other illicit drugs, with tertiary amino moieties or nonamino compounds, other reaction pathways for evaluating the efficiency of chlorine treatment to remove them need to be considered. For these compounds, chlorine reactions will be controlled by the affinity of chlorine with its oxygenated moieties, since unsaturated bonds or alkylic chains are poorly reactive to chlorine without catalyzation. Oxygenated moieties vary in reactivity, depending on their chemical nature and on the adjacent substituents. Acidic moieties are stable in the presence of chlorine, while aldehyde or ketone moieties show limited reactivity, which depends on the electron-donor or withdrawing properties of the substituents linked to the carbonyl group. Compounds such as cocaine, benzoylecgonine, norbenzoylecgonine, fentanyl, and methadone have carbonyl moieties linked to electron-donor groups. As these groups reduce the acidity of the α-carbon group, which is the one reactive to chlorine, low or negligible reactivity can be expected.

For instance, at real-scale experiments, cocaine and benzoylecgonine (Huerta-Fontela, 2008b) were hardly removed after chlorine treatment (13% and 9%, respectively) and methadone also had poor percentages (54%) (Boleda et al., 2009). For opiates, such as morphine, codeine, 6-acetylmorphine, and cannabinoids, the main chlorine reactivity can be expected through the alcohol moieties. In general, primary and secondary alcohol moieties react with chlorine by an oxidation step leading to the formation of carbonyl moieties (Caserio et al., 1960). In real-scale experiments, the reactivity of chlorine with some of these compounds has already been shown (Boleda et al., 2009). For heroin, an electrophilic substitution reaction with chlorine can be expected, because of the presence of electron-donor groups attached to the aromatic ring (Reaction 11.2).

Reaction 11.2 Suggested chlorination for heroin according to an electrophilic aromatic substitution mechanism.

Finally, compounds such as PCP or EDDP do not exhibit active moieties to chlorine treatment, suggesting negligible removals through this treatment, as observed for EDDP at real-scale experiments.

11.2.2.2 Chlorine Dioxide Oxidation

Chlorine dioxide (ClO_2) is an oxidant used for disinfection of high-quality water, such as groundwater or treated surface water. Its reactivity is because of free chlorine and ozone. Chemically, ClO_2 is a very selective oxidant for specific functional groups of organic compounds, such as phenolic moieties, tertiary amino groups, or thiol groups (Hoigne and Bader, 1994), leading to its reduction to chlorite through a one-electron transfer reaction.

Chlorine dioxide cleaves one of the N-C bonds of the tertiary amines (Buxton, 1998), leading to dealkylation to a secondary amine and then to their respective aldehydes. Therefore, illicit drugs containing tertiary amines, such as cocaine, benzoylecgonine, methadone or some opiates, among others, could be removed by chlorine dioxide treatment.

11.2.2.3 Ozonation Ozone is used in water treatment as both disinfectant and oxidant and reacts with a large number of organic and inorganic compounds. Ozone decays rapidly after a few minutes and its decomposition leads to a major secondary oxidant, the hydroxyl radical. Oxidation then occurs by means of the molecular ozone (O_3) or of the OH radical (OH·) (Hoigne et al., 1983). The OH radical, a powerful, but less selective, oxidant, reacts with high second-order rate constants with most organic compounds, but these reactions are less efficient, because a large fraction of the radical is scavenged by the water matrix (Buxton et al., 1998). Additional oxidation processes are the advanced oxidation processes (AOPs), which use OH radicals as the main oxidants. Of the oxidation processes, ozone reacts more readily with organic compounds than chlorine does and is particularly reactive toward unsaturated double bonds or moieties with electron-donating properties (Weyer and Riley, 2001; Xu et al., 2002). However, it must be taken into account that the ozone reaction rate with a specific contaminant is moiety-specific and pH-dependent. For instance, neutral tertiary amine moieties show high reactivity with ozone, but, at low pH, the protonation of the amines prevents their oxidation. In addition to pH, other water properties, such as temperature or organic carbon content, affect the reactivity of a compound with ozone (Huber et al., 2005; Dodd and Huang, 2007; Zwiener et al., 2007).

A general and easy approach to ozone reactivity is its reaction with unsaturated moieties, which occurs through the well-established Crigee mechanism. An electrophilic addition occurs over the double bond, leading to the cleavage of the bond and to the formation of aldehydes or ketones (Bailey and Ferrell, 1978; Dowideit et al., 1998). The presence of accessible double bonds in the structures of opiates and cannabinoids suggests a rapid reaction of these compounds with ozone (Reaction 11.3).

Reaction 11.3 Suggested chlorination for THC according to Crigee mechanism.

Real-scale experiments confirmed the reactivity of ozone with some opiates such as morphine, codeine, and norcodeine (Boleda et al., 2009). EDDP behaved differently: despite the presence of a double bond in its structure, it survived this treatment. This can be explained by the imine−enamine tautomeric equilibrium, which decreases the electron density of the C−C double bond, thus lowering its reactivity toward ozone.

Reactions of ozone with amine moieties are more complex, as previously mentioned. In general, amines, and especially tertiary amines, react with ozone by means of electron transfer reactions, leading to the formation of secondary amines and aldehydes (Muñoz and von Sonntag 2000). According to this chemical reactivity, amino-containing drugs, such as ATS, cocainelike compounds, methadone or LSD, among others, should react efficiently under ozone treatment. However, real-scale experiments demonstrated that MDMA, cocaine, benzoylecgonine, and norbenzoylecgonine were resistant to treatment, with removals lower than 30% (Huerta-Fontela et al., 2008b). The lower reaction rates of these compounds, together with short ozone contact times, explain this unexpected behavior.

11.2.2.4 Activated Carbon Sorption
Activated carbon, such as powdered activated carbon (PAC) or granular activated carbon (GAC), is a commonly used adsorbent for the removal of several water contaminants. Adsorption on activated carbon depends on the intrinsic properties of the activated carbon sorbent (surface area and charge, pore size distribution, and oxygen content) and on its solute properties (shape, size, charge, and hydrophobicity).

Removal of organic compounds is mainly controlled by hydrophobic interactions. GAC and PAC have proved efficient in removing hydrophobic compounds (log $K_{OW} > 3-5$), despite the competitive effects with natural organic matter, which tends to decrease the adsorption rates. Therefore, activated carbon treatment can be effective in removing illicit drugs, such as fentanyl or cannabinoids with octanol−water constants higher than 4.

However, since these compounds have relatively high hydrophobic properties, a percentage of these compounds could be found in the sludge rather than in the water phase.

Real-scale experiments showed the efficiency of GAC treatment to completely remove fentanyl (Boleda et al., 2009), as predicted. However, methadone and EDDP, which have an octanol−water constant higher than 3, were poorly removed, and codeine and norcodeine were completely removed, despite their poor hydrophobic properties (Boleda et al., 2009). These results suggested that the sorption of different substances on activated carbon is often regulated by factors other than their own physicochemical properties.

11.3 DISINFECTION BY-PRODUCTS

An additional problem related to conventional treatments of drinking water should also be considered, namely, the presence in final treated waters of new disinfection by-products (DBPs) produced during the treatment processes.

The formation of DBPs has been widely described since they were first reported in 1974 in two publications (Rook, 1974; Bellar et al., 1974) that demonstrated that conventional drinking water treatments could not completely guarantee tap water quality and safety. Rook (1974) reported that chloroform and other trihalomethanes (THMs) were found at higher concentrations in chlorinated drinking water than in raw waters and proposed that THMs were produced by reactions between chlorine and natural organic matter (NOM) in water. The discovery of these new water contaminants and their potential carcinogenicity for humans required the introduction of new regulations to control the presence of THMs and haloacetic acids (HAAs) in tap waters.

The establishment of regulatory levels and guidelines enhanced the study of the effects of conventional drinking water treatments in terms of DBPs. Therefore, in the almost 40 years since THMs were identified, several new DBPs have been reported in the literature (Richardson, 1998). This number is increasing exponentially with the discovery of new and potentially reactive contaminants in raw waters. Illicit drugs can be included in this group, since it has been demonstrated that they can enter DWTPs and may be reactive to some conventional treatments. For instance, several illicit drugs, such as amphetamine-type stimulants (ATSs) with amine moieties, have structures suited to generating N-nitrosodimethylamine (NDMA)-related compounds (Reaction 11.4).

Reaction 11.4   Suggested formation of NDMA and N-alkylamines from ATSs.

It has been demonstrated that the presence of primary or secondary amines, together with reactive chlorine, such as monochloramine or chlorine and ammonia, leads to the formation of NDMA. This compound belongs to the chemical class of the N-nitrosoamines and is a potential human carcinogen more active than trihalomethanes and chlorination DBPs (Mitch et al., 2003). In 1989, NDMA was first detected at high concentrations (up to 0.3 μg/L) in treated drinking water from Ohsweken (Ontario, Canada). This finding led to the survey of 145 Ontario DWTPs (Jobb et al., 1995; Graham et al., 1995), but the concentrations of NDMA detected in the treated water were lower than 5 ng/L.

In addition, several illicit drugs have aromatic rings that react readily with oxidants, such as chlorine and ozone, generating by-products, as has been demonstrated for other anthropogenic water contaminants, such as pesticides, pharmaceuticals, or personal care products. For instance, chlorination of amphetamine-type stimulants, led to the formation of two chlorination ring products, 4-chloro-1,3-benzodioxole and 1-chloro-3,4-dihidroxybenzene, in laboratory experiments (Huerta-Fontela et al., 2009; Huerta-Fontela et al., 2010). However, the presence of nitrogen-containing moieties, linked to aromatic rings, enhances the halogenation of the nitrogen moiety, leading to ring-chlorinated products, like those found for sulfamethoxazole. This reaction pathway was also that of some illicit drugs, such as ketamine or LSD (Dodd and Huang, 2004).

11.4   ANALYSIS OF DRINKING WATER SAMPLES: A REAL CASE STUDY

As previously mentioned, only two studies (Huerta-Fontela et al., 2008b; Boleda et al., 2009) in the literature deal with the analysis of illicit drugs in drinking-water treatment. Both these studies reported the analysis of stimulatory and depressor drugs and suggested that some of these compounds survive the treatment. In view of these results, our research group decided to perform an additional investigation with several illicit drugs and metabolites that, because of their structure, might survive the treatments. The study ran for an entire month, in order to compare and to evaluate removals under similar water and treatment conditions, thus excluding possible seasonal variations in drug concentrations. The results confirmed those obtained in our previous study and revealed that some other drug metabolites are also present in final treated waters. The results are summarized in Fig. 11.1.

Raw water collected at the intake of a DWTP facility in April 2008 showed the frequent occurrence of several illicit drugs at relatively high concentrations. However, ketamine, LSD, PCP, MDEA, heroin, 6-acetylmorphine, morphine glucuronide, and THC were not found in the samples collected. Amphetamine-type stimulants were found in raw waters at concentrations ranging from 2 to 155 ng/L. During the first treatment, which consisted of prechlorination, clarification, and sand filtration, all ATS (except MDMA), were efficiently removed at percentages higher than 99% (Fig. 11.1). These high removal percentages were related to the high reactivity of chlorine and chlorine dioxide with the primary and secondary amines described by other authors (Westerhoff et al., 2005; Chamberlain and Adams, 2006). Nevertheless, the removal of MDMA after the first treatments was only 23%, even though this compound contains a secondary amino moiety. A postchlorination step was, therefore, necessary to completely remove it from raw water. Opioid concentrations ranged from 4 to 14 ng/L at the DWTP intake and their removal in treatment was similar to that observed for ATS. Morphine, codeine, and norcodeine were completely removed during the first chlorination step, whereas normorphine persisted through the first chlorination treatment and was totally removed at the last chlorination step. THC was not detected in raw waters, but its main metabolites (THC-COOH and THC-OH) were detected at 8 and 5 ng/L, respectively. Both the compounds were completely removed during the first chlorination step, probably because of the specific reactivity of chlorine with carbon bond moieties present in their structures. For cocainelike compounds, a different behavior was observed, depending on the different molecules and their concentrations in raw water. For instance, norcocaine and cocaethylene were detected at low concentrations in raw water (0.6 and 0.2 ng/L, respectively) and were completely removed (>99%) during the first treatment step. However, cocaine, benzoylecgonine, and norbenzoylecgonine were poorly removed during this first chlorination step (9−14%) and unexpectedly were not removed by ozonation, either, even though they had aromatic rings, which have been described as reacting with ozone (Hoigne and Bader, 1983). The poor removal rates observed for cocaine, benzoylecgonine, and norbenzoylecgonine (24%, 43% and 20%, respectively) were probably related to the presence of electron-withdrawing groups in the aromatic rings, which led to lower reactivity with ozone (Ikehata et al., 2006). The water treated with ozone was successively passed through granulated activated carbon filters (GAC), which efficiently removed cocaine (>99%). Benzoylecgonine and norbenzoylecgonine were also removed at high percentages during this step (72% and 75%, respectively), and also survived postchlorination. They were, therefore,

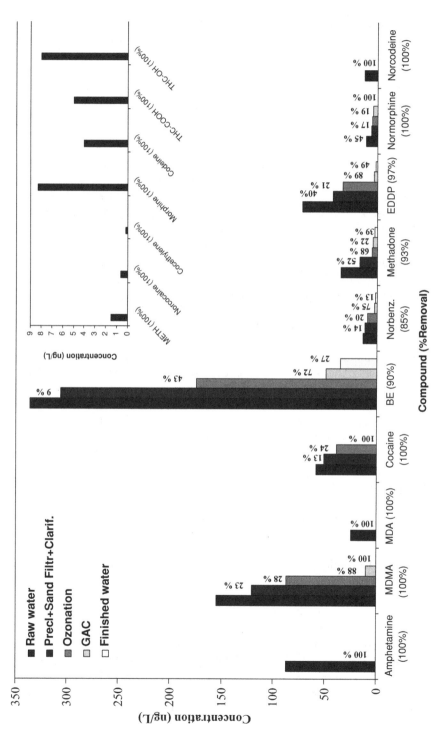

FIGURE 11.1 Concentrations of the studied drugs through a drinking water treatment consisting of prechlorination, clarification, sand filtration, ozonation, GAC filtration, and postchlorination (April 2008, $n = 3$). The elimination percentages of the compounds after each treatment are included on each bar and the total removal percentages are included in parenthesis.

measured in treated waters at concentrations of 36 ng/L for benzoylecgonine and 2 ng/L for norbenzoylecgonine. Methadone and EDDP also lasted throughout drinking water treatment. Neither the post- nor prechlorination treatments were effective in removing these compounds, probably because of the lack of reactive sites to chlorine. Ozonation and GAC filtration did not remove these compounds completely, found in the final treated waters at concentrations of 2.6 ng/L for methadone and 1.9 ng/L for EDDP.

These results showed that drinking water treatment was effective in removing 13 out of 17 target compounds found in raw water.

Since 2009, the process of this DWTP has been modified. After sand filtration and dilution with groundwater, the filtered water is split in two treatment lines. Around 30% of the water is subjected to ultrafiltration/reverse osmosis (UF/RO), whereas the remaining 70% is treated by ozonation and GAC filtration. The permeated water and the carbon-filtered waters are then blended, chlorinated, and distributed.

The behavior of illicit drugs in the new UF/RO process was studied. Cocaine, benzoylecgonine, EDDP, methadone, codeine, norcodeine, and ketamine, which were the only compounds identified after sand filtration, were fully eliminated. Thus, UF/RO treatment was able to remove all the target illicit drugs.

In conclusion, the results obtained reveal that methadone, EDDP, benzoylecgonine, and norbenzoylecgonine are not fully removed during conventional drinking water treatment and can be measured in the treated water. It was also demonstrated that improvements in water treatment technology, such as the introduction of new drinking water steps (that is, UF/RO), enable these recalcitrant illicit drugs to be removed and "better water quality" to be achieved. However, it must be taken into account that few DWTPs implement these new advanced technologies. The presence of these psychoactive substances in tap waters must be checked, since no studies evaluating the potential effects of these drugs on humans have yet been performed.

11.5 CONCLUDING REMARKS

The presence of illicit drugs in the water cycle, far from being a temporary or a national problem, has been a constant problem of global concern since it was first reported by researchers from the Mario Negri Institute. The illicit drugs used by humans reach the environment by means of sewage waters, which, in some cases, are discharged into rivers, lakes, or seas and may then be used for drinking water production. At this point, the crucial question is whether these drugs can return to human beings by means of our tap waters.

This chapter has conducted a review of the drinking water treatment technologies most commonly used to eliminate illicit drugs. Treatment efficiencies to remove these compounds were evaluated theoretically through examination of the chemical properties of each compound. The estimated behavior was found to be in line with the results obtained during several surveys carried out at real-scale DWTPs. Chlorine oxidation was shown to be effective in removing ATSs, cannabinoids, and several opiates; GAC treatment removed fentanyl and cocaine; while ozonation removed morphine completely. Only four compounds survived these treatments, benzoylecgonine, norbenzoylecgonine, methadone, and EDDP, which have also been found in finished waters. However, the complete removal of these substances was achieved by an UF/RO treatment. The low concentrations measured in finished drinking water suggest low exposure of humans to these substances. More research is needed on the topic, because the potentially adverse effects on human health and the environment that might be posed by exposure to chronic, subtherapeutic levels of these substances and/or their transformation products are still unknown. In addition, the possible formation of new disinfection by-products, such as chlorinated precursors or NDMA, should also be taken into account when evaluating the impact of illicit drugs' entrance into conventional drinking water treatment plants.

REFERENCES

Abia, L., X. L. Armesto, M. L. Canle, M. V. García, and J. A. Santaballa. 1998 Oxidation of aliphatic amines by aqueous chlorine. *Tetrahedron* **54**:521–530.

Bailey P. S., and T. M. Ferrell. 1978. Mechanism of ozonolysis. A more flexible stereochemical concept. *J. Am. Chem. Soc.* **100**:899–905.

Bellar, T. A., J. J. Lichtenberg, and R. C. Kroner. 1974. The occurrence of organohalides in chlorinated drinking waters. *J. Am. Waterworks Assoc.* **66**:703–706.

Bijlsma, L., J. V. Sancho, E. Pitarch, M. Ibañez, F. Hernandez. 2009. Simultaneous ultra-high-pressure liquid chromatography—tandem mass spectrometry determination of amphetamine and amphetamine-like stimulants, cocaine and its metabolites, and a cannabis metabolite in surface water and urban wastewater. *J. Chromatogr. A* **1216**:3078–3089.

Boleda, M. R., M. T. Galceran, and F. Ventura. 2007. Trace determination of cannabinoids and opiates in wastewater and surface waters by ultra-performance liquid chromatography tandem mass spectrometry. *J. Chromatogr. A* **1175**:38–48.

Boleda, M. R., M. T. Galceran, and F. Ventura. 2009. Monitoring of opiates, cannabinoids and their metabolites in wastewater, surface water and finished water in Catalonia, Spain. *Water Res.* **43**:1126–1136.

Bones, J., K. V. Thomas, and B. Brett. 2007. Using environmental analytical data to estimate levels of community consumption of illicit drugs and abused pharmaceuticals. *J. Environ. Monitor.* **9**:701–707.

Buxton G. V. 1998. The Montreal protocol on substances that deplete the ozone layer. *Assessment report of the technology and economic assessment panel. UNEP.*

Chamberlain, E., and C. Adams 2006. Oxidation of sulfonamides, macrolides, and carbadox with free chlorine and monochloramine. *Water Res.* **40**:2517.

Caserio M. C., W. H. Graham, and J. D. Roberts. 1960. Small-ring compounds—XXIX. A reinvestigation of the solvolysis of cyclopropylcarbinyl chloride in aqueous ethanol. Isomerization of cyclopropylcarbinol. *Tetrahedron* **11**:171–182.

Castiglioni, S., E. Zuccato, E. Crisci, C. Chiabrando, P. Grassi, and R. Fanelli. 2006. Identification and measurement of illicit drugs and their metabolites in urban wastewater by liquid chromatography-tandem mass spectrometry. *Anal. Chem.* **78**:8421–8429.

Dodd, M. C., and C. H. Huang. 2004. Transformation of the antibacterial agent sulfamethoxazole in reactions with chlorine: kinetics, mechanisms, and pathways. *Environ. Sci. Technol.* **38**:5607–5615.

Dodd, M. C., and C. H. Huang. 2007. Interactions of fluoroquinolone antibacterial agents with aqueous chlorine: reaction kinetics, mechanisms, and transformation pathways. *Water Res.* **41**:647–656.

Dowideit P., and C. von Sonntag. 1998. Reaction of ozone with ethene and its methyl- and chlorine-substituted derivatives in aqueous solution. *Environ. Sci. Technol.* **32**:1112–1119.

Gheorghe. A., B. van Nuijs, L. Pecceu, P. G. Bervoets, P. G. Jorens, and R. Blust R. 2008. Analysis of cocaine and its principal metabolites in waste and surface water using solid-phase extraction and liquid chromatography–ion trap tandem mass spectrometry. *Anal. Bioanal. Chem.* **391**:1309–1319.

Graham, J. E., O. Meresz, G. J. Farquahar, and S. A. Andrews 1995. Factors affecting NDMA formation during drinking water treatment. In *Proceedings of American Water Works Assocociation Water Quality Technology Conference*, Denver, CO.

Hoigne, J., and H. Bader. 1983. Constants of reactions of ozone with organic and inorganic compounds in water. *Water Res.* **17**:185–194.

Hoigne, J., and H. Bader. 1994. Kinetics of reactions of chlorine dioxide (OClO) in water–I. Rate constants for inorganic and organic compounds. *Water Res.* **28**:45–52.

Huber, M. M., S. Korhonen, T. A. Ternes, and U. von Gunten. 2005. Oxidation of pharmaceuticals during water treatment with chlorine dioxide. *Water Res.* **39**:3607–3615.

Huerta-Fontela M., and F. Ventura. 2010, Illicit drugs: Disinfection-by-products. Alliance Project He0607. Internal Report.

Huerta-Fontela, M., M. T. Galceran, F. Ventura. 2007 Ultraperformance liquid chromatography–tandem mass spectrometry analysis of stimulatory drugs of abuse in wastewater and surface waters. *Anal. Chem.* **79**:3821–3829.

Huerta-Fontela, M., M. T. Galceran, J. Martin, and F. Ventura. 2008a. Occurrence of psychoactive stimulatory drugs in wastewaters in north-eastern Spain. *Sci. Total Environ.* **397**:31–40.

Huerta-Fontela, M., M. T. Galceran, and F. Ventura. 2008b. Stimulatory drugs of abuse in surface waters and their removal in a conventional drinking water treatment plant. *Environ. Sci. Technol.* **42**:6809–6816.

Huerta-Fontela, M., M. T. Galceran, and F. Ventura. 2009. Illicit drugs through water treatment: generation of new disinfection-by-products. In *Spanish Society Chromatography Conference*, Donostia Spain.

Hummel, D., D. Loeffler, G. Fink, and T. A. Ternes. 2006. Simultaneous determination of psychoactive drugs and their metabolites in aqueous matrices by liquid chromatography mass spectrometry. *Environ. Sci. Technol.* **40**:7321–7328.

Ikehata, K., N. J. Naghashkar, and M. G. El-Din. 2006. Degradation of aqueous pharmaceuticals by ozonation and advanced oxidation processes: a review ozone. *Sci. Eng.* **28**:353–364.

Jobb, D. B., R. B. Hunsinger, O. Meresz, and V. Y. Taguchi. 1995. Removal of N-nitrosodimethylamine. In *Proceedings of American Water Works Association Water Quality Technology Conference*, Denver, CO.

Jones-Lepp T. L., D. A. Alvarez. J. D. Petty, and J. N. Huckins. 2004. Organic chemical integrative sampling and liquid chromatography electrospray/ion-trap mass spectrometry for assessing selected prescription and illicit drugs in treated sewage effluents. *Arch. Environ. Contamination Toxicol.* **47**:427–439.

Kasprzyk-Hordern, B., R. M. Dinsdale, and A. J. Guwy. 2008. The occurrence of pharmaceuticals, personal care products, endocrine disruptors and illicit drugs in surface water in South Wales, UK. *Anal. Bioanal. Chem.* **391**:1293–1308.

Kasprzyk-Hordern B., R. M. Dinsdale, and A. J. Guwy. 2009. The removal of pharmaceuticals, personal care products, endocrine disruptors and illicit drugs during wastewater treatment and its impact on the quality of receiving. *Water Res.* **42**:3498–3518.

Mitch, W. A., J. O. Sharp, R. R. Trussell, D. L. Alvarez-Cohen, Sedlak, D. 2003. N-nitrosodimethylamine (NDMA) as a drinking water contaminant: a review. *Environ. Eng. Sci.* **20**:389–404.

Muñoz F., and C. von Sonntag. 2000. The reactions of ozone with tertiary amines including the complexing agents nitrilotriacetic acid (NTA) and ethylenediaminetetraacetic acid (EDTA) in aqueous solution. *Chem. Soc. Perkin Trans.* **2**:2029–2041.

Nakada, N., H. Shinohara, A. Murata, K. Kiri, S. Managaki, N. Sato, and H. Takada. 2007. Removal of selected pharmaceuticals and personal care products (PPCPs) and endocrine-disrupting chemicals (EDCs) during sand filtration and ozonation at a municipal sewage treatment plant. *Water Res.* **41**:4373–4382.

Petrovic, M., M. J. Lopez de Alda, S. Diaz-Cruz, C. Postigo, J. Radjenovic, M. Gros, and D. Barceló. 2009. Fate and removal of pharmaceuticals and illicit drugs in conventional and membrane bioreactor wastewater treatment plants and by riverbank filtration. *Philos. Trans. R Soc. A* **367**:3979–4003.

Postigo, C., M. J. López de Alda, and D. Barceló. 2010. Drugs of abuse and their metabolites in the Ebro River basin: occurrence in sewage and surface water, sewage treatment plants removal efficiency, and collective drug usage estimation. *Environ. Int.* **36**:75–84.

Richardson, S. 1998. The Encyclopedia of Environmental Analysis and Remediation, p. 1398. Hoboken, New Jersey: John Wiley & Sons.

Rook, J. J. 1974. Formation of haloforms during chlorination of natural waters. *Water Treatment Exam.* **23**:234–240.

Rosenblatt, D. H., L. A. Hull, D. C. De Luca, G. T. Davis, R. C. Weglein, and H. K. R. Williams. 1967. Oxidations of amines. II. substituent effects in chlorine dioxide oxidations. *J. Am. Chem. Soc.* **89**:1158–1165.

Stackelberg, P. E., J. Gibs, E. T. Furlong, M. T. Meyer, S. D. Zaugg, and L. Lippincott. 2007. Efficiency of conventional drinking-water-treatment processes in removal of pharmaceuticals and other organic compounds. *Sci. Total Environ.* **377**:255–263.

van Nuijs, A. L. N., L. Pecceu, N. Theunis, C. Dubois, P. G. Charlier, L. Jorens, R. Bervoets, H. Blust, A. Neels, and A. Covaci. 2009. Cocaine and metabolites in waste and surface water across. *Belgium Environ. Pollut.* **157**:123–129.

Westerhoff, P., Y. Yoon, S. Snyder, and E. Wert. 2005. Fate of endocrine-disruptor, pharmaceutical, and personal care product chemicals during simulated drinking. *Water Treatment Processes Environ. Sci. Technol.* **36**:6649–6659.

Weyer, P., and D. Riley. 2001. Endocrine disruptors and pharmaceuticals in drinking water. *Am. Water Works Assoc.*

Xu P, Janex M-L, Savoye P, Cockx A, Lazarova V. 2002. Determination of nicotine and cotinine in human plasma by liquid chromatography-tandem mass spectrometry with atmospheric-pressure chemical ionization interface. *Water Res* 2002;**4**:1043–1055.

Zuccato E, Castiglioni S. 2009. Illicit drugs in the environment. *Phil Trans R Soc A* 2009;**367**:3965–3978.

Zuccato, E., C. Chiabrando, S. Castiglioni, D. Calamari, R. Bagnati, S. Schiarea, and R. Fanelli. 2005. Cocaine in surface waters: a new evidence-based tool to monitor community drug abuse. *Environ. Health A Global Access Sci Source* **4**:1–7.

Zuccato, E., S. Castiglioni, R. Bagnati, C. Chiabrando, P. Grassi, and R. Fanelli. 2008. Illicit drugs, a novel group of environmental contaminants. *Water Res.* **42**:961–968.

Zwiener, C., S. D. Richardson, D. M. DeMarini, T. Grummt, T. Glauner, and F. H. Frimmel. 2007. Occurrence and analysis of pharmaceuticals and their transformation products in drinking water treatment. *Environ. Sci. Technol.* **41**:363–371.

CHAPTER 12

ANALYSIS OF ILLICIT DRUGS IN WATER USING DIRECT-INJECTION LIQUID CHROMATOGRAPHY-TANDEM MASS SPECTROMETRY

REBECCA A. TRENHOLM and SHANE A. SNYDER

12.1 INTRODUCTION

Liquid chromatography—tandem mass spectrometry (LC—MS/MS) is becoming a powerful tool in the identification and quantitation of illicit drugs (Rook et al., 2005; Concheiro et al., 2007; Maralikova and Weinmann, 2004; Apollonio et al., 2006; Scheidweiler and Huestis, 2004). LC—MS/MS has many advantages over more traditional method, such as GC, GC—MS and LC—MS. Typically, compounds do not require derivatization prior to instrumental analysis and larger volumes can be injected. With advances in instrumentation and methodology, better sensitivity can be achieved, with reporting limits at or less than nanogram per liter concentrations (Huerta-Fontela et al., 2007; Barcelo et al., 2008). Because of the utmost importance of correct identification, LC—MS/MS can add the benefits of selectivity with additional confirmation techniques, such as multiple-reaction monitoring (MRM), in which two to three MS transitions are commonly monitored for each compound. Also simultaneous product ion scans can be collected, where each compound is monitored by MRM while collecting a product ion scan over a larger mass (m/z) range using a triple-quadrupole linear ion-trap hybrid (Herrin et al., 2005; de Castro et al., 2007; Apollonio et al., 2006).

Typically, in order to achieve low detection limits (nanogram per liter) some sample preparation may be required, such as sample clean-up and/or sample extraction and concentration. Depending on the matrix, sample clean-up may be vital to remove interferences from complex matrices (that is, wastewater, urine, blood, and hair), that could result in chromatography interferences or matrix suppression during

Illicit Drugs in the Environment: Occurrence, Analysis, and Fate Using Mass Spectrometry, Edited by Sara Castiglioni, Ettore Zuccato, and Roberto Fanelli
Copyright © 2011 John Wiley & Sons, Inc.

LC−MS/MS analysis (Concheiro et al., 2006; Cheze et al., 2007; Johansen and Bhatia, 2007; Kronstrand et al., 2004; Scheidweiler and Huestis, 2004), especially if the sample has been further extracted and concentrated. Solid-phase extraction (SPE), on-line SPE, and liquid−liquid extractions are commonly used methods for illicit drugs in water or liquid matrices (de Castro et al., 2007; Barcelo et al., 2008; van Nuijs et al., 2009; Deventer et al., 2006; Concheiro et al., 2007). However, with the improvements in technology and more sensitive instrumentation, some analytes can be measured directly with little or no sample preparation, known as direct injection (Chiaia et al., 2008). A direct-injection method is described here, which was used for screening of illicit drugs in wastewater effluent and for monitoring of amphetamines in wastewater effluent, surface water, and finished drinking water.

12.2 EXPERIMENTAL STUDIES

12.2.1 Compound selection

This method was developed for five illicit drugs and nine of their major metabolites. Selected compounds (Table 12.1) include: methamphetamine and its metabolite amphetamine; cocaine and its metabolites ecgonine, ecgonine methyl ester, benzoylecgonine, and norcocaine; MDMA (ecstasy) and its metabolite MDA; heroin and its metabolites 6-acetylmorphine and morphine; and Δ^9-THC and its metabolite 11-hydroxy-Δ^9-THC.

12.2.2 Standards and Sample Collection

All standards and reagents used were of the highest purity commercially available. Concentrated stocks were prepared in methanol and stored at −20°C. From these concentrated stocks, calibration standards were prepared by appropriate dilutions in reagent water and stored at 4°C. Water samples were collected in amber glass bottles, which contained 1 g/L sodium azide for preservation and 50 mg/L ascorbic acid to quench any residual oxidant; samples were stored at 4°C until analysis. Grab samples were collected from a single tertiary treatment wastewater treatment plant (WWTP) effluent and also from the combined wastewater effluent of three tertiary treatment WWTPs, depending on the analyses. Samples were analyzed using direct injection; therefore, no additional sample preparation or concentration was required.

12.2.3 Instrumentation

Analyses were performed by liquid chromatography−tandem mass spectrometry (LC−MS/MS) using a CTC autosampler, an Agilent 1100 LC binary pump, and an ABSCIEX 4000 QTRAP triple-quadrupole linear ion-trap hybrid mass spectrometer (Foster City, CA). All analytes were monitored using positive-electrospray ionization (ESI) mode and multiple-reaction monitoring (MRM). One metabolite of Δ^9-THC (11-nor-9-carboxy-Δ^9-THC) does not sufficiently ionize in ESI positive and was not

TABLE 12.1 List of Illicit Drugs and Their Structures, with Their Corresponding Metabolites (Indented) and Molecular Weights

Compound	Molecular weight	Structure
Methamphetamine	149	
Amphetamine	*135*	
Cocaine	303	
Ecgonine	*185*	
Ecgonine methyl ester	*199*	
Benzoylecgonine	*289*	
Norcocaine	*289*	
MDMA (Ecstasy)	193	
MDA	*179*	
Heroin	369	
6-Acetylmorphine	*327*	
Morphine	*285*	
Δ^9-THC	314	
11-Hydroxy-Δ^9-THC	*330*	

included in this method; however, it was able to ionize in ESI negative. A 100-μL sample loop (sample volume) was used for each injection. Separation was performed on a 150 × 4.6 mm Luna C18(2) with a 5-μm particle size (Phenomenex). A 0.1% formic acid in reagent water solution (A) and methanol (B) gradient was used for LC mobile phases at a constant 0.8 mL/min flow rate.

Table 12.2 lists the MRM transition ions (*m/z*) for each compound, along with their corresponding declustering potentials (DP), collision energies (CE), and exit

TABLE 12.2 Method Reporting Limits (MRL), MRM Transitions, Declustering Potentials (DP), Collision Energies (CE) and Exit Potentials (CXP) for Target Compounds

Compound	MRL (ng/L)	Q1 (m/z)	Q3 (m/z)	DP (V)	CE (V)	CXP (V)
Methamphetamine	5	150	119	41	15	20
Amphetamine	5	136	119	36	13	20
Cocaine	10	304	182	46	29	18
Ecgonine	25	186	168	61	27	26
Ecgonine methyl ester	25	200	82	61	35	14
Benzoylecgonine	10	290	168	71	75	14
Norcocaine	10	290	168	51	23	10
MDMA	10	194	163	41	19	28
MDA	10	180	163	31	15	8
Heroin	25	370	165	71	73	28
6-Acetylmorphine	10	328	165	76	53	8
Morphine	25	286	152	81	87	8
Δ^9-THC	100	315	193	56	31	34
11-Hydroxy-Δ^9-THC	100	331	313	41	21	22

potentials (CXP). Since the method does not require any extraction or sample preparation procedures, method reporting limits (MRL) were based on the lowest instrument sensitivity with a signal-to-noise ratio greater than 10. MRLs for each compound are also listed in Table 12.2. Calibration curves ranged from 5 to 500 ng/L.

12.2.4 Amphetamines in Water Samples

Amphetamine and methamphetamine were monitored in wastewater effluent samples, surface (source) water, and finished drinking water. The wastewater effluent is discharged into a larger body of water, which also serves as the source water for a drinking water treatment plant (DWTP), although there is mixing and dilution prior to the source water intake. The first sample tested was collected directly from tertiary-treated effluent leaving a wastewater treatment plant (WWTP), which employs UV disinfection at the end of the treatment process. This wastewater effluent then enters a wash system, which is comprised of a combination of tertiary treated effluent from two additional WWTPs and a small amount of urban and storm water run-off. A second sample was collected after the combined flow of all three WWTPs, just before the wash system converges into a larger lake/river system. Each of the three WWTPs uses slightly different treatment processes for disinfection. While the WWTP mentioned previously uses UV, a second WWTP uses chlorine, and the last uses chloramines for disinfection.

Samples were then collected at the source, or raw, water intake and finished drinking water from a DWTP, which utilizes the lake/river system for its source water. At the DWTP, the raw water is first treated using ozonation and then enters a multistage filtration system, followed by chlorination. Finished drinking water samples were collected just as the water exits the DWTP and prior to entering the distribution system.

Matrix spikes were performed on each water matrix to ensure that analyte recoveries, and the resulting data, were not compromised because of processes such as adsorption or matrix suppression, which is commonly observed using LC−MS/MS for complex matrices (Vanderford et al., 2003; Trenholm et al., 2008). Each matrix was spiked with 100 ng/L of amphetamine and methamphetamine. In addition, isotopically labeled analogs for amphetamine and methamphetamine (d_5-amphetamine and d_5-methamphetamine) were used during analyses for isotope dilution quantitation, which has been shown to be an effective approach to correct for matrix suppression (Vanderford and Snyder, 2006). Each compound was monitored using a primary MRM transition, along with additional confirmation transitions to ensure that the analytes were correctly identified. Amphetamine was monitored using the MRM listed in Table 12.2 for quantitation, as well as MRM transitions 136–65 m/z and 136–91 m/z for confirmation. Methamphetamine was monitored with two additional confirmation transitions as well (150–65 m/z and 150–91 m/z).

12.2.5 Screening of Illicit Drugs in Wastewater Effluent

Grab samples of tertiary treated wastewater effluent was analyzed for the entire list of illicit drugs listed in Table 12.1 to show how LC−MS/MS can be a useful tool for selective target analyte screening in a complex matrix. Although the LC−MS/MS method was essentially the same as used for the monitoring of amphetamine and methamphetamine in wastewater, the screening method contained MRMs (including a second confirmation transition) for all 14 compounds. Compounds were analyzed according to the instruments parameters described previously. Because of the selectivity of MRM, chromatographic separation becomes less important. Compound identification does not rely heavily on analyte retention time since each compound has a unique fragmentation pattern and precursor-to-product ion instrument parameters (that is, declustering potentials and collision energies). This can allow for faster LC programs and sample analyses.

12.3 RESULTS

12.3.1 Amphetamines in Water

Methamphetamine was detected in the tertiary treated wastewater effluent at 48 ng/L, but amphetamine was below the reporting limit (Table 12.3). Both methamphetamine and amphetamine were detected at higher concentrations in the combined wastewater effluent from the three WWTPs. This suggests that the contribution of the analytes to the wash system is higher for one or both of the other two plants. This could be because of the fact that the treatment processes employed by each treatment plant differ on their removal efficiency for these compounds or the flow of wastewater effluent entering the wash system is greater and, therefore, contributes more volume

TABLE 12.3 Occurrence of Methamphetamine and Amphetamine in Various Water Matrices (ng/L) and Matrix Spike Recoveries

Compound	WWTP 1 Tertiary Effluent	Combined Effluent[a]	Source Water DWTP 1[a]	Finished Water DWTP 1[a]	Wastewater Effluent Spike Recovery (%)	Source Water Spike Recovery (%)	Finished Water Spike Recovery (%)
Methamphetamine	**48**	**94**	<5.0	<5.0	109	87	110
Amphetamine	<5.0	**27**	<5.0	<5.0	97	95	97

[a]Average value of duplicate samples.

and analyte mass. Both compounds were less than their reporting limit in the source and finished drinking water suggesting that there is a substantial amount of dilution or removal before entering the drinking water treatment plant. Spike recoveries were between 87% and 110% for all matrices indicating that the use of isotope dilution corrected for any compound loss because of matrix suppression.

12.3.2 Screening of Illicit Drugs

Using LC–MS/MS with direct injection, wastewater effluent was screened for illicit drugs. Figure 12.1 shows an overlapping chromatogram of illicit drugs detected in a representative wastewater effluent sample. Six compounds can be qualitatively identified by their unique MRM transitions using an extracted-ion chromatogram (XIC) even though they all elute from the LC column very close to each other. Compounds detected in the wastewater effluent include methamphetamine, amphetamine, cocaine, benzoylecgonine, MDMA (ecstasy), and MDA. Initial screening analyses by LC–MS/MS can be a useful method for identifying illicit drugs in complex matrices for further quantitative monitoring.

12.3.3 Advances in LC-MS/MS Instrumentation for Compound Confirmation

There are a few circumstances where retention time and MRM cannot thoroughly determine an analyte correctly. One example is with the identification between benzoylecgonine and norcocaine. Without adequate chromatographic separation, it can be difficult to correctly identify these compounds, since they share some of the major MRM transitions (that is, 290–168 m/z) for each compound. New LC–MS/MS instrumentation techniques include scanning modes that will allow for the quantitation of target compounds using MRM, while simultaneously collecting additional confirmation information. Using a triple-quadrupole linear-ion trap hybrid LC–MS/MS system (4000 QTRAP, ABSCIEX), an enhanced product ion scan (EPI) can be collected along with the usual MRM method. For the example of benzoylecgonine and norcocaine, the peaks coelute close together and share many of the same MRM

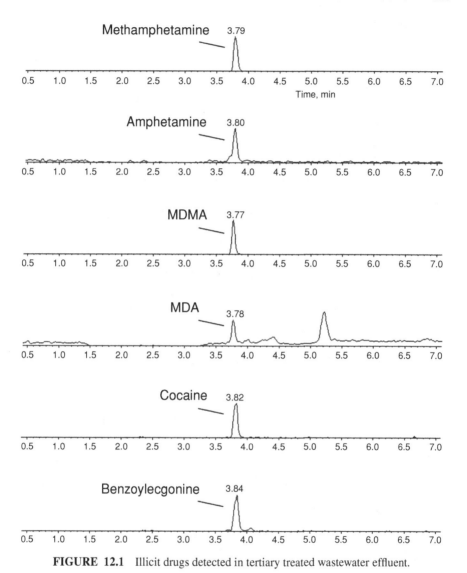

FIGURE 12.1 Illicit drugs detected in tertiary treated wastewater effluent.

transitions (Figure 12.2), which could result in misidentification and false positives. However, it is possible to correctly identify each analyte based on their EPI scans, as compared with a known standard, as there are clear differences between the two compounds. Although both compounds have the same precursor ion (290 m/z) and similar product ions (68, 77, 105, and 168 m/z), the ratios and intensities of their product ions differ. Norcocaine (Fig. 12.2b) also has a significant product ion at 136 m/z, that benzoylecgonine (Fig. 12.2a) does not have. This new technique can be a helpful tool in further confirming the identity of an analyte.

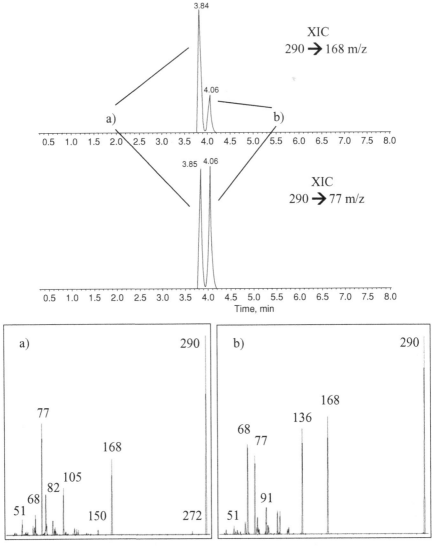

FIGURE 12.2 XIC chromatograms of quantitation (290–168 *m/z*) and confirmation (290–77) MRM transitions and product ion scans for (a) benzoylecgonine and (b) norcocaine.

12.4 CONCLUSIONS

Liquid chromatography−tandem mass spectrometry (LC−MS/MS) is emerging as the preferred method for the identification and quantitation of illicit drugs with many advantages over previous methods (GC and GC−MS). With increases in technology and instrumentation, samples can be measured directly with little or no sample preparation, such as clean up or extraction and concentrations. Using direct injection, water matrices (wastewater effluents and surface waters) can be monitored and screened for a variety of illicit drugs. LC−MS/MS with direct injection is also a fast and reliable method for monitoring the removal of illicit drugs through wastewater and finished drinking water treatment plants. Additional analyte confirmation can be obtained with new techniques using hybrid mass spectrometry, which can greatly minimize interferences from coeluting compounds and complex matrices, and which could result in false positives or incorrect quantitation. With LC−MS/MS, illicit drugs can be measured in a variety of matrices at or below nanogram per liter levels.

REFERENCES

Apollonio, L. G., I. R. Whittall, D. J. Pianca, J. M. Kyd, and W. A. Maher. 2006. Product ion mass spectra of amphetamine-type substances, designer analogues, and ketamine using ultra-performance liquid chromatography/tandem mass spectrometry. *Rapid Commun. Mass Spectrom.* **20**:2259–2264.

Barcelo, D, C. Postigo, and M. J. L. de Alda 2008. Fully automated determination in the low nanogram per liter level of different classes of drugs of abuse in sewage water by on-line solid-phase extraction liquid chromatography electrospray tandem mass spectrometry. *Anal. Chem.* **80**:3123–3134.

Cheze, M, M. Deveaux, C, Martin, M. Lhermitte, and G. Pepin, 2007. Simultaneous analysis of six amphetamines and analogues in hair, blood and urine by LC-ESI-MS/MS. Application to the determination of MDMA after low ecstasy intake. *Forensic Sci Int.* **170**:100–104.

Chiaia, A. C., C. Banta-Green, and J. Field, 2008. Eliminating solid phase extraction with large-volume injection LC/MS/MS: analysis of illicit and legal drugs and human urine indicators in U.S. wastewaters. *Environ. Sci. Technol.* **42**:8841–8848.

Concheiro, M., A. de Castro, O. Quintela, M. Lopez-Rivadulla, A. Cruz. 2006. Determination of drugs of abuse and their metabolites in human plasma by liquid chromatography-mass spectrometry. An application to 156 road fatalities. *J. Chromatogr. B Anal. Technol. Biomed. Life Sci.* **832**:81–89.

Concheiro, M., S. M. Simoes, O. Quintela, A. de Castro, M. J. Dias, A. Cruz, and M. Lopez-Rivadulla. 2007. Fast LC-MS/MS method for the determination of amphetamine, methamphetamine, MDA, MDMA, MDEA, MBDB, and PMA in urine. *Forensic Sci. Int.* **171**:44–51.

de Castro, A., M. Ramirez Fernandez Mdel, M. Laloup, N. Samyn, G. De Boeck, M. Wood, V. Maes, and M. Lopez-Rivadulla, M. 2007. High-throughput on-line solid-phase extraction-liquid chromatography-tandem mass spectrometry method for the simultaneous analysis of 14 antidepressants and their metabolites in plasma. *J. Chromatogr. A.* **1160**:3–12.

Deventer, K., P. Van Eenoo, and F. T. Delbeke. 2006. Screening for amphetamine and amphetamine-type drugs in doping analysis by liquid chromatography/mass spectrometry. *Rapid Commun. Mass Spectrom.* **20**:877–882.

Herrin, G. L., H. H. McCurdy, and W. H. Wall. 2005. Investigation of an LC-MS-MS (QTrap) method for the rapid screening and identification of drugs in postmortem toxicology whole blood samples. *J. Anal. Toxicol.* **29**:599–606.

Huerta-Fontela, M., M. T. Galceran, and F. Ventura 2007. Ultraperformance liquid chromatography-tandem mass spectrometry analysis of stimulatory drugs of abuse in wastewater and surface waters. *Anal. Chem.* **79**:3821–3829.

Johansen, S. S., and H. M. Bhatia. 2007. Quantitative analysis of cocaine and its metabolites in whole blood and urine by high-performance liquid chromatography coupled with tandem mass spectrometry. *J. Chromatogr. B.* **852**:338–344.

Kronstrand, R., I. Nystrom, J. Strandberg, and H. Druid. 2004. Screening for drugs of abuse in hair with ion spray LC-MS-MS. *Forensic Sci. Int.* **145**:183–190.

Maralikova, B., and W. Weinmann. 2004. Confirmatory analysis for drugs of abuse in plasma and urine by high-performance liquid chromatography-tandem mass spectrometry with respect to criteria for compound identification. *J. Chromatogr. B* **811**:21–30.

Rook, E. J., M. J. Hillebrand, H. Rosing, J. M. van Ree, and J. H. Beijnen. 2005. The quantitative analysis of heroin, methadone and their metabolites and the simultaneous detection of cocaine, acetylcodeine and their metabolites in human plasma by high-performance liquid chromatography coupled with tandem mass spectrometry. *J. Chromatogr. B* **824**:213–221.

Scheidweiler, K. B., and M. A. Huestis. 2004. Simultaneous quantification of opiates, cocaine, and metabolites in hair by LC-APCI-MS/MS. *Anal. Chem.* **76**:4358–4563.

Trenholm, R. A., B. J. Vanderford, J. E. Drewes, and S. A. Snyder. 2008. Determination of household chemicals using gas chromatography and liquid chromatography with tandem mass spectrometry. *J. Chromatogr. A* **1190**:253–262.

van Nuijs, A. L., B. Pecceu, L. Theunis, N. Dubois, C. Charlier, P. G. Jorens, L. Bervoets, R. Blust, H. Meulemans, H. Neels, and A. Covaci. 2009. Can cocaine use be evaluated through analysis of wastewater? A nation-wide approach conducted in Belgium. *Addiction* **104**:734–741.

Vanderford, B. J., and S. A. Snyder. 2006. Analysis of pharmaceuticals in water by isotope dilution liquid chromatography/tandem mass spectrometry. *Environ. Sci. Technol.* **40**:7312–7320.

Vanderford, B. J., R. A. Pearson, D. J. Rexing, and S. A. Snyder. 2003. Analysis of endocrine disruptors, pharmaceuticals, and personal care products in water using liquid chromatography-tandem mass spectrometry. *Anal. Chem.* **75**:6265–6274.

SECTION IVC

MASS SPECTROMETRIC ANALYSIS OF ILLICIT DRUGS IN THE ENVIRONMENT

Presence in Air and Suspended Particulate Matter

CHAPTER 13

PSYCHOTROPIC SUBSTANCES IN URBAN AIRBORNE PARTICULATES

ANGELO CECINATO and CATIA BALDUCCI

13.1 INTRODUCTION

Psychotropic compounds, such as caffeine and nicotine, can exist in air as a result of their release in tobacco smoke or when certain drugs or essences are inhaled as vapors. These psychotropic substances belong to various classes of organic compounds with different physico chemical properties and different routes of release into the environment, so they may exist in the gaseous or asparticulates and as native compounds or derivatives. For instance, nicotine is mostly gaseous when it is a free base, but combines with tobacco smoke particles when in acidic form (Liang and Pankow, 1996). Cocaine and heroin in the atmosphere presumably exist mainly as solid particulates (Dindal et al., 2000; Cecinato and Balducci, 2007).

These substances may be hydrophilic or hydrophobic, depending on their polarity and acid properties. Similar to nicotine, caffeine is fairly water soluble (22 g/L at 25°C and 180 g/L at 80°C). Cocaine chlorhydrate is hundreds of times more soluble than its base (1800 and 1.7 g/L), and heroin is poorly soluble in water (0.6 g/L). Vapor pressures of cannabinoids mean they can exist only when particle associated (McPartland and Russo, 2001) and their solubility in water is low (less than 2 mg/L for Δ^9-tetrahydrocannabinol (THC); Jarho et al., 1998). Moreover, these compounds decompose so rapidly in the environment that they can be detected only in trace amounts (a few pigogram per cubic meter), in spite of the fact that marijuana is the most abused drug worldwide.

Dedicated instruments and procedures exist for the detection and quantification of illicit substances in vapors of ambient or artificial atmospheres (Ziegler et al., 1996; Stubbs et al., 2003), but because of their low sensitivity they need "high" burdens of the compounds in air or require derivatization, e.g., with methylbenzoate

Illicit Drugs in the Environment: Occurrence, Analysis, and Fate Using Mass Spectrometry, Edited by Sara Castiglioni, Ettore Zuccato, and Roberto Fanelli
Copyright © 2011 John Wiley & Sons, Inc.

for detection by solid-phase microextraction (SPME) of crack smoke components (Lai et al., 2008).

The scientific literature is rich with papers dealing with tobacco smoke and its constituents, nicotine, in particular. However, most of them refer only to the health effects of active and passive smoking and to mainstream and sidestream (or "environmental") effects of smoke in air (Bacsik et al., 2007; Charles et al., 2007; Pieraccini et al., 2008). Composition studies in ambient air are less frequent, probably because of the difficulties of measuring the composition of both the particulate and the gaseous phase (Häger and Niesser, 1997; Thielen et al., 2008). Most investigations have aimed at constructing diffusive samplers or dedicated sensors, or identifying and assessing the contribution of tobacco smoke to environmental pollution. Much less frequent are studies on caffeine, the third recreational drug of importance after alcohol and tobacco. Just a handful of papers deal with this substance in rain, river, sea, and ground water (Weigel et al., 2002), although measures of caffeine have been proposed as a means of tracing the impact of domestic sewage on contamination (Verenitch and Mazumder, 2008). Caffeine is present in plants and beverages, including coffee, tea, cocoa, and tobacco and is teratogenic, enhancing the potency of some carcinogenic chemicals. It can promote infarction, convulsions, epilepsy, and neuromuscular disorders. However, it also has protective properties against Parkinsonism and has adjuvant activity in some analgesic preparations (Harinath and Sikdar, 2005; Eteng et al., 1997; Kiefer and Wiebel, 1998).

In recent years, illicit substances have attracted attention as a consequence of the accidental finding of cocaine in the air over Los Angeles, California (Hannigan et al., 1998) and Rome, Italy (Cecinato and Balducci, 2007). Cocaine, heroin, cannabinoids, amphetamines, ecstasy and their metabolites, pyro products and adulterants/excipients have been measured more often in environmental or man-related media, including rivers, wastes, urine, sweat, hair, and blood (Sleeman et al., 2000; Gentili et al., 2004; Lewis et al., 2004; Cognard et al., 2005; Castiglioni et al., 2006; Boleda et al., 2007; Mari et al., 2009; van Nuijs et al., 2009; Postigo et al., 2009). Since the presence of illicit compounds in these media was related to their consumption and the prevalence of some of these compounds have been calculated, based on the concentrations in waters (Zuccato et al., 2008a), identifying other indicators of drug consumption through their content in ambient air is challenging and deserves to be explored.

13.2   CHEMICAL APPROACH TO MONITORING PSYCHOTROPIC SUBSTANCES

The first observation of illicit drugs in the atmosphere was unexpected and occurred in a study with quite different goals. Thus, a preliminary objective of a subsequent study was to confirm the presence of these substances in air. Afterward, dedicated procedures were optimized to monitor psychotropic substances and drug residues in air with good precision and accuracy. The determination of psychotropic compounds in air is quite complex. Both air and airborne particles comprise thousands of components, including drugs, but at very low concentrations; thus, high sensitivity and selectivity are needed for reliable determination.

The methods aimed at monitoring drugs in airborne particulates consist of: (1) collecting samples by passing air through quartz filter membranes; (2) spiking samples with perdeuterated compounds as internal standards (namely nicotine-d4, cocaine-d3, and carbazole-d7 for caffeine); (3) solvent extraction in a Soxhlet apparatus with dichloromethane/acetone (4:1 in volume); (4) clean-up step using column chromatography on neutral alumina and elution with iso-octane, iso-octane/dichloromethane (60:40), and acetone, in sequence; (5) separation by capillary gas chromatography (CGC); and (6) instrumental detection by electron-impact, selected ion-monitoring mass spectrometry [(EI)MSD].

Some compounds can be easily detected by both CGC and high-pressure liquid chromatography (HPLC), for instance, cocaine, which has characteristic electron-impact mass and UV fluorescence spectra (Fernandez et al., 2009). Inert nonpolar or low-polar silicone phases have chromatographic features suitable for separation and analysis by CGC. In addition, dedicated base-treated stationary phases are now used to improve the identification and quantification of compounds. HPLC analyses are done by reversed-phase elution with buffered water/organic solvent mixtures. In contrast, numerous native drugs and almost all metabolites require chemical derivatization before analysis [e.g., with trifluoroacetic anhydride, N-methyl-N-trimethylsilyltrifluoroacetamide, or N-(t-butyldimethylsilyl)-N-methyltrifluoroacetamide, which is usually used for cannabinoids]; this improves the analytical performance of the method and drastically reduces the risk of misinterpretation and misidentification of the compounds.

Figure 13.1 shows the GC-(EI)MSD chromatogram of a sample of suspended particulate matter collected in Milan in December 2006. The analysis was repeated in scan and selected ion-monitoring (SIM) mode, and less than 40 min were necessary to scan drugs from amphetamine to heroin. The figure shows both total-ion currents and signals corresponding to molecular ion traces of nicotine, caffeine, cocaine-d3 (adopted as internal standard for cocaine determination), and native cocaine.

Illicit substances usually have high boiling and melting temperatures and low pressures in room conditions, although their partial volatility is exploited for determination in aqueous samples by head-space SPME. However, water solutions must be conditioned at $60° \sim 90°C$ to determine these compounds. By contrast, thus far, illicit drugs in ambient air have only been investigated in the particulate phase. Very few field experiments have been conducted to study the presence of cocaine vapor in the open atmosphere, or its distribution among the airborne particulate size fractions (PM1, PM2.5, PM10). Cocaine seems to accumulate in particles, with over 80% in PM2.5. Usually 5–50 mg of particulate are necessary for chemical characterization, corresponding to 150 to 1500 m^3 of air, to reach the limits of quantification of the analytical methods, as concentrations of illicit substances in PM2.5 are usually in the range $10 \sim 500$ pg/m^3.

Except for cocaine (Cecinato et al., 2009a), the stability of drugs in particulates stored in the dark and at ambient temperature has thus far not been investigated. This is a limitation of the studies to date, because the amounts of illicit substances were

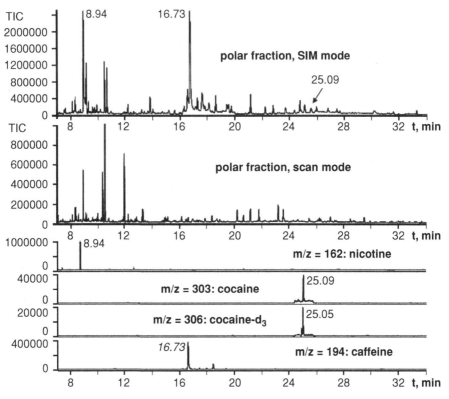

FIGURE 13.1   GC–MSD profile of the "neutral highly polar fraction" in airborne particulate collected in Milan, Italy, in December, 2006.

probably underestimated. There was no relation between the estimated amounts of some psychotropic substances consumed in various locations, and the corresponding drug levels measured in the aerosol form; for instance, THC was often found at much lower concentrations than cocaine. The use of ultraclean inert membranes to filter air and pooling short-period samplings instead of using a single medium-period collection seems important to correctly measure airborne psychotropic substances.

13.3   CONCENTRATIONS AND SOURCES OF PSYCHOTROPIC SUBSTANCES IN AIR

As of November 2009, only two research groups investigated the occurrence of psychotropic substances in the atmosphere, namely, the IDAEA-CSIC in Spain and the CNR-IIA in Italy, although cocaine was first identified in air over Los Angeles, California (Hannigan et al. 1998). Most measurements have been made in Europe, particularly in Italy, with sketchy data for South America and Algeria.

TABLE 13.1 Average Concentrations of Psychotropic Substances in Airborne Particulates of Three Italian Cities (ng m^{-3}) with the Cocaine/Benzo[a]pyrene Ratio (CO/BaP)[a]

	Milan, 2006–2007							Bari	Taranto
	Jun	Aug	Sep	Dec	Jan	Feb	Chr.	Oct '08	Mar '05
Nicotine	7.52	7.44	34.69	10.40	5.77	53.5	6.64	21.6	3.1
Caffeine	0.17	0.07	0.40	4.91	0.69	1.00	1.92	0.13	0.14
Cocaine	0.073	0.097	0.216	0.080	0.252	0.266	0.470	0.014	0.010
Cannabidiol	<0.005	<0.005	<0.005	0.009	0.037	<0.005	0.038	0.002	na
Δ^9-THC	0.027	0.027	<0.005	<0.005	<0.005	<0.005	0.100	0.039	na
Cannabinol	0.074	0.098	<0.005	<0.005	<0.005	<0.005	0.310	0.032	na
CO/BaP	1.13	1.51	0.76	0.06	0.12	0.19	0.09	0.16	0.01

[a]Chr.=Christmas time (late December to early January); na, not assessed.

No investigations have been done in the countries where coca, cannabis or opium poppy are grown, which would help clarify how cultivation conditions and/or preliminary manufacturing affects drug levels in the atmosphere.

Table 13.1 shows the average atmospheric concentrations of cocaine, THC, cannabinol, cannabidiol, nicotine, and caffeine in samples collected in Milan, Bari, and Taranto. Milan is in northern Italy, while Bari and Taranto are in the south, the three cities having different drug consumption patterns (Cecinato et al., 2009a). In Milan, measurements were repeated from June 2006 to February 2007 in order to explore the variability over time. The ratio of cocaine to benzo[a]pyrene (CO/BaP) is also reported to illustrate the relationship between illicit substances and other anthropogenic pollutants.

Table 13.2 reports drug concentrations in particulate samples collected in Pančevo (Serbia), Oporto (Portugal), and Madrid and Barcelona (Spain) (Postigo et al., 2009). Measurements were taken at sites indicative of the background air pollution, except in Pančevo; there, the investigation was carried out in three residential districts in the city center (data averaged) and in the suburb, during the four seasons of the year.

Table 13.3 shows results from some field experiments in three cities in Brazil, Ouro Preto, Sao Paulo, and Piracicaba, and in Santiago del Chile. Except for Ouro Preto (2002), samples were collected in 2006–2007 in a university campus, far from other emission sources like traffic, boilers, and kitchens. Ouro Preto is a small/medium-size town with extensive air and water pollution because of mining, while Sao Paulo and Santiago are two very dynamic cities with economic and government affairs. Other samples were collected in Araraquara (Brazil), where the cocaine contents were similar to those in Piracicaba.

In Algiers (Algeria), investigations were done in the downtown area and in two suburbs, Bab-el-Oued and Ben Akroun (Table 13.4). The presence of drugs in the atmosphere over the year in a social, religious, and productive context different from the western countries was studied. Preliminary samplings for cocaine were taken in 2004 and 2005 in the town center; a more extensive study followed at the two suburban locations in 2006–2007 (Ladji et al., 2009).

TABLE 13.2 Average Concentrations (ng m^{-3}) of Psychotropic Substances in Airborne Particulates of Pančevo (Serbia), Oporto (Portugal), Madrid (Spain), and Barcelona (Spain)[a,b]

	Pančevo downtown, 2006–2007				Pančevo outskirts, 2006–2007				Oporto	Madrid	Barcelona
	Jul	Oct	Jan	Mar	Jul	Oct	Jan	Mar	Jun '07	Wi. '07	Au. '07
Nicotine	10.3	42.2	30.4	16.8	6.4	10.9	10.5	6.0	5.1	na	na
Caffeine	4.83	4.99	11.80	22.8	0.17	0.87	1.10	2.39	10.4	na	na
Cocaine	<0.002	<0.002	<0.002	<0.002	<0.002	<0.002	<0.002	<0.002	0.148	0.48	0.204
Cannabidiol	<0.003	0.012	0.107	0.026	<0.003	0.016	0.058	0.010	0.003	na	na
Δ^9-THC	<0.003	<0.003	<0.003	<0.003	0.111	<0.003	0.073	<0.003	<0.003	0.044	0.027
Cannabinol	<0.003	<0.003	0.064	0.022	<0.003	<0.003	<0.003	<0.003	0.010	na	na
CO/BaP	–	–	–	–	–	–	–	–	0.40	–	–

[a]The cocaine/benzo[a]pyrene ratio (CO/BaP) is reported only for Oporto, where both compounds were measured.
[a]na, not assessed; Wi, winter; Au, autumn.

TABLE 13.3 Average Concentrations (ng m^{-3}) of Psychotropic Substances in Airborne Particulates of Three Brazilian Cities and Santiago, Chile with the Cocaine/Benzo[a]pyrene Ratio (CO/BaP)

	Ouro Preto		Sao Paulo				Piracicaba				Santiago	
	Au '02	Wi '02	Wi '06		Au '07		Wi '06		Wi '07		Wi '06	
Nicotine	2.86	10.15	6.2	16.0	14.9	17.1	2.85	2.84	4.54	2.92	14.3	21.6
Caffeine	0.03	0.03	1.88	12.3	0.37	0.13	0.62	1.06	0.14	0.13	13.0	8.2
Cocaine	0.047	0.101	0.333	0.101	0.206	0.346	0.086	0.079	0.110	0.069	2.44	4.11
Cannabidiol	0.152	<0.001	0.040	0.008	0.001	0.003	0.126	0.089	0.009	<0.001	0.001	<0.001
Δ^9-THC	<0.001	<0.001	<0.001	<0.001	<0.001	0.013	0.050	<0.001	<0.001	<0.001	<0.001	<0.001
Cannabinol	<0.001	<0.001	<0.001	<0.001	0.071	0.262	<0.001	<0.001	0.055	0.019	<0.001	<0.001
CO/BaP	≪0.01	≪0.01	0.46	0.05	0.57	1.23	0.07	0.07	0.09	0.21	0.58	0.77

[a] Au, autumn; Wi, winter.

TABLE 13.4   Average Concentrations (ng m^{-3}) of Psychotropic Substances in Airborne Particulates Collected in Algiers, Algeria, in 20062007[a,b]

	Algiers Suburb				Algiers Downtown			
	Su	Au	Wi	Sp	Su	Au	Wi	Sp
Nicotine	0.76	1.04	2.39	1.71	2.56	0.90	3.30	3.97
Caffeine	0.50	0.38	4.17	2.6	2.28	2.34	1.58	0.71
Cocaine	< 0.001	< 0.001	< 0.001	< 0.001	< 0.001	< 0.001	< 0.001	< 0.001
Cannabidiol	< 0.001	< 0.001	< 0.001	< 0.001	Na	< 0.001	< 0.001	< 0.001
Δ^9-THC	< 0.001	< 0.001	< 0.001	< 0.001	Na	< 0.001	< 0.001	< 0.001
Cannabinol	< 0.001	< 0.001	< 0.001	< 0.001	Na	0.008	0.012	0.009

[a] na, not assessed.
[b] Au, autumn; Wi, winter; Sp, spring.

To our knowledge, the most studied metropolitan area from this point of view is Rome, Italy, where two investigations were done. First, short-field experiments, each lasting two weeks, mostly in winter or summer, were carried out at sites with different anthropogenic emissions. In the city, they were in viale Regina Elena (university district, area with heavy traffic), via Bissolati (city center, business and touristic area), piazza Fermi (car park), via Belloni (residential area, with a school and a street market), and the garden of Villa Ada (urban area); in the suburbs, Malagrotta (~20 km SW of the city center, a suburban area with houses, small industry, a landfill, and an incinerator), and Montelibretti (semirural area, ~30 km NE of the city center). Second, long campaigns were done at Villa Ada (September 2006–2007) and Montelibretti (January 2006–September 2007). Table 13.5 and Fig. 13.2 summarize these studies.

The concentrations of nicotine and caffeine varied widely, depending on the site and time of year of the sampling. Monthly average concentrations of nicotine were from 0.7 to 50 ng m^{-3}, with peaks exceeding 200 ng m^{-3} (data not reported). Caffeine

TABLE 13.5   Average Concentrations (ng m^{-3}) of Cocaine, Nicotine, and Caffeine in the Air over Rome, Italy (Autumn 2005–Summer 2006), Expressed as ng m^{-3a}

	RE	Via Bissolati		Piazza Fermi		Via Belloni		VA		Montelibretti		MG
	Au[c]	Su	Wi	Su	Wi	Su	Wi	Su	Wi	Su	Wi	Sp
Nicotine	2.68	15.6	27.5	10.1	22.4	4.29	6.22	1.50	1.15	0.72	1.74	1.67
Caffeine	0.18	1.00	1.80	3.38	7.72	5.38	8.06	0.34	0.75	0.24	0.21	1.04
Cocaine	0.084	0.030	0.036	0.038	0.069	0.053	0.069	0.048	0.070	0.015	0.016	0.047
CO/BaP[b]	0.06	0.13	0.07	0.07	0.05	0.75	0.27	0.30	0.08	0.29	0.03	0.35

[a] RE= viale Regina Elena; VA=Villa Ada garden; MG=Malagrotta.
[b] CO/BaP=cocaine to benzo[a]pyrene ratio.
[c] Su, summer; Au, autumn; Wi, winter; Sp, spring.

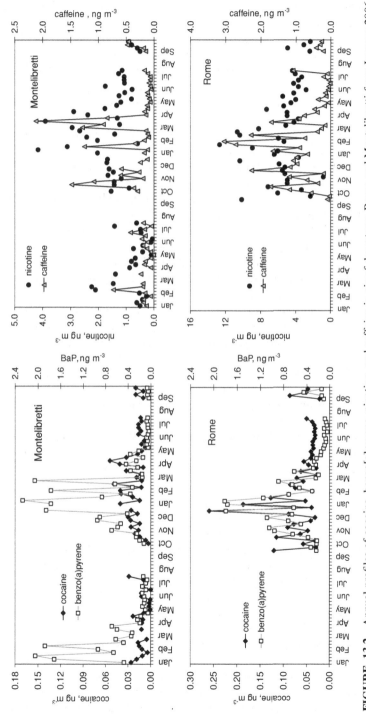

FIGURE 13.2 Annual profiles of cocaine, benzo[a]pyrene, nicotine, and caffeine in air of downtown Rome and Montelibretti from January 2006 to September 2007.

levels ranged from less than 0.1 to over 10 ng m^{-3}, with values exceeding 23 ng m^{-3}. The caffeine/nicotine ratio was also very variable from site to site and season of the year. Nicotine is semivolatile, so its partition between the gaseous and particulate phases is affected by ambient conditions. At room temperatures, from 10% to 90% of the total nicotine, depending on pH, would be in particulates and the percentage would decrease with any rise in the temperature. This is in agreement with the observed behavior of this drug, whose average content in particulates is lower in summer than winter.

Cocaine is the most widely studied illicit substance in air and also seems to reach the highest concentrations. Values as high as 3000 pg m^{-3} were recorded in Santiago, Chile and 500 pg m^{-3} were measured in Barcelona, Spain and Milan, Italy. Urban sites are generally more contaminated than suburban and rural areas. For instance, in two areas in the Taranto countryside, cocaine was close to (Statte, residential area) or below (Palagiano, rural area) the quantification limit of 0.002 ng m^{-3}. Large differences were found in different cities. Very high concentrations were detected in Milan, while in the urban areas of Taranto and Bari, levels were lower than in the suburban areas of Malagrotta and Montelibretti. In Rome, the cocaine values were highest in the university district (which also hosts highly populated areas and a large hospital complex), while levels in the city center were lower than in residential zones.

The case of Villa Ada is noteworthy: it is the second largest park in Rome, and the sampling station was in a no-entry zone, that frequently hosts events, exhibitions, and meetings. The cocaine level in this area was sometimes higher than in any other part of the city. The principal metabolite of cocaine, benzoylecgonine, was monitored in air only in Barcelona and Madrid, where it reached, respectively, 0.029 and 0.014 ng m^{-3}, and in Rome, where it was virtually absent.

Cannabinoids were measured less frequently than cocaine. Attention was mostly paid to THC, cannabinol (CBL), and cannabidiol (CBD), all associated with marijuana smoking; studies were mainly restricted to detection, but no quantitative assessments were made. The contents of THC, CBL, and CBD in air were usually lower than those of cocaine, ranging, respectively, from <0.001 to 0.11 ng m^{-3}, <0.001 to 0.31 ng m^{-3}, and <0.001 to 0.15 ng m^{-3}. Nevertheless, the compounds were measurable in air samples from Algiers and Pančevo (where CBD was absent) and THC levels exceeded those of cocaine in Bari (0.039 versus 0.014 ng m^{-3}) and Montelibretti (0.058 versus 0.050 ng m^{-3}, in October 2008).

Heroin was detected only in Oporto (~0.07 ng m^{-3}) and Madrid (0.084 ng m^{-3}), but not in Italy or Barcelona; methadone was never found. Amphetamine and methamphetamine were detected in Barcelona and Madrid at very low levels (~0.002 ng m^{-3}); they have not been measured in Italy. However, these two substances are semivolatile and measurements focused on particulates and did not include the vapor phase. Ecstasy or MDMA (0.003 ng m^{-3}) and acetylmorphine, a metabolite of heroin (0.023 ng m^{-3}) were measurable only in Barcelona and Madrid, respectively.

In general, concentrations of psychotropic substances in air have characteristic profiles during the year, with lowest levels in summer and highest in winter. This pattern is probably more related to atmospheric conditions than to the release of the substances in the environment, as drug consumption in the population does not seem to follow this trend. Differences are seen between weekdays and weekends. A typical case was in September 2007, when cocaine levels in air were about 1.3 and 2.1 times higher at weekends than on weekdays both in downtown Rome (0.048 versus 0.037 ng m^{-3}) and in Montelibretti (0.021 and 0.010 ng m^{-3}). The drug levels in Villa Ada peaked up to 0.087 ng m^{-3} during the White Night celebrations, when the CO/BaP ratio reached 0.6, while on weekdays in the same period this ratio was only 0.2 (Cecinato et al., 2009b).

Like other pollutants, the high concentrations of drugs in air also reflect the atmospheric conditions. For instance, cocaine levels were 2–4 times higher in winter than in summer both in Rome and Milan and this may be related to the weather, which in winter promotes atmospheric stability with accumulation of pollutants (Perrino et al., 2001; ARPA Lombardia, 2005; Vecchi et al., 2007). Particularly high concentrations of cocaine, benzo[a]pyrene (the only particulate pollutant regulated in Italy) (IME, 1994), n-alkanes, and PM10 were detected in Santiago, which lies in an area surrounded by high mountains.

The CO/BaP ratio generally shows a lack of correlation between the concentrations of these substances and the same is true for cannabinoids. For instance, CO/BaP was ≤0.1 in most areas in winter, but exceeded 1.0 in Milan in summer. In Rome, both low and high cocaine concentrations were measured at sampling stations with high BaP levels.

According to the limited data available, cocaine and cannabinoid concentrations in air do not seem correlated either between themselves, or with those of nicotine and caffeine. Looking at sources of well-known pollutants, BaP and fine particulates are produced by combustion of organic fuels (traffic, domestic heating, industry, waste and wood combustion), while n-alkanes are associated with both anthropogenic and natural emissions (petroleum and vegetation, respectively). Thus, cocaine's behavior in the atmosphere seems to call for some source independent of the main anthropogenic and biogenic emitters.

As coca and cannabis crops are not, or are only occasionally grown in Europe, this source cannot explain the drug residues in air, which are more likely to be related to the consumption/abuse of psychotropic substances. This is similar to that for nicotine and caffeine, whose emission sources are mainly tobacco smoking and coffee drinking or roasting. However, the sources of illicit drugs in the environment remain to be confirmed. Hypothetically, residues of drugs of abuse in air might also originate from sales or disposal of these substances. In the former case, production, refinery, cutting, transport, dose preparation, street pushing and consumption, and excretion in biological fluids or sweat can contribute. A contribution from the clinical uses of some of these substances cannot be excluded and, in this case, the incineration of substances seized or disposed of in hospitals or special incinerators merits investigation. As of November 2009, only one attempt has been made to compare the amounts of cocaine seized by the police in Italy (Italian Police Annual Drug Report, 2007; UNODC, 2008) and drug concentrations in air. Cocaine was found in air only in areas of "high" consumption and sales and never where consumption and sales were low. A rough correspondence was observed by plotting the drug levels in air against the corresponding abuse rates in Milan (maximum), Rome (medium) and Taranto or

Bari (both minimum consumption and air concentrations) (Presidency of the Italian Council of Ministers, 2008).

13.4   FUTURE NEEDS AND FIELDS OF INVESTIGATION

Various activities in the field of illicit drugs in the environment appear important and promising in the future. One of the first topics is monitoring. Only a few countries have been studied up to now and only with "explorative" aims; nothing is known about regions where drug crops are extensively grown (e.g., Colombia and Bolivia for coca, Morocco for cannabis, and Afghanistan for opium) or consumed (United States, northern Europe, eastern Asia) (Central Intelligence Agency, 2008). Nothing is known about the presence in air of psychotropic substances from recreational or illicit use consumed in highly populated regions like China, India, Japan, and Africa. Second, the true origin of the illicit substances in the atmosphere still has to be identified in order to address a number of objectives: (1) to understand whether it is the market or the disposal of these substances that is responsible for their occurrence in air; (2) to profit from lessons that promote reduction of drug consumption; (3) to learn about possible health consequences, for which very little is known. It has been calculated that one lifetime is not enough to inhale an amount of cocaine from the air equivalent to one dose. Nevertheless, considering the size of the exposed population, some effects cannot be excluded. Finally, (4) the level of contamination of the atmosphere might serve as an easy, cheap, rapid, and nonintrusive way to estimate drug consumption at national, regional, or local levels. This will be even more useful if this drug-related index can be combined with other parameters derived from measurements in parallel in surface and waste waters (Bones et al., 2007; Zuccato et al., 2008b).

ACKNOWLEDGMENTS

Financial contributions were received from The Italian Ministry of University and Research (MIUR) through the Basic Research Program 2001. The kind cooperation of many colleagues (Prof. C. Pio from Aveiro University, Prof. P. Vasconcellos De Castro from Sao Paulo University, Dr. R. Ladji from USTHB University of Algiers, Prof. E. Bolzacchini from Bicocca University, Milan, Prof. G. De Gennaro from University of Bari, and Dr. P. Romagnoli from CNR-IIA) was valuable for investigations in cities other than Rome.

REFERENCES

ARPA Lombardia, Rapporto sulla qualità dell'aria di Milano e provincia anno 2005. http://www.arpalombardia.it/qaria/pdf/RQA-2005/RQA%20MI%202005.pdf

Bacsik, Z., J. McGregor, and J. Mink. 2007. FTIR analysis of gaseous compounds in the mainstream smoke of regular and light cigarettes. *Food Chem. Toxicol.* **45**:266–271.

Boleda, M. R., M. T. Galceran, and F. Ventura. 2007 Trace determination of cannabinoids and opiates in wastewater and surface waters by ultra-performance liquid chromatography–tandem mass spectrometry. *J. Chromatogr. A* **1175**:38–48.

Bones, J., K. V. Thomas, and B. Paull. 2007. Using environmental analytical data to estimate levels of community consumption of illicit drugs and abused pharmaceuticals. *J. Environ. Monit.* **9**:701–707.

Castiglioni, S., E. Zuccato, E. Crisci, C. Chiabrando, R. Fanelli, R. Bagnati. 2006. Identification and measurement of illicit drugs and their metabolites in urban wastewater by liquid chromatography-tandem mass spectrometry. *Anal. Chem.* **78**:8421–8429.

Cecinato, A., and C. Balducci. 2007. Detection of cocaine in the airborne particles of the Italian cities Rome and Taranto. *J. Sep. Sci.* **30**:1930–1935.

Cecinato, A., C. Balducci, and G. Nervegna. 2009a. Occurrence of cocaine in the air of the world's cities: an emerging problem? A new tool to investigate the social incidence of drugs? *Sci. Total Environ.* **407**:1683–1690.

Cecinato, A., C. Balducci, G. Nervegna, G. Tagliacozzo, and I. Allegrini 2009b. Ambient air quality and drug aftermaths of the Notte Bianca (White Night) holydays in Rome. *J. Environ. Monit.* **11**:200–204.

Central Intelligence Agency. 2008. The World Factbook. https://www.cia.gov/library/ publications /the-world-factbook/fields/2086.html

Charles, S. M., S. A. Batterman, and C. Jia. 2007. Composition and emissions of VOCs in main- and side-stream smoke of research cigarettes. *Atm. Environ.* **41**:5371–5384.

Cognard, E., S. Rudaz, S. Bouchonnet, and C. Staub. 2005. Analysis of cocaine and three of its metabolites in hair by gas chromatography-mass spectrometry using ion-trap detection for CI/MS/MS. *J. Chromatogr. B* **826**:17–25.

Dindal, A. B., M. V. Buchanan, R. A. Jenkins, and C. K. Bayne. 2000. Determination of cocaine and heroin vapour pressures using commercial and illicit samples. *Analyst* **125**:1393–1396.

Eteng, M. U., E. U. Eyong, E. O. Akpanyung, M. A. Agiang, and C. Y. Aremu. 1997. Recent advances in caffeine and theobromine toxicities: A review. *Plant Foods Hum. Nutr.* **51**:231–243.

Fernandez, P., M. Lago, R. A. Lorenzo, A. M. Carro, A. M. Bermela, and M. J. Tabernero. 2009. Optimization of a rapid microwave-assisted extraction method for the simultaneous determination of opiates, cocaine and their metabolites in human hair. *J. Chromatogr. B* **877**:1743–1750.

Gentili, S., M. Cornetta, and T. Macchia 2004. Rapid screening procedure based on headspace solid-phase microextraction and gas chromatography–mass spectrometry for the detection of many recreational drugs in hair. *J. Chromatogr. B* **801**:289–296.

Häger, B., and R. Niessner. 1997. On the distribution of nicotine between the gas and particle phase and its measurement. *Aerosol Sci. Technol.* **26**:163–174.

Hannigan, M. P., G. R. Cass, W. B. Penman, C. L. Crespi, A. L. Lafleur, W. F. Busby, Jr., W. G. Thilly, and B. R. T. Simoneit. 1998. Bioassay-directed chemical analysis of Los Angeles airborne particulate matter using a human cell mutagenicity assay. *Environ. Sci. Technol.* **32**:3502–3514.

Harinath, S., and S. K. Sikdar. 2005. Inhibition of human TREK-1 channels by caffeine and theophylline. *Epilepsy Res.* **64**:127–135.

Italian Ministry of the Environment IME 1994. "Updating of the technical rules concerning the concentration limits and the precautional and warning levels regarding the atmospheric pollutants in the urban areas and indications for the measure of some pollutants referred in the Ministerial Decree released on April 15th, 1994", released on November 25th, 1994 and published in the Gazzetta Ufficiale Italiana, Suppl. No. 290, 1994 13, December.

Italian Police Annual Drug Report 2007. http://www.poliziadistato.it/pds/online/ antidroga/dati_sic_antidrog_2007.htm.

Jarho, P., D. W. Pate, R. Brenneisen, and T. Järvinen. 1998. Hydroxypropyl-β-ciclodextrin and its combination with hydroxypropyl-methylcellulose increases aqueous solubility of Δ9-tetrahydrocannabinol. *Life Sci.* **63**:381–384.

Kiefer, F., and F. J. Wiebel. 1998. Caffeine potentiates the formation of micronuclei caused by environmental chemical carcinogens in V79 Chinese hamster cells. *Toxicol. Lett.* **96**:131–136.

Ladji, R., N. Yassaa, C. Balducci, A. Cecinato, and B. Y. Meklati. Annual variation of particulate organic compounds in PM10 in the urban atmosphere of Algiers. *Atm. Res.* **92**:258–269.

Lai, H., I. Corbin, and J. R. Almirall. 2008. Headspace sampling and detection of cocaine, MDMA, and marijuana via volatile markers in the presence of potential interferences by solid phase microextraction–ion mobility spectrometry (SPME-IMS). *Anal. Bioanal. Chem.* **392**:105–113.

Lewis, R. J., R. D. Johnson, M. K. Angier, and R. M. Ritter RM. Determination of cocaine, its metabolites, pyrolysis product, and ethanol adducts in post-mortem fluids and tissues using Zymark® automated solid phase extraction and gas chromatography-mass spectrometry. *J. Chromatogr. B* **806**:141–150.

Liang, C., and J. F. Pankow. 1996. Gas/particle partitioning of organic compounds to environmental tobacco smoke: partition coefficient measurements by desorption and comparison to urban particulate material. *Environ. Sci. Technol.* **30**:2800–2805.

Mari, F., L. Politi, A. Biggeri, G. Accetta, C. Trignano, M. Di Padua, and E. Bertol. 2009. Cocaine and heroin in waste water plants: A 1-year study in the city of Florence, Italy. *Foren. Sci. Int.* **189**:88–92.

McPartland, J. M., and E. B. Russo. 2001. Cannabis and cannabis extracts: greater than the sum of their parts? *J. Cannabis Therap.* **1**:103–132;

Perrino, C., A. Pietrodangelo, and A. Febo. An atmospheric stability index based on radon progeny measurements for the evaluation of primary urban pollution *Atm. Environ.* **35**:5235–5244.

Pieraccini, G., S. Furlanetto, S. Orlandini, G. Bartolucci, I. Giannini, S. Pinzauti, and G. Moneti. 2008. Identification and determination of mainstream and sidestream smoke components in different brands and types of cigarettes by means of solid-phase microextraction–gas chromatography–mass spectrometry. *J. Chromatogr. A* **1180**:138–150.

Postigo, C., M. J. L. de Alda, M. Viana, X. Querol, A. Alastuey, B. Artiñano, and D. Barcelò. 2009. Determination of drugs of abuse in airborne particles by pressurized liquid extraction and liquid chromatography-electrospray-tandem mass spectrometry. *Anal.Chem.* **81**:4382–4388.

Presidency of the Italian Council of Ministers. 2008. Drugs addictions. Annual Report on the State of Drugs Addiction in Italy in 2007. http://www.governo.it/GovernoInforma/ Dossier/relazione _droga_2007/.

Sleeman, R., F. Burton, J. Carter, D. Roberts, and P. Hulmston. 2000. Drugs on money. *Anal. Chem.* **72**:397 A–403 A.

Stubbs, D. D., S. –H. Lee, and W. D. Hunt. 2003. Investigation of cocaine plumes using surface acoustic wave immunoassay sensors. *Anal. Chem.* **75**:6231–6235

Thielen, A., H. Klus, and L. Müller 2008 Tobacco smoke: Unraveling a controversial subject. *Exp. Toxicol. Pathol.* **60**:141–156.

UNODC. United Nation Office on Drug And Crime; 2008. On-line database https://www.unodc.org/unodc/en/data-and-analysis/Research-Database.html.

van Nuijs, A. L. N., B. Pecceu, L. Theunis, N. Dubois, C. Charlier, P. G. Jorens, L. Bervoets, R. Blust, H. Neels, and A. Covaci. 2009. Cocaine and metabolites in waste and surface water across Belgium. *Environ. Pollut.* **157**:123–129.

Vecchi, R., G. Marcazzan, and G. Valli. A study on nighttime–daytime PM10 concentration and elemental composition in relation to atmospheric dispersion in the urban area of Milan (Italy). *Atm. Environ.* **41**:2136–2144.

Verenitch, S. S., and A. Mazumder. 2008. Development of a methodology utilizing gas chromatography ion-trap tandem mass spectrometry for the determination of low levels of caffeine in surface marine and freshwater samples. *Anal. Bioanal. Chem.* **391**:2635–2646.

Weigel, S., J. Kuhlmann, and H. Hühnerfuss. 2002. Drugs and personal care products as ubiquitous pollutants: occurrence and distribution of clofibric acid, caffeine and DEET in the North Sea. *Sci. Total Environ.* **295**:131–141.

Ziegler, T., O. Eikenberg, U. Bilitewski, and M. Grol M. 1996. Gas phase detection of cocaine by means of immunoanalysis. *Analyst* **121**:119–125.

Zuccato, E., C. Chiabrando, S. Castiglioni, R. Bagnati, and R. Fanelli R. 2008a. Estimating community drug abuse by wastewater analysis. *Environ. Health Perspect.* **116**:1027–1032.

Zuccato, E., S. Castiglioni, R. Bagnati, C. Chiabrando, P. Grassi, R. Fanelli. 2008b. Illicit drugs, a novel group of environmental contaminants. *Water Res.* **42**:961–968.

APPLICATIONS OF ILLICIT DRUG ANALYSIS IN THE ENVIRONMENT

CHAPTER 14

ILLICIT DRUGS IN THE ENVIRONMENT: IMPLICATION FOR ECOTOXICOLOGY

GUIDO DOMINGO, KRISTIN SCHIRMER, MARCELLA BRACALE, and FRANCESCO POMATI

14.1 LITERATURE REVIEW

14.1.1 Effects on Humans

Acute and long-term effects of what we call "illicit drugs" have been mostly observed on human subjects. Here we review the effects of cocaine, tetrahydrocannabinol (THC), 3,4-methylenedioxymethamphetamine (MDMA), and morphine as a reference for further discussion of potential consequences on nontarget organisms.

14.1.1.1 Acute Effects Acute cocaine and amphetamine consumption produce similar signs: psychiatric disturbance (agitation, hallucinations), neurological effects (mydriasis, convulsions), cardiovascular problems (tachycardia, raise in blood pressure, arrhythmia, and acute coronary insufficiency) and respiratory difficulties (cardiorespiratory arrest). Headaches may occur because of stroke or transient ischemic attack or to intracerebral or subarachnoid hemorrhage. Spontaneous cerebral hemorrhage can occur in normotensive human subjects (Gilman et al., 1985). Although the oral lethal dose of cocaine is usually given as 1.2 g, death has been reported to follow with as little as 20 mg parenterally (not intestinally) administered cocaine in humans (Haddad, 1990).

Users exposed to MDMA or other amphetamines experience psychological difficulties including confusion, depression, sleep problems, drug craving, severe anxiety, and paranoia; physical symptoms such as muscle tension, involuntary teeth clenching, nausea, blurred vision, rapid eye movement, faintness, and chills or sweating;

Illicit Drugs in the Environment: Occurrence, Analysis, and Fate Using Mass Spectrometry, Edited by Sara Castiglioni, Ettore Zuccato, and Roberto Fanelli
Copyright © 2011 John Wiley & Sons, Inc.

increases in heart rate and blood pressure, is a special risk for people with circulatory or heart disease (Mas et al., 1999). The toxic dose of amphetamines as MDMA is variable, with near fatal and fatal ingestions having been reported with blood levels between 0.11 to 2.1 mg/L. Survival has also been reported after MDMA blood levels of 4.3 mg/L drawn 13 h after ingestion.

Acute overdose of morphine is manifested by respiratory depression, somnolence progressing to stupor or coma, skeletal muscle flaccidity, cold and clammy skin, constricted pupils, and, sometimes, pulmonary edema, bradycardia, hypotension, and death. Marked mydriasis rather than miosis may be seen because of severe hypoxia in overdose situations (Thomson Healthcare, Inc. 2006). Morphine leads to death in amounts of 0.15 to 0.2 g (subcutaneous) or 0.3 to 0.4 g (oral) in adults. Babies and young children are much more susceptible and death has been observed at doses of 30 mg (Harvey, 1993). Toxic morphine blood concentration is 0.1−1 mg/L, whereas the lethal morphine blood has concentration higher then 4 mg/L (Gossel and Bricker, 1994).

THC produces effects on mood, memory, motor coordination, cognitive ability, sensorium, time sense, and self-perception. Most commonly there is an increased sense of well-being or euphoria, accompanied by feelings of relaxation and sleepiness when subjects are alone; where users can interact, sleepiness is less pronounced and there is often spontaneous laughter. Short-term memory is impaired and there is deterioration in capacity to carry out tasks requiring multiple mental steps to reach a specific goal (Gilman et al., 1985). There are few data on acute effects of THC in humans, since the main exposure route is inhalation by smoking and it is generally considered to be impossible to achieve lethal doses of THC by smoking cannabis.

14.1.1.2 *Chronic Effects*

Chronic cocaine, but not chronic amphetamine, use is associated with perseverative responding in humans, which is linked with memory and learning capabilities (Ersche et al., 2008). In chronic high-dose cocaine or amphetamine abuse, energy and euphoria, induced by active drug administration, is replaced in withdrawal by rebound dysphoric and anergic symptoms that appear to occur whether or not the stimulant abuser meets the diagnostic criteria for a mood disorder. The sustained increase in the sensitivity to the psychosis-inducing properties of stimulants suggests that chronic consumption of central stimulants induces a permanent alteration in the functional organization of the central nervous system, especially dopaminergic systems. This hypothesis has been presented in reference to amphetamines and, indeed, similar considerations have been raised for cocaine (Ellinwood et al., 1998). The effects of cocaine on kidneys have recently become more apparent more, whereas the acute and chronic effects of cocaine use on other vasculatures are reasonably well known (Van der Woude, 2000). Long-term use of THC can lead to squamous metaplasia of tracheal mucosa (Braude et al., 1984). Regular use may result in reproductive effects in males including decreased testicular size and testosterone levels. Reproductive effects have also been reported in animals (Braude et al., 1984).

14.1.2 Effects on Mammalian Model Species and Other Terrestrial Organisms

Effects reported in the literature on illicit drugs in model organisms can be considered similar to those reported above for humans. According to Bedford et al. (1982), the lethal dose for 50% (LD_{50}) of cocaine when administered intraperitoneal (IP) in mice was 95.1 mg/kg of weight.

The LD_{50} for morphine sulfate anhydrous (oral dose) is 461 mg/kg in rat and 600 mg/kg in mouse. The estimated oral LD_{50} for MDMA in rat is 325 mg/kg. Intraperitoneal LD_{50} for methylendioxyamphetamine (MDA) and MDMA were determined in mice by Davis et al. (1987). The LD_{50} of MDMA and MDA were basically the same, with MDA equaling 90.0 and MDMA 106.5 mg/kg, respectively.

The LD_{50} of rats tested by inhalation of THC is 42 mg/kg of body weight (Budavari, 1996 [Merck Index, 12th edition, 1996). For oral consumption, the LD_{50} for rats is 1270 and 730 mg/kg for males and females, respectively. In female rats intravenous LD_{50} is 29 mg/kg (Budavari, 1996 [Merck Index, 12th edition, 1996]). THC is more soluble in fat than in water, which may account for the significant fluctuations in LD_{50} values between injection and oral administration.

Little is known about toxicity of illicit drugs on plants. Mach and Livingston (1922) showed that lupine roots treated for 24 h with a 2.04% (v/v) cocaine solution causes complete growth inhibition. The lowest concentration of morphine showing toxicity to ladino clover seedlings in a complete nutrient culture was reported as 500 mg/kg (McCalla and Haskins, 1964).

14.1.3 Effects on Aquatic Organisms

Most of the toxicological data available for illicit drugs are on acute toxicity in humans or mammalian model organisms, mostly rats and mice. Little is known regarding the risk toward aquatic species for which exposure has never been intended or predicted. Darland and Dowling (2001) have shown that cocaine has specific effects on the behavior of zebrafish (*Danio rerio*). The lowest observed effect concentration (LOEC) determined by measuring change in place preference was 5 mg/L. The acute responses of the fish to cocaine and another local anaesthetic, lidocaine, were different. When treated with cocaine, the fish typically displayed slow circling, low in the water column, with fins more or less extended—indicating arousal. Cocaine also partially inhibited dark adaptation (Darland and Dowling, 2001). Both codeine and morphine had similar EC_{50} values when tested in 24-h immobilization tests using *Daphnia pulex*, 83.5 and 88.3 mg/L respectively (Morrow et al., 2001). LC_{50} of amphetamine sulpfate determined in *Daphnia magna* acute toxicity tests after 24-h exposure ranged from 60.4 to 265.3 mg/L (Guilhermino et al., 2000; Lilius et al., 1995).

Zebrafish embryos at late high blastula exposed to THC for 24 h showed a statistically significant reduction of spontaneous tail muscle twitch (at the normal time of its development) and subsequent embryo death above 2.0 mg/L THC levels, while no effect on spontaneous muscle twitch or any subsequent death of embryos was observed at 2.0 mg/L THC, a dose at which 37% distal trunk anomalies (curved spine or bulbous-tipped tail) were, however, evident. No identifiable effects could be detected at doses below 2.0 mg/L THC (Thomas, 1975). As reported in Cerilliant Material Safety Data Sheet for THC in methanol (http://www.cerilliant.com), the LC_{50} for rainbow trout (*Onchorhynuchus mykiss*, 96-h test) and carp (*Cyprinus carpio*, 48-h test) were 19 and 36 mg/L, respectively, while EC_{50} (48-h test) for *Daphnia magna* has been reported as 24.5 mg/L.

Suzuki et al. (2000, 2004) have shown how cocaine-derived synthetic local anesthetics enhance the growth of several algae, including cyanobacteria. Procaine, procainamide, tetracaine, lidocaine, and dibucaine stimulated the growth of *Anabaena cylindrica, Anabaena variabilis, Dunaliella primolecta,* and *D. parva* at levels of exposure in the range of 0.01 to 1000 mg/L (Suzuki et al., 2000). Lidocaine, in particular, appeared to be the most effective growth enhancer in cyanobacteria, being effective at concentration levels of 0.1 to 1 μM on *Synechococcus leopoliensis* (Suzuki et al., 2004). In the saxitoxin-producing cyanobacterium *Cylindrospermopsis raciborskii* T3 (Pomati et al., 2004), lidocaine was effective in stimulating both growth and the intracellular accumulation of the potent neurotoxin saxitoxin at 1-μM doses, while inducing the expression of genes related to bloom formation in another saxitoxin-producing cyanobacterium, *Anabaena circinalis* (Pomati et al., 2006). Lidocaine and synthetic analogs of cocaine, in general, seem to have the capability of interacting with cyanobacterial ion fluxes or channels, modulating membrane potential and other homeostasis-related activities, which can impact algal growth and production of secondary metabolites in aquatic environments (Suzuki et al., 2004; Pomati et al., 2004; Pomati et al., 2006). Excessive proliferation of algae and, in particular, harmful cyanobacterial blooms are a continuous concern in the management of aquatic ecosystems, especially in relation to climate change (Paerl and Huisman, 2008).

Considering the scarcity of data on aquatic organisms, we include in this article a brief review of the effects of two other plant-derived alkaloids that commonly contaminate surface waters, caffeine and nicotine, which may serve as an additional example of possible effects on fish and aquatic invertebrates. Material Safety Data Sheet (ERM AC802) for nicotine reported an EC_{50} (immobilization) for *Daphnia pulex* of 0.24 mg/L and a LC_{50} for rainbow trout of 4 mg/L, both after 48 h of acute exposure. Zebrafish treated with 100 mg/L nicotine for 3 min by immersion before testing caused a significant decrease in diving throughout the session, while 50 mg/L was effective during the first minute when the greatest bottom dwelling was seen in controls (Levin et al., 2007). Chronic exposure of larval zebrafish to nicotine concentrations of 5, 10, or 20 mg/L beginning at 1-day postfertilization significantly reduced notochord length and eye diameter (growth). Exposure also adversely affected the startle responses (behavior) and decreased fish survival (Levin et al., 2007).

After 24-h exposure to caffeine, the LC_{50} for *Daphnia magna* was recorded as 683.7 mg/L (Guilhermino et al., 2000) and an EC_{50} (immobilization) as 161.18 mg/L (Lilius et al., 1995). Freshwater invertebrate species, such as *Ceriodaphnia dubia, Pimephales promelas*, and *Chironomus dilutus*, commonly used in water monitoring tests, were exposed to aqueous caffeine solutions under static exposure for 48 h and then daily for 7 days. Averaged responses for 48-h acute endpoints indicated that *C. dubia* was more sensitive to caffeine exposures ($LC_{50} = 60$ mg/L) than either *P. promelas* ($LC_{50} = 100$ mg/L) or *C. dilutus* ($LC_{50} = 1230$ mg/L) (Moore et al., 2008).

14.1.4   Natural Biological Function of Illicit Drugs

Chemistry and chemical signals can play a significant role in determining the outcome of species interactions and coexistence in natural communities. Many organisms can produce a wide variety of so-called "secondary metabolites," which include signaling molecules and toxins. For many years, secondary metabolites were thought to be waste products of plants, without any apparent function in the producing organism. Gradually, however, recognition of the important role of secondary metabolites has increased, for example, in terms of resistance to pests and disease. Many secondary metabolites can be involved in offensive and defensive mechanisms utilized to inhibit the growth of competitor plants and to deter potential predators, grazers, or pathogens (Bell and Charlwood, 1980). They can be toxic or repellent directly to insects and microorganisms or they can be produced and released into the air to attract parasites and predators that kill grazers (Bell and Charlwood, 1980). Recent research suggests further primary roles for some secondary metabolites in plants as antioxidants (Campos et al., 2006).

Most pharmaceuticals are based on plant chemical compounds and secondary metabolites are widely used for recreation and stimulation (the alkaloids nicotine and cocaine; the terpene THC). Among plant secondary metabolites, a few have been commercially developed and processed into an insecticide product used for pest control. The best known and most widely used botanical insecticide is pyrethrum, a mixture of different insecticidal compounds found in the flower head of the pyrethrum daisy, *Chrysanthemum cinerariaefolium*. A study demonstrated that cocaine exerts insecticidal effects at concentrations, which occur naturally in coca leaves (Nathanson et al., 1993). Unlike its known action on dopamine reuptake in mammals, cocaine's pesticide effects have been shown to result from a strengthening of insect octopaminergic neurotransmission. In insects, cocaine can interfere with the neurohormone octopamine (OA) functions as neuromodulator and neurotransmitter (Orchard, 1982). Monoamine neurotransmitters in insects include OA, serotonin, histamine, and possibly tyramine (Osborne, 1996). Amine-reuptake blockers of other structural classes also exert pesticide activity with a rank order of potency distinct from that known to affect vertebrate amine transporters. These findings suggest that cocaine may function in plants as a natural insecticide and that octopamine transporters may be selective target sites of interest in the risk assessment of illicit drugs on aquatic invertebrates.

As in the case of nicotine and caffeine, THC's most likely function in *Cannabis* is to protect the plant from herbivores or pathogens (Taiz and Zeiger, 1991). THC also possesses high UV-B (280–315 nm) absorption properties, protecting the plant from potentially harmful radiation (Lydon et al., 1987). MDMA is a synthetic molecule derived from piperonal, isosafrole, or safrole (Kamdem and Gage, 1995). *Sitophilus zeamais motschulsky* (*Coleoptera, Curculionidae*) and *Tribolium castaneum* (*Coleoptera, Tenebrionidae*) adults were equally susceptible to contact toxicity of either safrole or isosafrole (Huang et al., 1999).

14.1.5   Molecular Targets of Illicit Drugs

There are four principal protein targets with which illicit drugs can interact: enzymes, membrane carriers, ion channels, and receptors (Lambert, 2004). Receptors are macromolecules involved in chemical signaling between and within cells and activated receptors directly or indirectly regulate cellular biochemical processes (e.g., ion conductance, protein phosphorylation, DNA transcription, and enzymatic activity), which, in turn, can affect the homoeostasis of cells and organisms.

Few drugs are absolutely specific for one receptor or subtype, but most have relative selectivity. Many recreational drugs, such as cocaine and amphetamines, alter the functionality of the dopamine transporter (DAT), the protein responsible for removing dopamine from the neural synapse. When DAT activity is blocked, the synapse floods with dopamine and increases dopaminergic signaling, inducing euphoric effects (Di Chiara et al., 2004; Hummel and Unterwald, 2002). Targets of cocaine can also be κ-type opioid receptors, the sodium-dependent serotonin transporter, and the sodium-dependent noradrenaline transporter (Lee et al., 2006; Imming et al., 2006; Overington et al., 2006). γ-Aminobutyric acid-C (GABA [C]) receptors can also be targeted and are the least studied of the three major classes of GABA receptors (Johnston et al., 2003). The physiological role and pharmacology of GABA(C) receptors are still being unraveled (Johnston et al., 2003). Little is known about binding characteristics and downstream effects on cell physiology of the octopamine transporter (OAT), a target site of cocaine in insects. Cocaine might be acting biochemically through a blockade of octopamine reuptake, thereby augmenting octopamine neurotransmission and functionally acting as octopamine agonists (Nathanson et al., 1993).

Amphetamines stimulate the release of norepinephrine from central adrenergic receptors. At higher dosages, they cause the release of dopamine. Amphetamine may also act as a direct agonist on central 5-HT receptors and may inhibit monoamine oxidase (MAO) activity impairing reuptake of catecholamines from the synaptic cleft (Banerjee et al., 1978; Juhila et al., 2005).

Morphine is the principal alkaloid in opium and the prototype opiate analgesic and narcotic, with widespread effects in the central nervous system (CNS) and on smooth muscle. The precise mechanism of the analgesic action of morphine is unknown. However, specific CNS opiate receptors have been identified and likely play a role in the expression of analgesic effects. Opioids interact with so-called opioid receptors, a group of G protein-coupled receptors. There are three types of opioid receptors, μ, δ and κ (Suzuki and Misawa, 1997). Recent evidence suggests the existence of two functional duplicates of the δ-opioid receptor gene in zebrafish (Pinal-Seoane et al., 2006). The δ-opioid receptor inhibits neurotransmitter release by reducing calcium ion currents and increasing potassium ion conductance (Knapp et al., 1994).

Cannabinoid receptors are a class of cell membrane receptors under the G protein-coupled receptor superfamily (Howlett, 2002; Mackie, 2008; Graham et al., 2009).

There are currently two known subtypes of cannabinoid receptors, termed CB1 and CB2 (Matsuda et al., 1990; Gérard et al., 1991), both activated by cannabinoid lipid ligands, which may be either endogenous or exogenous. The CB1 receptor is expressed mainly in the brain (central nervous system, CNS), but also in the lungs, liver, and kidneys. The CB2 receptor is mainly expressed in the immune system and in hematopoietic cells. Different articles (Járai et al., 1999; Ho and Hiley, 2003; McHugh et al., 2008) suggest the existence of an additional cannabinoid receptor. A recent study has shown that the protein GPR55 and a G protein-coupled receptor (GPCR) in the brain should, in fact, be characterized as cannabinoid receptors (Ryberg et al., 2007).

14.2 RISK ASSESSMENT

14.2.1 Risk Characterization

The occurrence of illicit drugs in the environment can potentially lead to adverse effects on a number of nontarget aquatic organisms. Although adverse effects of illicit drugs on animals have been described, very little is known about the toxicity of these compounds in an environmentally relevant scenario. A range of aquatic organisms are unintentionally exposed to a large number of illegal drug residues in their natural habitats (Zuccato and Castiglioni, 2009). Since many bioactive compounds are capable of affecting specific protein targets at relatively low doses, illicit drugs can potentially represent a hazard, even at environmentally relevant exposure levels. The risk assessment of existing illicit drugs is, however, hampered by the limited availability of experimental data.

Currently, the environmental risk assessment of chemicals is based on ecotoxicological data from standard assays. However, acute toxic effects in nontarget organisms are not expected or have hardly been observed for environmentally relevant concentrations. Given that illicit drugs are designed to elicit specific biological action, potentially, long-term effect information is more relevant for the assessment of unwanted effects. Predictions of possible impacts upon aquatic biodiversity should be based on comparative toxicity tests with species of different feeding preferences, habitats, physiology, and size to determine a toxicant's effects (Rodgers et al., 1997). The prediction of long-term effects in nontarget organisms may provide a way forward in order to allow a tailored risk assessment of illicit drugs. Even if the concentration of an individual drug is sufficiently low, there is perhaps need for concern if other drug sharing the same mode of action can combine to reach a threshold level (Silva et al., 2002).

The exposure potential (namely, the extent to which aquatic life might be exposed to these compounds) and the effects potential (that is, whether the compounds are present and at what levels they may affect aquatic life) need to be assessed to estimate the potential impacts to aquatic life. One method is to estimate the predicted or measured environmental concentration (PEC or MEC) and the predicted no-effect concentration (PNEC). PNECs are derived from experimentally generated effect concentrations multiplied by an assessment factor, which shall compensate for

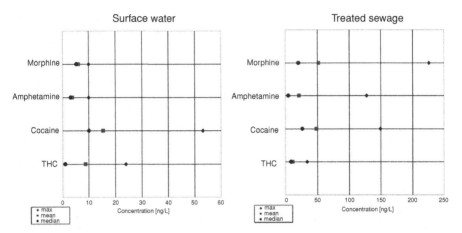

FIGURE 14.1   Concentration of illicit drugs in treated wastewater (a) and surface water (b) (grey rhombus = maximum value; grey square = mean; black circle = median) (Huerta-Fontela et al., 2008; Postigo et al., 2008; Castiglioni et al., 2006; Zuccato et al., 2005; Kasprzyk-Hordern et al., 2008; Zuccato et al., 2008; Gheorghe et al., 2008; Hummel et al., 2006; Bones et al., 2007; Boleda et al., 2007).

uncertainties based on interspecies and interlaboratory variance and the extrapolation from acute to chronic toxicity. PECs and PNECs are then used for the calculation of a "risk quotient" (RQ). In general, if PEC or MEC is less than PNEC (MEC/PNEC < 1), the environmental risk is deemed acceptable. In case of RQ >1 (PNEC are lower than PEC or MEC indicating the likelihood of an adverse effect), appropriate mitigation measures are considered.

MECs of illicit drugs have been mainly reported in sewage treatment plant (STP) effluents and in surface water in many countries, often at locations near STPs. Figure 14.1 summarizes the concentrations of most frequently assessed illicit drugs in wastewater and surface water reported so far. In STP effluents, a number of different drugs occur at concentrations generally in the nanogram to microgram per liter range. In rivers, lakes, and seawaters, they are in the nanogram per liter range (Huerta-Fontela et al., 2008; Postigo et al., 2008; Castiglioni et al., 2006; Zuccato et al., 2005; Kasprzyk-Hordern et al., 2008; Zuccato et al., 2008; Gheorge et al., 2008; Hummel et al., 2006; Bones et al., 2007; Boleda et al., 2007).

In Fig. 14.2, we summarized PNEC values for illicit drugs based on the available literature for different aquatic organisms (*Danio rerio, Oncorhynchus mykiss, Cyprinus carpio, Daphnia magna, D. pulex*), applying a safety factor of 10,000 (1000 accounting for interspecies, interlaboratory and acute to chronic extrapolation plus an additional safety factor of 10 to account for the availability of a very reduced toxicity data set) (Thomas, 1975; Martins et al., 2007; Guilhermino et al., 2000; Lilius et al., 1995; Morrow et al., 2001; Darland and Dowling, 2001).

According to our preliminary analysis of the literature, however, PNEC values of illicit drugs are about one order of magnitude higher than maximum concentrations

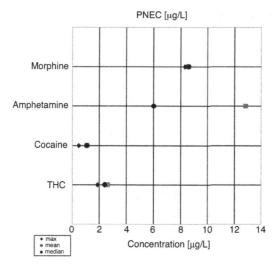

FIGURE 14.2 Predicted no-effect concentration of illicit drugs for aquatic organisms) (grey rhombus = maximum value; grey square = mean; black circle = median) (Thomas, 1975; Guilhermino et al., 2000; Lilius et al., 1995; Morrow et al., 2001; Darland and Dowling, 2001).

reported in either surface waters or STP effluents (Fig. 14.1 and 14.2). This consideration would suggest that no toxicity concerns apply for aquatic organisms. The highest PNEC for illicit drugs is 13 µg/L reported for amphetamines (Fig. 14.2) and the RQ does not exceed the level of one for any illicit drug we considered. For cocaine, morphine, amphetamines, and THC, the margins of safety appear to be, therefore, rather large. Concentration and toxicity data on illicit drugs in the environment are, however, currently insufficient to draw definitive conclusions. The RQ practice relies largely on standard toxicity tests, which have not been designed to target the specific effects elicited by illicit drugs, such as altered behavior. In addition, no information on illicit drugs' toxicity has been derived by quantitative structure–activity relationships (QSAR) predictions (Jones et al., 2002; Sanderson et al., 2004). This highlights the inadequacy of the RQ approach for the environmental risk assessment of illicit drugs.

On one side, more experimental data on chronic toxicity of illicit drugs as well as more information about MEC are needed to fully judge the environmental risk posed by these micropollutants. Many illicit drugs and their metabolites co-occur in surface waters and may show additive or synergistic effects on aquatic organisms, rendering their low levels of exposure more hazardous than expected. Aquatic species can also be exposed to illicit drugs at different stages of growth, including embryo development, for which no exposure or side effects are available. On the other hand, we suggest the necessity of new approaches in the initial desk-top risk assessment for illicit drugs in the environment, including *in silico* evaluation of potential drug targets.

14.2.2 Conservation of Drug Targets

In this study, we explore the application of an emerging approach in order to help improve our understanding of biological effects potentially elicited by illicit drugs, aiding to our current risk-assessment procedure. The information about illicit drug-target sites summarized in this review can be utilized to hypothesize and assess biological effects on a set of aquatic model organisms.

A molecular approach for the prediction of long-term effects has already been recognized as a valuable tool in regulatory applications (OECD, 2006; USEPA, 2007). Genomics and phylogenomics, which advocate an evolutionary view of genomic data, can be useful tools in the prediction of protein function, based on significant sequence and structural similarity, and of protein interactions, including biological processes. Gunnarsson et al. (2008), studying water-borne pharmaceuticals, recently suggested that orthology predictions can be used as a guide to prioritize test species for a certain drug, to interpret the relevance of existing ecotoxicity data, or to deduce which drugs may pose an increased risk to a certain group of organisms. Environmental effects of certain drugs can be associated with binding and interactions of the bioactive compound with conserved or convergent drug targets in wild species (Gunnarsson et al., 2008). These interactions can be anticipated by analysis of evolutionary or structural conservation of the drug targets in different organisms (Scholz et al., 2010). High homology may imply an evolutionary relationship between two sequences. Because products of homologous genes should tend to keep their original functions, targets with high homology may have a high probability of a functional interaction with the same illicit drug. Gunnarsson and colleagues suggested that the presence of human drug orthologs (and possible homolog receptors) in aquatic species may highlight possible effects of concern. In addition, the probability of a toxic effect in species with many human illicit drug target homologs and high degree of similarity will be great. The value of orthologous prediction has already been widely recognized in the field of pharmacology (Searls, 2003).

In this work, we seek to find the potential human illicit drug targets in a set of model aquatic species to further assess if some organisms may be potentially sensitive to these compounds and identify potential biological processes in model species that can be affected by illicit drugs in the environment. Sequence alignment tools such as BLAST and PSI-BLAST can be utilized for homology detection, aided by the increasing wealth of genome sequence databases (Altschul et al., 1990, 1997). If one or more model sequences of known function are found and exhibit a sufficiently high level of similarity to the query sequence, hypotheses may be possible on the function of the query sequence. A homology above 40% with conserved functional motifs is considered sufficient for hypothesizing that the query sequence has similar function to that of the model sequence (Lengauer and Zimmer, 2000). As the level of similarity decreases, the conclusions on function that can be drawn from sequence similarity become increasingly less reliable for prediction.

As a case study, a set of human illicit drug-target sequences were downloaded from DrugBank (http://www.drugbank.ca). We included primary and alternative targets.

We compiled homology predictions for these illicit drug targets for a group of model aquatic species commonly used in environmental risk assessment. They comprised *Danio rerio, Daphnia magna, Xenopus laevis and Chlamydomonas reinhardtii*, plus *Synechococcus elongatus* and *Escherichia coli* as prokaryotic representatives for comparison and *Drosophila melanogaster* to account for effects on insects (Table 14.1). We were not able to include any bentic macroinvertebrate in our selection, since no fully sequenced and annotated genomes are currently available in the databases for model aquatic species.

Homology data have been retrieved from the National Center for Biotechnology Information (NCBI) database (www.ncbi.nlm.nih.gov) using PSI-BLAST analysis based on our set of human drug-target protein sequences. Each human drug target was aligned with potential homolog from the translated genome sequences of our selection of model organisms.

In the aquatic vertebrate *D. rerio* all homologous sequences were predicted with relatively high similarity (above 48%) and E-score (threshold value $= 0.005$) (Table 14.1). In *X. leavis* the similarity was less, but still above 34%. Homologous sequences for all drug targets were also identified for *D. melanogaster, C. Reinhardtii,* and *S. elongatus* with, respectively, a similarity less then 57%, 44%, and 37% and E-scores lower then the threshold value. The total number of homologous sequences in *D. magna* and *E. coli* were fewer than in the vertebrates with a similarity prediction less than 44% (Table 14.1). The presence of a drug target ortholog in a species does not guarantee that a functional interaction with the drug can occur. Vice versa, functional interactions between a drug and other nonorthologous proteins are also possible (Gunnarssonn et al., 2003).

According to our preliminary assessment and among the evaluated species, aquatic vertebrates as expected appear to have the greatest number of human drug-target homologous sequences with the highest degree of similarity.

14.2.3   Identification of Possible Biological Processes at Risk

The classifications of amino acid sequences into families that form clusters of structurally or functionally related proteins can aid in the prediction of protein functions and related biological processes. To predict protein function from sequence similarity, a large database of protein sequences is screened for "model sequences," which exhibit a high level of similarity to the query protein sequence. The Gene Ontology (GO) annotation project aims to provide standard gene ontology annotations for genes and proteins (Gene Ontology Consortium, 2001). GO provides for each gene product a series of attributes across different species and databases (terms). GO terms describe gene product characteristics, such as the cellular component of destination and the molecular function within the cell, or general and specific biological processes in which genes/proteins are involved.

Our set of human and homolog illicit-drug targets were classified into different functional categories based on GO-associated terms. GO assignments for drug targets were achieved using Blast2GO, a research tool for sequence annotation (Conesa et al., 2005). For each organism of our preliminary selection (Table 14.1), we annotated

TABLE 14.1 Homology Prediction for Human Illicit Drug Targets in a Set of Model Aquatic Species[a]

Human Target	Drug or Class of Drugs	D. rerio	X. leavis	D. melanogaster	D. magna	C. reinhardtii	S. elongatus	E. coli
Dopamine receptor D1	Cocaine	73(9e-175)	80(0,0)	40(6e-63)	29(3.5)	21(2.7)	28(5.1)	Nd
Sodium channel protein type 10 subunit alpha	Cocaine	55(0,0)	48(5e-139)	48(0.0)	33 (2.3)	31(2e-100)	25(1e-04)	Nd
Sodium channel protein type 11	Cocaine	48(0,0)	40(4e-106)	45(0.0)	22(0.93)	29(5e-96)	26(1e-05)	32(6.2)
Sodium channel protein type 5 subunit alpha	Cocaine	64(0,0)	52(5e-160)	44(0.0)	27(1.8)	31(4e-101)	30(7e-05)	27(0.45)
Sodium- and chloride-dependent GABA transporter 1	Cocaine	84(0,0)	87(0.0)	57(0.0)	30(1.8)	32(6.9)	22(8e-19)	nd
Sodium-dependent noradrenaline transporter	Cocaine	71(0,0)	100(0,0)	55(0.0)	29(1.8)	21(0.36)	23(1e-14)	29(0.059)
D(3) dopamine receptor	Cocaine	57(1e-125)	46(1e-146)	37(2e-66)	31(1.4)	44(2.4)	23(2.9)	34(7.5)
κ-Type opioid receptor	Cocaine/morphine	69(1e-150)	48(2e-109)	36(8e-51)	31(3.0)	25(0.72)	31(0.76)	nd
Sodium-dependent dopamine transporter	Cocaine	71(24e-34)	64(3e-116)	41(5e-16)	29(1.8)	29(.7)	37(4.9)	nd
δ Opioid receptor	Morphine	68(2e-136)	43(5e-16)	33(9e-51)	20(3.6)	26(2.2)	22(0.55)	34(2.9)
μ Opiate receptor	Morphine	83(3e-154)	56(8e-112)	35(5e-51)	34(6.3)	32(1.9)	31(0.76)	26(6.7)
α-2 Adrenergic receptor	Amphetamine	56(1e-127)	34(1e-62)	46(3e-50)	44(2.7)	30(0.13)	27(2.9)	nd
α1C Adrenergic receptor	Amphetamine	59(1e-156)	36(1e-61)	25(2e-56)	34(6.3)	32(0.88)	31(3.2)	26(1.5)
Sodium-dependent dopamine transporter	Amphetamine	71(4e-34)	43(5e-16)	41(5e-16)	24(6.2)	29(3.7)	37(4.9)	nd
Cocaine- and amphetamine-regulated transcript protein precursor	Amphetamine/cocaine	55(1e-30)	73(5e-26)	35(3.9)	22(0.52)	40(2.2)	31(0.33)	26(9.9)
Trace amine receptor 1	Amphetamine	48(2e-90)	34(1e-44)	36(1e-44)	34(5.4)	31(1.8)	30(2.7)	22(4.8)
Cannabinoid receptor	THC	70(0,0)	83(0.0)	27(4e-20)	28(1.1)	27(1.8)	31(4.3)	19(3.2)

[a] For each entry a similarity (conservation of amino acids) percentage is reported together with the E-score (in brackets). E-values represent the expected number of similar alignment similarity score occurring by chance (the lower the E-value, the more significant the similarity score).

all possible biological processes associated with our identified set of illicit drug-target homologous sequences. Blast2GO was unable to predict potential biological processes in the *X. leavis* and *E. coli* sets of sequences because of poor sequence similarity or to low availability of annotation terms. Figure 14.3 gives reports on a ranking of the biological processes potentially affected by illicit drugs in our list of model species. Our preliminary analysis suggests that most drugs tend to interfere with processes like behavior, ion transport, signal transduction, homoeostasis, and oxidative stress. Many drugs may interact with more than one target with convergence of different proteins in the same biological process.

Illicit drug targets in humans, *D. rerio*, and *D. melanogaster* involve several common processes, such as altered behavior, and functions, such as altered adenylate cyclase activity, neurotransmitter transport, and sodium ion transport (Fig. 14.3). Adenylate cyclase signaling is an illicit drug-target process that appears to be common in humans, fish, *D. melanogaster*, and *Daphnia*. Targeted processes predicted in *D. rerio* included behavior, nervous system development, sensory perception, and immune response. Zebrafish CNS, although perhaps less complex, is essentially organized like the mammalian (Rink and Wullimann, 2001).

Several recent articles support our inferred drug-target processes and, in particular, altered behavioral responses in zebrafish as a consequence of exposure to illicit drugs. When treated with cocaine, zebrafish have displayed slow circling, lower water column position, and extended fins – as in mating behavior. In small groups of fish, cocaine induced a striking increase in aggressive behavior marked by dominance displays and chasing. Cocaine also induced a decrease in visual sensitivity (Darland and Dowling, 2001).

A recent study (Rihel et al., 2010) shows relationships between drugs and their targets, demonstrating a conserved vertebrate neuropharmacology. Behavioral profiling of psychoactive drugs in zebrafish reveal new factors implicated in responses such as an ether-a-go-go-related gene (ERG), encoding a potassium voltage-gated channel, and immunomodulators in control of rest and locomotor activity. Behavioral effects can also be linked to amphetamines and have been mainly attributed to changes in dopamine transmission. In the knifefish *Gymnotus carapo*, amphetamine treatment significantly decreased the intensity of aggressive behavior typical of this species, while it increases the intensity and probability of occurrence of jamming avoidance response (JAR) (Capurro et al., 1997).

Biological processes affected by illicit drugs in *D. melanogaster* mainly pertain to behavioral aspects. *Drosophila melanogaster* responds, after repeated cocaine exposures, with an increase in the locomotor response and with many of the same stereotypic motor behaviors seen in vertebrates (McClung and Hirsh, 1998). Psychostimulants, like cocaine and amphetamines, are potent inhibitors of all three monoamine transporters (Ritz et al., 1987) which help to maintain low extracellular amine concentrations. Disruption of the monoamine transporter genes in mice results in profound behavioral (Giros et al., 1996) and neurochemical (Bengel et al., 1998; Xu et al., 2000) changes. In *D. magna* the most likely processes that could be affected by the presence of illicit drugs appeared to be behavior, oxidative stress regulation, female meiosis, and the regulation of cell proliferation. Our analysis of

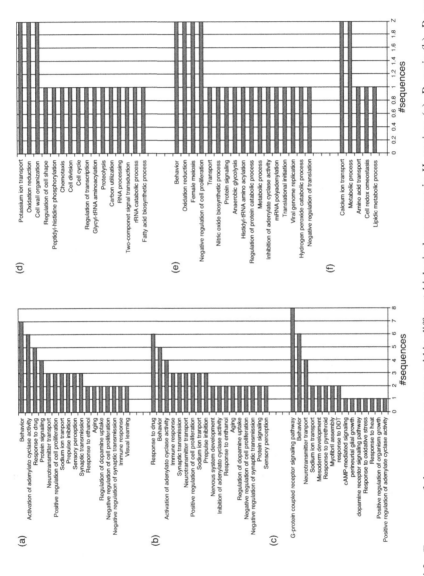

FIGURE 14.3 Frequency of drug target sequences within different biological processes in *H. sapiens* (a), *D. rerio* (b), *D. melanogaster* (c), *S. elongatus* (d), *D. magna* (e), and *C. reinhardtii* (f). The most frequent biological processes reported for each organism are the ones that may potentially be more affected by the presence of illicit drugs.

GO terms also showed an over-representation of enzymes involved in the biosynthesis of nitric oxide, a molecule that can be involved in a number of signaling processes.

The reduced number of available illicit drug-target homologs in the green alga *Chlamydomonas reinhardtii* resulted in a limited range of possible affected biological processes. Prediction of possible effects in this species is, therefore, very restrictive. Cellular redox homoeostasis and ion transport are, however, common illicit drug-target processes that have been highlighted by our analysis in all our model organisms. In *Chlamydomonas*, calcium transport is also involved in cell–cell interactions leading to cell fusion and mating (Goodenough et al., 1993).

Voltage-dependent channels are present in virtually all organisms from bacteria to animals and the voltage-gated Na^+ channels were predicted by our analysis to be rather conserved as illicit drug targets. For example cocaine, which acts on voltage-gated Na^+ channels in humans, also increases intracellular sodium concentration via the same mechanism in pond snail neurons (Onizuka et al., 2004). In our preliminary analysis homologous sequences for this human target were predicted in *D. rerio, C. reinhardtii,* and *S. elongatus* and annotated, respectively, as voltage-gated Na^+, voltage-gated Ca^{2+} channel, and K^+ channel proteins. As mentioned above, cyanobacteria may actually be very sensitive to sodium channel modulators. Chemotaxis also appeared to be a biological process potentially affected by illicit drugs in cyanobacteria. There is evidence that channel blockers also interfere with chemotaxis, motility, and metabolism in heterotrophic bacteria (Tisa et al., 2000; Pomati et al., 2003). Other important cyanobacterial processes that emerge include cell wall organization and carbon utilization, which may further indicate detrimental effects of illegal drugs on algal metabolism. To further emphasize metabolic and growth effects, hystidyl-tRNA aminoacylation, mRNA polyadenylation, glycyl-tRNA aminoacylation, and RNA processing, in general, appeared to represent illicit drug-target processes in both *S. elongates* and *D. magna*.

14.3 GENERAL CONCLUSIONS AND FUTURE DIRECTIONS

Illicit drugs are secondary metabolites often produced to defend plants against herbivores, pathogens, insects, or competitors and so there would be a reason to hypothesize that these substances once excreted in the environment may have some toxic effects on nontarget organisms. Most of toxicological data of illicit drugs found in the literature are, however, on acute toxicity in humans or model organisms and are, therefore, insufficient for a complete ecotoxicological assessment.

Our preliminary risk characterization performed with the scarce data found in literature suggests that no toxicity concerns apply for aquatic organisms. Yet, even if the concentration of an individual drug is sufficiently low, there is need to identify synergistic or antagonistic effects, as well long-term effects, to fully judge the environmental risk, especially in nontarget organisms. Aquatic species can in fact be continually exposed to illicit drugs and at different stages of growth, including embryo development.

Besides the need for more comprehensive experimental data of illegal drugs, there is the necessity of new approaches in the initial desk-top risk assessment. An emerging and useful approach for the illicit drug risk assessment is the orthology and homology predictions. As a case study, we compared the conservation of drug targets and predicted hypothesis regarding possible biological processes affected by illicit drugs in a representative set of organisms. This approach can be a guide to prioritize tests, to understand the biological effects potentially elicited by illicit drugs or to deduce which drugs may pose an increased risk to a certain group of organisms (Gunnarsson et al., 2008).

In our proof-of-principle assessment, PSI-BLAST analysis of each human drug target showed that homology decreases with increasing evolutionary distance between studied species and man. Aquatic vertebrates (represented by *D. rerio*) appear to have the greatest number of human drug-target homologous sequences with the highest degree of similarity (above 48%). All possible biological processes associated with our identified set of illicit drug-target homologous sequences were annotated except for *X. leavis* and *E. coli*. This tentative analysis of biological processes involved suggests that many illicit drugs tend to interfere with processes like behavior, ion transport, signal transduction, homoeostasis, and oxidative stress.

Several recent articles support our inferred drug-target processes, especially for *D. rerio* and *D. melanogaster*. Cocaine and other local anesthetics are already known to have specific effects on zebrafish behavior (Darland and Dowling, 2001) and psychostimulants are also known as potent inhibitors of all three monoamine transporters (Ritz et al., 1987) involved in behavioral (Giros et al., 1996) and neurochemical (Bengel et al., 1998; Xu et al., 2000) changes in insects. A negative influence on the behavior of species, such as zebrafish, may suggest possible impacts on the ecology of fish and, therefore, we propose that aquatic environmental risk assessments for illicit drugs should always include comprehensive studies on aquatic vertebrates including reproduction and behavioral responses.

Suzuki et al. (2000, 2004) have shown how synthetic local anesthetics (cocaine-derived) enhance the growth of several algae, including cyanobacteria. We highlight in this study that illicit drugs can target processes like hystidyl-tRNA aminoacylation, mRNA polyadenylation, glycyl-tRNA aminoacylation, and RNA processing in *Synechococcus*, which emphasize metabolic and growth effects. Lidocaine has been shown to stimulate growth and intracellular accumulation of the potent neurotoxin saxitoxin in cyanobacteria (Pomati et al., 2004, 2006) at the microgram per liter concentration range; cocainelike compounds may, therefore, influence the development of harmful algal blooms (HABs). HABs can cause negative impacts to both the ecosystem and the services that aquatic ecosystem can provide, including drinking water and recreation. Cyanobacteria should also be considered in detail for an environmental risk assessments of human drugs. Pollutant absorption by microalgae may also have repercussions throughout the food chain because of biomagnification.

Genomic and proteomic approaches may be useful to provide information on subtle effects of illicit drugs on aquatic organisms. In addition, with the study of metabolic changes in organisms in response to environmental challenges (environmental metabolomics), hundreds of metabolites can be monitored simultaneously, providing a much more comprehensive snapshot of the effects that a particular chemical has on a living organism (Bundy et al., 2009). These approaches may also be useful to identify biomarkers, sensitive bioindicators of effect and exposure of illicit drugs, which can provide early warning of an environmental damage.

REFERENCES

Altschul, S. F., W. Gish, W. Miller, E. W. Myers, and D. J. Lipman. 1990. Basic local alignment search tool. *J. Mol. Biol.* **215**:403–410.

Altschul, S. F., T. L. Madden, A. A. Schaffer, J. Zhang, Z. Zhang, W. Miller, and D. J. Lipman. 1997. Gapped BLAST and PSI-BLAST: A new generation of protein database search programs. *Nucl. Acids Res.* **25**:3389–3402.

Banerjee, S. P., V. K. Sharma, Kung, and S. Lily. 1978. Amphetamine induces β-adrenergic receptor supersensitivity. *Nature (London)* **271**(5643):380–381.

Bedford, J. A., C. E. Turner, and H. N. Elsohly. 1982. Comparative lethality of coca and cocaine. *Pharmacol. Biochem. Behav.* **17**(5):1087–1088.

Bell, E. A., and B. V. Charlwood. 1980. *Encyclopedia of Plant Physiology: Secondary Plant Products*. Berlin: Springer-Verlag.

Bengel, D., D. L. Murphy, A. M. Andrews, C. H. Wichems, D. Feltner, A. Heils, R. Mossner, H. Westphal, and K. P. Lesch. 1998. Altered brain serotonin homeostasis and locomotor insensitivity to 3,4-methylenedioxymethamphetamine ("Ecstasy") in serotonin transporter-deficient mice. *Mol. Pharmacol.* **53**:649–655.

Boleda, M. R., M. T. Galceran, and F. Ventura. 2007. Trace determination of cannabinoids and opiates in wastewater and surface waters by ultra-performance liquid chromatography–tandem mass spectrometry. *J. Chromatogr. A* **1175**:38–48.

Bones, J., K. V. Thomas, and P. Brett. 2007. Using environmental analytical data to estimate levels of community consumption of illicit drugs and abused pharmaceuticals. *J. Environ. Monit.* **9**:701–707.

Braude, M. C., and J. P. Ludford (editors).1984. *Marijuana: Effects on the Endocrine and Reproductive Systems*, pp. 82–96. NIDA, Rockville.

Budavari, S. (editor) 1996. *The Merck Index—An Encyclopedia of Chemicals, Drugs, and Biologicals*, p. 631. Whitehouse Station, NJ: Merck and Co.

Bundy, J. B., M. P. Davey, and M. R. Viant. Environmental metabolomics: a critical review and future perspectives. *Metabolomics* **5**(1):3–21.

Campos, D., G. Noratto, R. Chirinos, C. Arbizu, W. Roca, and L. Cisneros-Zevallos. 2006. Antioxidant capacity and secondary metabolites in four species of Andean tuber crops: native potato (*Solanum* sp.), mashua (*Tropaeolum tuberosum* Ruiz & Pavon), Oca (*Oxalis tuberosa* Molina) and ulluco (*Ullucus tuberosus* Caldas). *J. Sci. Food Agr.* **86**:1481–1488.

Capurro, A., M. Reyes-Paroda, D. Olazabal, R. Perrone, R. Silveira, and O. Macadar. 1997. Aggressive behaviour and jamming avoidance response in weakly electric fish *Gymnotus carapo*: effects of methylenedioxymethamphetamine (MDMA). *Comp. Biochem. Physiol.* **118A**:831–840.

Castiglioni, S., E. Zuccato, E. Crisci, C. Chiabrando, R. Fanelli, and R. Bagnati. 2006. Identification and measurement of illicit drugs and their metabolites in urban wastewater by liquid chromatography-tandem mass spectrometry. *Anal. Chem.* **78**:8421–8429.

Conesa, A., S. Götz, J. M. Garcia-Gomez et al. 2005. Blast2GO: A universal tool for annotation, visualization and analysis in functional genomics research. *Bioinformatics* **21**:3674–3676.

Darland, T., and J. E. Dowling. 2001. Behavioral screening for cocaine sensitivity in mutagenized zebrafish. *Proc. Natl. Acad. Sci. U.S.* **98**(20):11691–11696.

Davis, W. A., H. T. Hatoum, and L. W. Waters. 1987. Toxicity of MDA (3,4-methylenedioxy-amphetamine,l considered for relevance to hazards of MDMA (Ecstasy) abuse. *Alcohol Drug Res.* **7**:123–134.

Di Chiara, G., V. Bassareo, S. Fenu, M. A. De Luca, L. Spina, C. Cadoni, E. Acquas, E. Carboni, V. Valentini, and D. Lecca D. Dopamine and drug addiction: the nucleus accumbens shell connection. *Neuropharmacology* **47**(Suppl. 1):227–241.

Ellinwood, E. H., G. R. King, and T. H. Lee. 1998. Chronic amphetamine use and abuse. In Psychopharmacology: The Fourth Generation of Progress edited by F. E. Bloom and D. J. Kupfer, fourth ed. CD Rom Version. Philadelphia, PA: Lippincott, Williams and Wilkins; 1998.

Ersche, K. D., Roiser J. P., Robbins T. W., Sahakian B. J. Chronic cocaine but not chronic amphetamine use is associated with perseverative responding in humans. *Psychopharmacology* 2008; 24 Jan, 2008 (e-pub ahead of print)

Gene Ontology Consortium Creating the gene ontology resource: design and implementation. *Genome Res* 2001; 1425–1433.

Gérard C. M., C. Mollereau, G. Vassart, M. Parmentier. Molecular cloning of a human cannabinoid receptor which is also expressed in testis. *Biochem J* 1991; **279**(1):129–134.

Gheorghe A., A. Van Nuijs, B. Pecceu, L. Bervoets, P. G. Jorens, R. Blust, H. Neels, A. Covaci. Analysis of cocaine and its principal metabolites in waste and surface water using solid-phase extraction and liquid chromatography-ion trap mass spectrometry. *Anal Bioanal Chem* 2008; **391**:1309–1319.

Gilman, A. G., Goodman L. S., Gilman A.. In: Goodman L. S., Gilman, A., editors. *The pharmacological basis of therapeutics.* 7th ed. New York: Macmillan; 1985. p 550.

Giros, B., M. Jaber, S. R. Jones, R. M. Wightman, and M. G. Caron. 1996. Hyperlocomotion and indifference to cocaine and amphetamine in mice lacking the dopamine transporter. *Nature (London)* **379**:606–612.

Goodenough, U. W., B. Shames, L. Small, T. Saito, R. C. Crain, M. A. Sanders, and J. L. Salisbury. 1993. The role of calcium in the *Chlamydomonas reinhardtii* mating reaction. *J Cell Biol.* **121**:365–374.

Gossel, T. A., and J. D. Bricker. 1994. Principles of Clinical Toxicology. third ed, p. 421. New York, NY: Raven Press.

Graham, E. S., J. C. Ashton, M. Glass. 2009. Cannabinoid receptors: a brief history and "what's hot." *Front Biosci* **14**:944–957.

Guilhermino, L., T. Diamantino, M. C. Silva, and A. M. V. M. Soares. 2000. Acute toxicity test with *Daphnia magna*: An alternative to mammals in the prescreening of chemical toxicity? *Ecotoxicol. Environ. Safety* **46**:357–362.

Gunnarsson, L., J. W. Thornton, E. Need, and D. Crews. 2003. Resurrecting the ancestral steroid receptor: ancient origin of estrogen signaling. *Science* **301**:1714–1717.

Gunnarsson, L., A. Jauhiainen, E. Kristiansson, O. Nerman, and D. G. J. Larsson. 2008. Evolutionary conservation of human drug targets in organisms used for environmental risk assessments. *Environ. Sci. Technol.* **42**:5807–5813.

Haddad, L.M. 1990. Clinical Management of Poisoning and Drug Overdose, second edition, p. 731. Philadelphia, PA: W.B. Saunders.

Harvey, A. L. 1993. Natural and Synthetic Neurotoxins, p. 268. London Academic Press.

Ho, W. S., and Hiley, C. R. 2003. Vasodilator actions of abnormal-cannabidiol in rat isolated small mesenteric artery. *Br. J. Pharmacol.* **138**(7):1320–1332.

Howlett, A. C. 2002. The cannabinoid receptors. *Prostaglandins Other Lipid Mediat.* **68–69**:619–631.

Huang, Y., S. H. Ho, and R. K. Manjunatha. 1999. Bioactivities of safrole and isosafrole on *Sitophilus zeamais* (Coleoptera: Curculionidae) and *Tribolium castaneum* (Coleoptera: Tenebrionidae). *J. Econ. Entomol.* **92**:676–683.

Huerta-Fontela, M., M. T. Galceran, and F. Ventura. 2008. Occurrence of psychoactive stimulatory drugs in wastewaters in north-eastern Spain. *Sci. Total Environ.* **397**:31–40.

Hummel, M., and E. Unterwald. 2002. D1 dopamine receptor: a putative neurochemical and behavioral link to cocaine action. *J. Cell. Physiol.* **191**(1):17–27.

Hummel, D., D. Loeffler, G. Fink, and T. A. Ternes. 2006. Simultaneous determination of psychoactive drugs and their metabolites in aqueous matrices by liquid chromatography mass spectrometry, *Environ. Sci. Technol.* **40**:7321–7328.

Imming, P., C. Sinning, and A. Meyer. 2006. Drugs, their targets and the nature and number of drug targets. *Nat. Rev. Drug Discov.* **5**(10):821–834.

Járai, Z., J. A. Wagner, K. Varga, K. D. Lake, D. R. Compton, B. R. Martin, A. M. Zimmer, T. I. Bonner, N. E. Buckley, E. Mezey, R. K. Razdan, A. Zimmer, and G. Kunos G. 1999. Cannabinoid-induced mesenteric vasodilation through an endothelial site distinct from CB1 or CB2 receptors. *Proc. Natl. Acad. Sci. US* **96**(24):14136–14141.

Johnston, G. A., M. Chebib, J. R. Hanrahan, and K. N. Mewett. 2003. GABAC receptors as drug targets. *Curr. Drug Targets CNS Neurol. Disord.* **2**:260–268.

Jones, O. A. H., N. Voulvoulis, and J. N. Lester. 2002. Aquatic environmental assessment of the top 25 English prescription pharmaceuticals. *Water Res.* **36**:5013–5022.

Juhila, J., A. Honkanen, J. Sallinen, A. Haapalinna, E. R. Korpi, and M. Scheinin. 2005. Alpha(2A)-Adrenoceptors regulate d-amphetamine-induced hyperactivity and behavioural sensitization in mice. *Eur. J. Pharmacol.* **517**(1–2):74–83.

Kamdem, D. P., and D. A. Gage. 1995. Chemical composition of essential oil from the root bark of *Sassafras albidum. J. Org. Chem.* **61**(6):574–575.

Kasprzyk-Hordern, B., R. M. Dinsdale, and A. J. Guwy. 2008. The occurrence of pharmaceuticals, personal care products, endocrine disruptors and illicit drugs in surface water on South Wales, UK. *Water Res.* **42**:3498–3518.

Knapp, R. J., E. Malatynska, L. Fang, X. Li, E. Babin, M. Nguyen, G. Santoro, E. V. Varga, V. J. Hruby, W. R. Roeske, et al. 1994. Identification of a human delta opioid receptor: cloning and expression. *Life Sci.* **54**(25):463–469.

Lambert, D. G. 2004. Drugs and receptors. *BJA - CEPD Rev.* **4**(6):181–184(4).

Lee, K. W., Y. Kim, A. M. Kim, K. Helmin, A. C. Nairn, and P. Greengard. 2006. Cocaine-induced dendritic spine formation in D1 and D2 dopamine receptor-containing medium spiny neurons in nucleus accumbens. *Proc. Natl. Acad. Sci. US* **28**;103(9):3399–3404.

Lengauer, T., and R. Zimmer. 2000. Protein structure prediction methods for drug design. *Briefings Bioinform.* **1**(3):275–288.

Levin, E. D., Z. Bencan, and D. T. Cerutti. 2007. Anxiolytic effects of nicotine in zebrafish. *Physiol. Behav.* **90**:54–58.

Lilius, H., T. Hastbacka, and B. Isomaa. 1995. A comparison of the toxicity of 30 reference chemicals to *Daphnia magna* and *Daphnia pulex. Environ. Toxicol. Chem.* **14**:2085–2088.

Lydon, J., A. H. Teramura, and C. B. Coffman. 1987. UV-B radiation effects on photosynthesis, growth and cannabinoid production of two Cannabis sativa chemotypes. *Photochem. Photobiol. A* **46**:201.

Mach, D. I., and M. B. Livingston. 1922. Effects of cocaine on the growth of *Lupinus albus*. A contribution to the comparative pharmacology of animal and plant protoplasm. *J. Gen. Physiol.* **4**:573–584.

Mackie, K. 2008. Cannabinoid receptors: where they are and what they do. *J. Neuroendocrinol.* **20**(Suppl 1):10–4.

Martins, J., O. L. Teles, and V. Vasconcelos 2007. Assays with *Daphnia magna* and *Danio rerio* as alert systems in aquatic toxicology. *Environ. Int.* **33**(3):414–425.

Mc Calla, T. M., and F. A. Haskins. 1964. Phytotoxic substances from soil microorganisms and crop residues. *Bacteriol. Rev.* **28**(2):181–207.

McClung, C., and J. Hirsh. 1998. Stereotypic behavioral responses to free-base cocaine and the development of behavioral sensitization in *Drosophila. Curr. Biol.* **8**:109–112.

Mc Hugh, D., C. Tanner, R. Mechoulam, R. G. Pertwee, and R. A. Ross. 2008. Inhibition of human neutrophil chemotaxis by endogenous cannabinoids and phytocannabinoids: evidence for a site distinct from CB1 and CB2. *Mol. Pharmacol.* **73**(2):441–450.

Mas, M., M. Farré, R. de la torre, et al. 1999. Cardiovascular and neuroendocrine effects and pharmacokinetics of 3,4-methylenedioxymethamphetamine in humans. *J. Pharmacol. Exp. Therap.* **290**:136–145.

Matsuda, L. A., S. J. Lolait, M. J. Brownstein, A. C. Young, and T. I. Bonner. 1990. Structure of a cannabinoid receptor and functional expression of the cloned cDNA. *Nature (London)* **346**(6284):561–564.

Moore, M. T., S. L. Greenway, J. L. Farris, and B. Guerra. 2008. Assessing caffeine as an emerging environmental concern using conventional approaches. *Arch. Environ. Contaminations Toxicol.* **54**(1):31–35.

Morrow, D., D. Corrigan, and S. Waldren. 2001. Development of a bioassay for phytochemicals using *Daphnia pulex. Planta Med.* **67**(9):843–846.

Nathanson, J. A., E. J. Hunnicutt, L. Kantham, and C. Scavone. 1993. Cocaine as a naturally occurring insecticide. *Proc. Natl. Acad. Sci. US.* **90**:9645–9648.

OECD. 2006. Report of the validation of the 21-day fish screening assay for the detection of endocrine substances (phase 1b), series on testing and assessment. Number 61, ENV/JM/MONO 29.

Onizuka, S., T. Kasaba, T. Hamakawa, S. Ibusuki, and M. Takasaki. 2004. Lidocaine increases intracellular sodium concentration through voltage-dependent sodium channels in an identified lymnaea neuron. *Anesthesiology* **101**:110–120.

Orchard, I. 1982. Octopamine in insects: neurotransmitter, neurohormone, and neuromodulator. *Can. J. Zool.* **60**:659–669.

Osborne, R. H. 1996. Insect neurotransmission: neurotransmitters and their receptors. *Pharmacol. Therap.* **69**:117–142 .

Overington, J. P., B. Al-Lazikani, and A. L. Hopkins. 2006. How many drug targets are there? *Nat. Rev. Drug Discov.* **5**(12):993–996.

Paerl, H., and J. Huisman. 2008. Blooms like it hot. *Science* **320**:57–58.

Pinal-Seoane, N., I. R. Martin, V. Gonzalez-Nunez, E. M. de Velasco, F. A. Alvarez, R. G. Sarmiento, and R. E. Rodriguez. 2006. Characterization of a new duplicate delta-opioid receptor from zebrafish. *J. Mol. Endocrinol.* **37**:391–403.

Pomati, F., C. Rossetti, D. Calamari, and B. A. Neilan. 2003. Effects of saxitoxin (STX) and veratridine on bacterial Na^+ -K^+ fluxes: a prokaryote-based STX bioassay. *Appl. Environ. Microbiol.* **69**(12):7371–7376.

Pomati, F., C. Rossetti, G. Manarolla, B. P. Burns, and B. A. Neilan. 2004. Interactions between intracellular Na^+ levels and saxitoxin production in *Cylindrospermopsis raciborskii* T3. *Microbiology* **150**(Pt 2):455–461.

Pomati, F., R. Kellmann, R. Cavalieri, B. P. Burns, and B. A. Neilan. 2006. Comparative gene expression of PSP-toxin producing and non-toxic *Anabaena circinalis* strains. *Environ Int.* **32**(6):743–748.

Postigo, C., M. J. Lopez de Alda, and D. Barceló. 2008. Fully automated determination in the low nanogram per litre level of different classes of drugs of abuse in sewage water by on-line solid-phase extraction–liquid chromatography–electrospray–tandem mass spectrometry. *Anal. Chem.* **80**:3123–3134.

Rihel, J., D. A. Prober, A. Arvanites, K. Lam, S. Zimmerman, S. Jang, S. J. Haggarty, D. Kokel, L. L. Rubin, R. T. Peterson, and A. F. Schier. 2010. Zebrafish behavioral profiling links drugs to biological targets and rest/wake regulation. *Science* **327**(5963):348–351.

Rink, E., and M. F. Wullimann. 2001. The teleostean (zebrafish) dopaminergic system ascending to the subpallium (striatum) is located in the basal diencephalon (posterior tuberculum). *Brain Res.* **889**:316–330.

Ritz, M. C., R. J. Lamb, S. R. Goldberg, and M. J. Kuhar. 1987. Cocaine receptors on dopamine transporters are related to self-administration of cocaine. *Science* **237**:1219–1223.

Rodgers, J. H., Jr, E. Deaver, B. C. Suedel, and P. L. Rogers. 1997. Comparative aqueous toxicity of silver compounds: laboratory studies with freshwater species. *Bull. Environ. Contamination Toxicol.* **58**:851–858.

Ryberg, E., N. Larsson, S. Sjögren, S. Hjorth, N. O. Hermansson, J. Leonova, T. Elebring, K. Nilsson, T. Drmota, and P. J. Greasley. 2007. The orphan receptor GPR55 is a novel cannabinoid receptor. *Br. J. Pharmacol.* **152**(7):1092–1101.

Sanderson, H., D. Johnson, T. Reitsma, C. Wilson, R. Brain, and K. R, Solomon. 2004. Ranking and prioritization of environmental risks of pharmaceuticals in surface waters. *Regul. Toxicol. Pharmacol.* **39**:158–183.

Searls, D. B. 2003. Pharmacophylogenomics: genes, evolution and drug targets. *Nat. Rev. Drug Discov.* **2**:613–623.

Scholz, S., K. Schirmer, and R. Altenburger. 2010. Pharmaceutical contaminants in urban water cycles—a discussion of novel concepts for environmental risk assessment. In: Xenobiotics in the Urban Water Cycle; Mass Flows, Environmental Process, Mitigation and Treatment Strategies, Environmental Pollution, edited by D. Fatta-Kassinos, K. Bester, and K. Kümmerer,, Vol. 16, DOI 10.1007/978-90-481-3509-7_12; Berlin: Springer.

Silva, E., N. Rajapakse, and A. Kortenkamp. 2002. Something from "nothing" − eight weak estrogenic chemicals combined at concentrations below NOECs produce significant mixture effects. *Environ. Sci. Technol.* **36**(8):1751–1756.

Suzuki, T., and M. Misawa. 1997. Opioid receptor types and dependence. *Nippon Yakurigaku Zasshi* **109**(4):165–174.

Suzuki, T., T. Ezure, T. Yamaguchi, H. Domen, M. Ishida, and W. Schmidt. 2000. Stimulatory effect of procaine on the growth of several microalgae and cyanobacteria. *Pharmaceut. Press* **52**:243–251.

Suzuki, T., K. Nakasato, S. Shapiro, F. Pomati, A. Brett, and B. A. Neilan. 2004. Effects of synthetic local anaesthetics on the growth of the cyanobacterium *Synechococcus leopoliensis*. *J. Appl. Phycol.* **16**:(2)145–152.

Taiz, L., and E. Zeiger. 1991. *Plant Physiology*, p. 559. Redwood, CA: Benjamin/Cummings; 1991.

Thomas, R. J. 1975. The toxicologic and teratologic effects of Δ 9-tetrahydrocannabinol in the zebrafish embryo. *Toxicol. Appl. Pharmacol.* **32**(1):184–190.

Thomson Healthcare, Inc. 2006. *Physicians' Desk Reference*, 60th edition, pp. 1175–1177 Montvale, NJ: Thomson PDR.

Tisa, L. S., J. J. Sekelsky, and J. Adler. 2000. Effects of organic antagonists of Ca^{2+}, Na^+, and K^+ on chemotaxis and motility of *Escherichia coli*. *J. Bacteriol.* **182**:4856–4861.

USEPA 2007. Validation of the fish short-term reproduction assay: integrated summary report, U.S. Environmental Protection Agency, Endocrine Disruptor Screening Program, Washington, DC, December 15, 2007.

Van Der Woude, F. J. 2000. Cocaine use and kidney damage. *Nephrol. Dialysis Transplant.* **15**(3):299–301.

Xu, F., R. R. Gainetdinov, W. C. Wetsel, S. R. Jones, L. M. Bohn, G. W. Miller, Y. M. Wang, and M. G. Caron. 2000. Mice lacking the norepinephrine transporter are supersensitive to psychostimulants. *Nat. Neurosci.* **3**:465–471.

Zuccato, E., and S. Castiglioni. 2009. Illicit drugs in the environment. *Philos. Trans. A Math. Phys. Eng. Sci.* **367**:3965–3978.

Zuccato, E., C. Chiabrando, S. Castiglioni, D. Calamari, R. Bagnati, S. Schiarea, and R. Fanelli. 2005. Cocaine in surface waters: a new evidence-based tool to monitor community drug abuse. *Environ. Health Global Access Sci. Source* **4**:1–7.

Zuccato, E., S. Castiglioni, R. Bagnati, C. Chiabrando, P. Grassi, and R. Fanelli. 2008. Illicit drugs, a novel group of environmental contaminants. *Water Res.* **42**:961–968.

CHAPTER 15

DRUG ADDICTION – POTENTIAL OF A NEW APPROACH TO MONITORING DRUG CONSUMPTION

NORBERT FROST

15.1 INTRODUCTION

This chapter is about drugrelated information, reliability of data, why these data are collected, and who takes responsibility in this field, which is a highly stigmatized behavior. This question will be discussed from changing viewpoints. Are there alternatives or complementary methods that exist to obtain such information?

Drug monitoring in urban wastewater systems will be the principal perspective here as a source to estimate overall consumption of dedicated substances within given populations and spatial context. The aim is not to compete with existing methods but to provide input, which could be helpful in developing new opinions to enrich existing knowledge. Wastewater drug monitoring can be particularly interesting, when there is pressure to react, e.g., where short-term information about new trends in drug consumption is required. Since other chapters are provide in-depth contributions regarding technical and methodological aspects, the discussion here will focus mainly on "philosophical" concepts around drug addiction and drug monitoring, including existing instruments for data collection.

15.1.1 Drug Addiction – Nothing Is "Normal"

How many people die directly or indirectly from taking drugs and how many drug users have been sentenced for committing a drug-related crime? How can supply routes be successfully identified and what would this mean in terms of countermeasures? How many people seek drug addiction treatment? Are there sufficient healthcare services and who winds up paying the bill? Monitoring illicit drug use

Illicit Drugs in the Environment: Occurrence, Analysis, and Fate Using Mass Spectrometry, Edited by Sara Castiglioni, Ettore Zuccato, and Roberto Fanelli
Copyright © 2011 John Wiley & Sons, Inc.

and its synergistic effects is a complex task where information is not always easy to obtain. The UNODC estimates that between 172–250 million persons worldwide have used illicit drugs at least once during the past year (World Drug Report, 2009).

Drug addiction is a disease, the result of a way of living, of conscious or unconscious decisions, the consequence of misconduct, your own fault. There are many explanations and definitions to classify substance dependence. Since this is costly for society and unhealthy for the individual, it is constantly surveyed. However, it is not just health consequences that are being looked at. It is also consumption of illicit drugsthat will leadto a a clinical picture of drug dependence. However, even if considered a disease, drug addiction is special in the canon of human suffering, as is its monitoring.

Drug monitoring trys to understand and to describe a cultural and societal phenomenon, which does not follow normative conventions. There are "intrinsic" difficulties, expressed in a variety of ideological readings and interpretations around the concept of drug use and drug addiction. Dependence on one or a variety of substances is the result of consumption habits, their long-term negative effects on the organism, and an increasing loss of control when seeking a stimulant. For a long time, drug addiction was regarded as a consequence of individual failure and misconduct. With new research results and increasing knowledge, derivedfrom the field of neuroscience, there has been a shift occurring where the scientific community began to understand and to describe drug addiction as a disease (Koob and Le Moal. 2005). Actually, there is an ambiguous situation where we are close to hermeneutics. Existing definitions are either, stressing the disease, or the guilt and criminal aspect or they are presenting mixed concepts, quite often in the realm of political ideologies. This ambiguity in perception is also manifested in the fact that in many countries drug addiction affairs are administratively settled between ministries of health and/or justice. The European Monitoring Centre for Drugs and Drug Addiction has strong connections to the Directory General for Justice, Freedom and Security, and DG-SANCO, which represents the European Commission's directorate for health related Questions and consumer protection. Apart from ideological nuances, there is common sense, of course, that drug addiction creates a large variety of problems, where it is a burden to any society. Public measures and drug programs are focused accordingly, depending on the ideological perspective and emphasis.

To provide a brief example: Diabetes Mellitus type II, which is also called "the metabolic syndrome," is a disease of civilization (Pschyrembel, 2007). As such, it is culturally embedded, with immense consequences at the individual level and costs for the society at large (Pschyrembel, 2007). The same applies for drug addiction. Both diseases depend, in part, on genetic disposition and on individual behavioral aspects, where external impulses play an essential, additional role, in lifestyle modulation. For either disease, the public health systems have to cope with the consequences of providing medical staff, treatment, pharmaceuticals, infrastructure, etc. There are insurance conditions for the treatment of diabetes type II, but there is big variation, regarding accessibility to drug addiction treatment. Thus, a website in the United States advertises for drug addiction treatment abroad.

In the United States, treatment of cocaine addiction can range from thousands to tens of thousands of dollars. Medical insurance does not cover the cost of drug addiction treatment plans, which may range in duration from weeks to months to years, depending on the individual (source: http://newbizshop.com/2009/11/alcohol-addiction-treatment-abroad).

It is also easier to obtain epidemiological background information on prevalence and incidence of diabetes than on drug addiction. Stigmatization and, in the case of usage of illicit drugs, criminalization because of the law, creates many problems, which do not occur in the context of disease. This fact, that there is a distinct connotation of drug addiction with guilt and individual failure is essential in many aspects. Is it the allocation of public funding for indicated therapy, or the reputation of those involved in treating drug addicts or the difficulties faced by investigators performing data collection on drug use? Data collection is harder once people enter treatment, but we know that, by far, not all who need drug addiction treatment are also receiving it or accepting adequate therapy because of existing thresholds. It is, likewise, easy to convince patients at risk to undergo glucose tolerance test in the case of diabetes, whereas drug-addicted patients would rather avoid diagnostic procedures for drug addiction.

The reasons for the observation of variation in epidemiological data quality are many (European Monitoring Centre for Drugs and Drug Addiction, 2009), but it is obvious that all of these circumstances in such a highly stigmatized area create specific problems when obtaining background information in order to improve understanding. People afraid to confess their "negative habits" will contribute to "recall biases" and underreporting artifacts in surveys. "Currently, accurate estimates of drug consumption are unavailable, as a consequence of the usually stigmatized nature of drug use, which often leads drug users to hide their habits" (Frost et al., 2008). This quote is still valid in 2010.

Looking at the reliability of drug-related information about consumption habits gives some reasons to think about alternatives or complements, to cope with the previously mentioned dilemma. Although it is well known, that there is drug use existing in upper and middle class society, it rarely happens that leading personalities will openly talk about their addiction problems (Hblatt, 2003; Forster, 1984). The existence of numerous treatment centers offering anonymously conducted detoxification and psychotherapy underlines this "confession problem." Drug addiction is nothing that anybody wants to be connected with, except that the disease has developed so far that is it no longer easily hidden. Until its consumption, related information is difficult to obtain.

A conclusion that can be drawn from this is that wastewater screening, or continuous surveillance, can be a useful complement to provide information on overall consumption of specific drugs. Such information will not address any particular group within the society, but will only provide quantities and show peaks over time. The advantage is that the coverage is across the entire population, without excluding hidden minorities or subpopulations. Another advantage, which will be discussed later in detail, is anonymity, since there is no link between the consumer and any personally filled-out questionnaire.

15.1.2 The Question of Responsibility

If there are potential new methods to collect data, maybe cheaper, faster, and more reliable to measure overall drug consumption, they need funding, as do all the existing conventional data-collection tools. Therefore, it is important to mention the underlying concept of responsibility, i.e., to discuss why governments are spending public money, why they are establishing drug policies, or why they develop versatile legislature and penal codes to tackle illegal drug consumption, supply, and the like. Responsibility could be seen as a shared virtue between the citizen and the state, closely related to the concept of freedom and the right of self-determination. How far can state paternalism go and where are its limits? Who has to take care of the damage done? And, finally, who has to pay? These are some aspects for the following discussion.

15.1.2.1 Individual Responsibility?

Is it the drug user acting most harmfully in an irresponsible way, being the target? Addicted people suffer from this disease as other people suffer from any other disease. In addition, they are losing their social grounding, jobs, and quality of life. Psychodynamic views of addiction have identified two contributory elements "...disordered self-esteem and disordered relationships, which have evolved into a modern self-medication hypothesis, where individuals with substance use disorders are hypothesized to take drugs as means to cope with painful and threatening emotions" (Koob and Le Moal 2005). What finally ends up as addiction may have originally been an impulse, a wish for self-help, and self-responsibility (Hart et al., 2008).

When looking at daily news from press and mass media it is obvious that there are other factors influencing individual consumptive behavior beyond free choice. Marketing strategies for legal and illegal substances and dealing methods are becoming more aggressive, contributing to victimization. Often, violence is involved pushing people into addiction, particularly in trafficking, e.g., women forced into prostitution. There is also significant variance in vulnerability, where self-protection works better with one person than it does with another. It is self-evident that there is no way to leave the individual alone with a manifest drug addiction.

How does peer pressure within societal systems influence teenage habits or adult behavior (Quensel, 2004)? How intensive are intercommunicative codes working along with consumption of legal or illegal substances? To what extent does advertising contribute and how large is the influence of transporting this type of behavior over the World Wide Web? Do we know the real size of medically initiated addiction by highly addictive pharmaceuticals, e.g., benzodiazepines? Medically prescribed drugs are seen to a large extent in environmental screening. Focused substance detection can help to better understand the size of the problem. Much has been done here over the last decade, as has been shown within this book.

15.1.2.2 Societal Responsibility?

Again the question arises—Is it the social community (the government), that must cope with drug-related damage because of the failure of educational systems or insufficient countermeasures to avoid trafficking? Of course, there is no societythat can guarantee the completeness of individual sanctity, but there are laws within constitutions in many countries of the world that assure security and freedom for citizens. Therefore, the state cannot escape taking the responsibility, to some extent. This implies measures to understand how effective public responses are, like supply reduction, drug addiction treatment, or prevention programs? Drug monitoring is a natural consequence to know more about the success or failure of interventions. Environmental surveillance of external fluids like waste- and surface waters are an option that needs to be further investigated. There are objectives existing in the new EU Action Plan on drugs 2008–2012, where such a concept could have a place (EU Action Plan on Drugs, 2009–2012).

State responsibility on a large scale is required when focusing on organized crime. This touches much more on international task force activities than merely just control at a national level. Are there sufficient means to establish international collaboration between states where national sovereignty ends (World Drug Report, 2009)? What about the role of international customs and task forces to combat drug trafficking and who will pay for them? Just looking at those few questions - and there are many more – that show that leaving the responsibility to the individual drug user would not address the complexity of the problem. For the most part, drug users are the victims rather than the offenders.

Self-responsibility of the individual is one aspect of this question, includingsocial stability within a society, acceptable living conditions, reasonable salaries and equal opportunities, equality in educational programs, and illiteracy. The UNODC states that drugs are not harmful, because they are controlled, but they are controlled because of their harmfulness (World Drug Report, 2009). Further there is the statement that UNODC favors the evaluation of the effectiveness of the current approach in controlling drugs and while changes would be needed, they should "be in favor of different means to protect society against drugs, rather than by pursuing the different goal of abandoning such protection" (World Drug Report, 2009). This touches on the kernel of the dilemma namely, to act responsibly, between protectionism and paternalism, on the one hand, and liberalization and decriminalization, on the other. Reality around the subject of drugs does not follow black and white drawings but is rather a gray-scaled field and any approach to expect completely satisfying answers will likely fail. This ambiguity between law and order approaches and liberalization of drugs explains why there are still so many nuances in existing drug policies. "... Modern society must, and can, protect both these assets with unmitigated determination" (World Drug Report, 2009). Governmental programs on national level or on the continent,like the EU Action Plan on Drugs, do exist around the world and have existed for quite a long time. It is obvious that, over time, fine tuning of these plans have occurred with a growing understanding of the problems. The degree of understanding again depends on the availability and reliability of collected information. Once more, wastewater and environmental screenings should be seriously taken into consideration as an additional option. The EU Action Plan foresees the development of new data collection methods (EU Action Plan 2009–2012).

15.1.3 Drug Policies

In 2009, "The most serious issue concerns organized crime" (World Drug Report, 2009). "First, law enforcement should shift its focus from drug users to drug traffickers. Drug addiction is a health condition: people who take drugs need medical help, not criminal retribution" World Drug Report, 2009). A more moderate position has now been adopted by the European Monitoring Centre for Drugs and Drug Addiction (EMCDDA). "Broadly speaking the European policy debate has moved towards the view that priority should be given to interdiction activities targeting the supply rather than the use of drugs" (EMCDDA, 2009). A position paper, drafted by the European Council, which "stressed the importance of public health as the first principle of the international drug control system" (EMCDDA, 2009) also underlines this shift of attention toward the "real" sources of drug addiction.

Day to day reality, however, still differs throughout Europe. "If millions of Europeans have at some stage taken drugs or are doing so right now, there is a gap between public policy and public behavior that no society can afford to ignore" (EU-Action Plan on Drugs 2009–2012). It should also be clear, that causal responses alone, like just fighting drug trafficking, will not reach what they are aiming at, because of the real world's nonlinearity. Anything in the context of drugs is highly dynamic, which demands continuously flexible and adapted responses at various levels. This also includes control systems to evaluate dedicated programs and measures. Wastewater mass flow quantification of consumed substances using the waterwater treatment plant profile can function as a screening instrument to control intervention effects over time.

Many countries around the world are collecting data on drug consumption and drug supply. Many of them also have accredited special governmental delegates to take responsibility and coordinate leadership in political programs. For the European Union, the Horizontal Drugs Group is a forum where national drug policies, law enforcement issues, and health programs are discussed between member states and EU Commission.

According to the 2009 World Drug Report issued by the United Nation's Office on Drugs and Crime in Vienna, 118 countries out of 192 have replied regarding the UN questionnaire on Drug Use Demand and 116 contributed to the questionnaire on Illicit Supply of Drugs with considerable variation across continents (World Drug Report, 2009). The overall coverage shows up in the World Drug Report 2009 (download at: http://www.unodc.org/documents/wdr/WDR_2009/WDR2009_eng_web.pdf)

Like the European Monitoring Centre for Drugs and Drug Addiction, the UNODC also states that data quality still greatly varies between countries, which presents problems regarding comparability (World Drug Report, 2009). Apart from such statements, where competitive aspects regarding data quality are predominantly concerned, the question remains, are other ways to get closer to realistic estimates about overall drug consumption? There are many countries where even surveys do not exist, let alone other sophisticated investigations. This is not just a question of reluctance but, particularly for third-world countries, a question of finances. Wastewater or even surface water drug screenings, when not established as a continuous program, but with periodic standardized sampling routines once or twice a year in the most important regions, could serve to create comparable profiles based on standardized methodological protocols. This is particularly interesting, if there are no other data-collection tools existing. From an economical viewpoint, this could be a challenging perspective toward reduction of public expenditure. The International Narcotics Control Board INCB monitors the implementation of United Nations drug control conventions. Agencies like NIDA (U.S. National Institute for Drugs and Alcohol) or EMCDDA monitor the North American/European situation. Institutions around the globe are aiming at a better understanding of the drug phenomenon. However, they are also somehow competing. Instead of developing common data collection protocols, they spend public money for the development of proprietary systems. There is a global alliance existing regarding drug control, but, thus far, there is no clear and common understanding. This should be the main target, like international agreements against uncivil society (World Drug Report, 2009).

The existence of antidrug programs indicates that policy and civil society, national and regional health systems, do see a responsibility resting on their shoulders and because of some of the reasons have been previously outlined. At least in the democratic tradition and following the principles of the charter of the United Nations – a state has to provide guarantees regarding the well-being and health of its citizens. This protective function requires counterbalancing the observation of human rights of freedom and inviolability of the individual. Environmental monitoring already demonstrates in many areas of routine surveillance that it does not violate individual freedom, but rather can contribute to this responsibility widening the understanding of otherwise hidden phenomena.

15.2 DRUG MONITORING IN EUROPE AND THE WORLD

The European Monitoring Centre for Drugs and Drug Addiction as the main supplier of drug-related information in Europe has set up five key epidemiological indicators to inform politicians, scientists, and the population about the drug situation (EMCDDA, 2009). Actually, there are new indicators that have been developed in response to drug use, where drug supply and counter measures are becoming increasingly important. The United Nations and WHO as global organizations are collecting epidemiological information by using their own networks. In Europe, for the last 15 years, drug consumption, patterns of drug use, habits in school populations, drug related crime and mortality, drug-related infections, and treatment requests on behalf of addicted patients have been part of EMCDDA's data collection through the REITOX network. On the response side, treatment availability, observation and evaluation of drug policy programs, prevention measures and, more recently, surveys on supply reduction have been carried out which is in line with the UNODC recommendations and the EU Action Plan on Drugs. There is also collaboration with customs and Europol. In its latest Annual Report 2009, the Agency provides estimates for drug use in Europe, relating to the adult population, that is, 15 to 64 years (EMCDDA, 2009). The worldwide epidemiological situation varies considerably, as stated in the latest UN World Drug

Report (2009). Comparison of the European coverage is by far the best, followed by Asia and the United States of America (World Drug Report, 2009). How are these data obtained and do they provide reliable results to understand the dimension of the problem? To date, the answer is that there is a great variance in exisiting data quality not only in Europe, but also around the world (EMCDDA, 2009; World Drug Report, 2009). Data to estimate prevalence of drug use are traditionally derived from cross-sectional studies, like large population surveys, which are performed periodically at national level with usually large sets of items where drug consumption habits and frequency play a rather marginal role. The British Crime Survey, which by default is a "victimization" survey, is one of the most famous among a large variety of country-specific national surveillance instruments, but clearly not dedicated to provide "hard" information on drug consumption. These surveys are all conducted under the auspices of national responsibility, which means that up until now, there are no common methodological standards existing throughout Europe. However, the EMCDDA are providing recommendations toward a common data collection protocol in EU member states. These are recommendations with no binding character and it depends on the good will of countries whether they implement such advice or continue with their own standards. In the early years of the drugdata collection, there were studies carried out to find common denominators across Europe regarding population survey variables (EMCDDA, 2009). Although improvements have been made, even 10 years later there is no unique general population survey dataset. In general population surveys, there are dedicated instruments for subpopulations existing, like the ESPAD questionnaire (European School Project on Alcohol and Other Drugs, 2007). This survey is actually considered to be the best European standard on drug consumption estimation. "... to reach the goal of providing data that are cross-nationally comparable, the methodology of the ESPAD project is strictly standardized. The standardization regards the target population, data collection instrument, field procedure, timing and the data processing. "(ESPAD, 2007).

15.2.1 Survey Pitfalls

Survey data result from telephone or face to face interviews. The data can also result from controlled or instructed filling in of questionnaires in classrooms where pupils are requested to provide answers regarding their existing (or nonexistent) drug consumption habits. Linked problems like recall bias, unwillingness to respond honestly, personal feelings of shame, etc., have been initially mentioned and are reason for concern. Such problems create a variety of biases, which can produce underreporting effects and deviating estimates. Information quality has been improved for the past 15 years, but, unfortunately, there are still considerable methodological gaps, as has been noted in the World Drug Report on behalf of the United Nations for many countries not only outside Europe (World Drug Report, 2009). Regarding the most important data source for drug prevalence in the general population, we face the problem that almost 30% of the population survey data do not match the proposed age ranges given by the EU drug agencies, which causes problems when trying to compare information between countries (EMCDDA, 2009). The explanation for this inconsistency is simple. Large surveys are expensive and usually have a long national tradition. Any abrupt change of essentials in data structure like age groups and age ranges would severely influence the overall evaluation of the national statistical system requiring limitless work to adapt historical data with new information resulting from a changed data collection protocol. Changes are, therefore, not easily to apply and it can take considerable time, before a common European surveillance system may be successfully established. Europe is a conglomerate of more than 30 independent countries with autonomous governments including health and judicial policies. The Lisbon treaty, which will amend the Treaty of the European Union and the Treaty establishing the European Community (Lisbon Treaty, 2007), common data collection protocols could become more feasible. With this Treaty, the EU member states will "...confer competences to attain objectives, they have in common" (Lisbon Treaty, 2007). This could mean, in practice, that new data collection protocols like a common wastewater screening approach or an embedding of such a procedure within a greater environmental screening framework could be more easily established in the future starting with an official recommendation. This statement is valid for any type of harmonized data collection in the European Union, regardless of whether the information will derive from surveys or wastewater analyses. A dedicated and decentralized agency like EMCDDA with a high level of insight and expertise, including the European data exchange network REITOX, could play a leading coordinative role as it currently does. Alternatively, if established predominantly as an environmental monitoring approach, there is also the European Environmental Agency EEA in Copenhagen that can play a key role. A collaboration of these agencies with the European Commission is another scenario that could work, which would depend on the character of a final approach, that is. whether the main aspect would be on just monitoring drug/illicit drug mass-flow or, even broader, it could be taken from the perspective of environmental toxicology.

Surveys, as has been previously shown, are usually quite large instruments. They also require costly and tiring procedures from the beginning of data collection until final dissemination of the evaluated statistical output. Only a few are performed annually, some biannually, and some every four years, like the ESPAD survey (ESPAD, 2007). Regarding general population surveys, the delay between dissemination at the national and European level takes even longer. This happens because the national statistical offices and evaluating institutions are, first, working on their own interests, making national use out of national datasets. Only then, is the data then passed for other purposes outside the country. Probably direct European funding or at least indirect contribution on behalf of the European Commission through institutions like EMCDDA or other agencies on these thematic issues will contribute to speeding up those procedures. In the long run, Eurostat, the official statistical unit of the European Union, will likely play a key role in data acquisition. For regularly collected drug prevalence information in Europe, this currently means a minimum delay of 2 years (EMCDDA, 2009). Again, the question is whether datasets and information coming from other sources, like the environment, could help to improve quality and speed. The answer regarding quality depends essentially on methodological questions, including applied technology and laboratory standards, and to what extent environmental monitoring data shall be enriched or cross-matched with overlay information from complementary sources like demography or epidemiology. The second regarding timeliness and speed could bring considerable advantages. Wastewater surveys carried out according to fixed protocols and standardized analytical procedures could give results, theoretically and in practice, in the very same day when the samples were taken. For identification of trends, such a reduced indicator can already give reliable information. At a later stage, the variable set can always be combined with overlay data from other statistical or bioanalytical sources when more comprehensive and exhaustive exploration is needed. This also provides an interesting option for plausibility controls among various datasets.

Measuring mass flow of drugs within wastewater treatment plants as part of a greater concept is similar to existing routines like data collection of weather conditions or earthquake surveillance. Detection of drugs in wastewater systems could become simply part of already existing analytical procedures, that are technically and analytically extended. At European level with the Council Directive 91/271/EEC, from 21 May c15+bib+0002 on urban wastewater treatment, there is already a fundamental framework existing for administrative decision making (CD 91/271 1991). Further investigation in the field of environmental toxicology will also probably underline the need for extending the existing practice in wastewater analyses. Currently, there is greater ". . .concern existing for long term health and environmental impacts of wastewater constituents" (Tchobanoglous et al., 2002).

15.2.2 Outlook

Political pressure to establish common sampling procedures and analytical protocols for wastewater drug detection commonly applied in Europe or, even worldwide, will most likely not emerge from the illicit drugs research community but from the environmental movement, where engineering expertise is more distributed including technical insight. However, powerful networks exist. In Europe, the Reitox network, in collaboration with the EMCDDA and the European Commission, or a joint venture between various agencies and the Commission, could promote further investigation, toward methodological symbioses between conventional and environmental data collection. Eventually, monitoring of illicit drugs in wastewater could become a potent add-on, established as a module within broader routine procedures. Even if not serving as a primary source for prevalence of drug use, environmental data can provide valuable material for cross-validation with conventional epidemiological data material.

This discussion might be novel grounds for the social scientific drug research community, but it is not within other areas. Environmental data have already helped to better understand how human behavior "echoes" our own living conditions. Concentrations of toxic substances, air-borne or water carried, weather and climate surveillance, satellite imaging to monitor erosion, or ocean dumping are well-established techniques, and there is little doubt regarding the usefulness of these types of investigations. Environmental drug monitoring can contribute toward a better understanding of drug consumption quantities without violating data protection issues. In the same way as public health investigation looks at population issues rather than observing individuals, analyses for mass flow of drugs can occur at a level where the focus is accumulated input. The advantage is, however, that the linked population is well defined in terms of demographic and infrastructural details. Thus, biases like over-reporting resulting from other input, like pharmaceutical industries or big hospitals, can be avoided.

Technically speaking, environmental water screening is not very different from drug screening at a personal level when saliva, blood, or urine is examined. The difference is that concentrations in biofluids, like surface water or wastewater, are highly diluted and that the "body" investigated is a public one – a technical plant, but with intrinsic system information that mirrors collective human behavior. One crucial point is that when we deliver individual samples for laboratory diagnostics, we do this with informed consent; environmental samples are taken without our knowledge and explicit permission. However, when we look at this in more detail, we have to acknowledge that contracts between households and municipalities are already shifting the responsibility from the dismissing party toward municipal services. Such regulations are also necessary in order to avoid problems of responsibility in the case of leakage in sewage systems or other accidentsthat can occur during the waste removal.

What remains is the question, whether it is ethically acceptable to investigate domestic waste once disposed of in public removal services or sewage systems. De facto, there is a lot statistical evaluation of data already taking place with all types of waste. Often, the individual consumer is not aware of this fact or does simply not very much interest. Thus, there are detailed statistics that exist about upcoming domestic or industrial discharge at the national, as well as at international level. The national "Agendas21" and their international equivalent, the Global Agenda21, are good examples about how far environmental policy has already entered the fine-tuned surveillance that bridges horizontally over the wide spectrum between economy, ecology, and social/human issues like health, genetics, biotechnology, population, and globalization. Thus, we clearly have to state, that environmental surveillance is neither exotic nor dramatic in terms of violating individual integrity, but has rather developed to a point where humankind may learn by applying measures to avoid further ecological damage. The crucial point seems to be where or when the analysis takes place. Thus, it makes a big difference to examine litter when it is still in the dustbin in front of a house or if discharged residues are investigated only after they have been mixed up with those from thousands of other households at a distant deposit, where former "owners" are no longer traceable. Applying these rules of caution, there is virtually no damage that could be done by wastewater screening as long as it is realized at the wastewater treatment site (Frost et al., 2008).

15.2.3 The Urban Space

Text in the introduction to Paul Manning's book (2007) on "Drugs and Popular Culture," reads: "The use of illegal drugs is so common that a number of commentators now refer to the 'normalization' of drug consumption." Furthermore, the author notes that "This collection of readings will apply an innovator, multi-disciplinary approach to this theme, combining some of the most recent research on 'the normalization thesis' with fresh work on the relationship between drug use and popular culture." (Manning, 2007) With the rapid technological revolution, since the beginning of industrialization in the 19th century, we are now far away from those times, where regional developments took a long time to spread to other regions of the earth. Today, thousands of existing internet channels, blogs, and chats can easily disseminate or give guidance to any new, fashionable drug culture. What developed yesterday in Sydney spreads around the globe within days, thus establishing itself in the same or a similar way. This is a new dimension, leaving behind the long lasting heritage procedures of ancient cultures. Hence, there is another argument for the potential benefits of environmental screening. One can apply them faster and they will produce overall consumption estimates in the present, as opposed to the two- or three year- delay as it was in the past. This is of particular interest with drug "waves" flooding big cities as short-term trends. The facts may initially derive from insider knowledge and spread through daily news. However, usually we do not really know what the dimensions are until we can see reliable estimates. Wastewater screening may be the appropriate tool to provide rapid answers. This is of particular interest in large cities where new trends are usually born. Quite often, infrastructures for new drug cultures in urban areas mirror the synergies of the "globalization" in drug trafficking. Diversification of international markets has also led to a streamlining in rapid drug designs, sacrificing purity of substances in favor of profit margins. Thus, substance stretching is often subject to additional, severe health risks. An increased interest in early warning systems has emerged during the last two decades. Here, the international community wishes to respond more rapidly in terms of joint action, Wastewater screening as part of such alert systems could be a useful tool.

15.3 THE WATER HAS KNOWLEDGE AND MEMORY

Regarding data collection and prevalence of illicit drug use, the EMCDDA organized a small conference in 2007, entitled "In Aqua Veritas" (Frost et al., 2008), where, for the first time, a few protagonists came together to discuss potentials and possibilities to measure overall drug consumption through applied wastewater sampling and subsequent analysis. The title for the meeting was not chosen by chance, but with the underlying assumption that it could be worthwhile, even necessary, to look for and at other potential data sources to estimate overall consumption of legal or illegal psychoactive substances. The entrance point was to look where individual behavior becomes public and adds to collective behavior, leaving accumulated traces in the environment. In environmental investigations, it is useful to bear in mind that this type of accumulated contribution to environmental side effects do not just occur with excreted drug metabolites, but with almost anything we are discharging as waste products from daily life.

Although we have known for quite some time that the environment mirrors our habits, we are reluctant to accept negative images from this mirror, particularly when the "echo" reflects the dark side of what we are providing. To some extent, we have accepted changes in the Ozone layer and climate, which are now constantly being monitored. We also begin to realize that the water is more than a consumer good, but also an information carrier at a micro level. In the same way as drug dependence manifests within a human organism, leaving traces in organs, tissues, and interneuron, the repeated and accumulated discharge of toxic substances accumulates in environmental fluids over time. If this process lasts for a long time, substances will not only remain in the water itself, but become also part of sedimentation, thus infiltrating deeper layers of the ground. In other words our own environment echoes back to us what we are doing in our daily life and since the water makes up an essential part of it, we can ask this fluid to relinquish its intrinsic knowledge. Instead of asking individual people about their drug consumption habits and hidden activities, wastewater sampling and analysis provides this possibility of looking into the environmental mirror of humankind.

The answers are there, whether we like them or not. Quite often, this reading from environmental carriers is not feasible for social Scientists (due to lack of professional training and technical equipment) but increasingly adapted by bio-engineering Science. This implies also a shift in focus, which means that the obtained answers are differently acquired and of a different type/character. There is a difference in output and evaluation, but it is not such a complicated one that it would create too many problems between the various scientific communities. Social scientists have developed their subject as have natural scientists. The most challenging part here is, that multidisciplinarity is required to make these groups capable of understanding each other's arguments to build a bridge between their different areas of knowledge. This is also what has occurred within this reference book and it was the aim of the first Lisbon meeting in 2007 where people started to communicate their ideas and what would be necessary to bring this type of analysis forward. It turned out, that wastewater screening for illicit drugs and their metabolites could give reasonable enlightenment. Enriched with epidemiological information and other meta-information it was even seen as a most useful complement in data collection around drug prevalence (Frost et al., 2008). This was the main message coming out of the meeting and this is still the message for further investigation in this field. The method, as such, is not very complicated, but needs further theory building. Once sufficiently elaborated, dissemination of acquired knowledge should spread toward the protagonists in political decision making, because this field is hardly one where private investors will step in. It is not a lucrative area with big profits but, as said initially, it is an area of public interest and responsibility.

The methods applied so far for data collection on prevalence of drug use are based on protocols that have been and still are individually developed by various institutions more or less competing round the world with a low degree of international standardization. The monitoring institutions are using different proprietary instruments to collect data. They belong to different governmental regimes, or are independent NGOs based on international funding systems. Although they share common questions and common knowledge, they are not yet collaborating in a way that would be desirable in order to develop worldwide-standardized data collection instruments. This is admitted by all of them and is especially crucial when aimed at data comparison, either at the local or at international level. Therefore, interinstitutional collaboration should aim at the development of common data collection protocols, including international standardization, in data interchange. Here again, natural science within the framework of environmental bioanalytics have a potential advantage. Laboratory techniques and analytical standards are already implemented worldwide providing recognized international acceptance. This would allow for an easier international implementation of new methods, if fine-tuning of methodological aspects could be initiated and agreed upon. The United Nations would be the most adequate international body to coordinate collaboration with continental partners worldwide.

A very last argument for wastewater screening, detection, and analysis for illegal drugs could be a monetary aspect. There is an increasing bottleneck to be observed which requires governmental administrations as well as municipal expenditure to become more tailored. The need for budgetary savings has become a concern, which extends across all societies. Wherever there is a possibility to reach a dedicated public task with less money compared to a more expensive alternative, the choice is often clearly made in favor of the economic one. If, as it was discussed before, wastewater detection for illicit drugs and their metabolites would become an integrated part of routine surveillance, this could be a cost-effective and appealing alternative to obtain relevant information. Computer-aided processing and evaluation has – at least – the potential to reduce costs. Wherever there are huge amounts of data to process nowadays, there is, at the same time, space for computational rationalizing.

15.4   SUMMARY: "WASTEWATER DATA" - THE WWTP AS A PUBLIC "BODY"

Even if wastewater sampling and detection of drug residues in the environment cannot lead to direct per capita consumption, it has the potential to design reliable profiles of total consumption, if monitored continuously, based on a solid and commonly applied methodology. Without additional information about user populations, their age structures, consumption habits, and patterns and frequencies of use, it will remain a subset to add to a more complex epidemiological indicator. The detection of total mass flow of substances under constant focus has some advantages, which conventional methods do not:

- Real time information, but at reduced complexity compared to survey information! Results from wastewater screenings may quickly serve to alert any rise or decrease of a particular substance within a surveyed area.
- Intervention control: The same approach may also serve as a fast controlling instrument to measure the effectiveness of programs before, during, and after the intervention. When looking at field studies that were done by various research groups publishing here, this is encouraging.

- A wastewater treatment plant can be seen as the collective societal body of a given population. Enriched with epidemiological information and demographic overlays, any investigative bioanalytical result from this figurative body may contribute creating a picture that could mirror the health status of the population in many ways.
- Environmental monitoring can vary depending on its focus. Thus, it can stress aspects of environmental toxicology, environmental forensics, or environmental epidemiology.

REFERENCES

CD 91/271/EEC 1991. Agenda 21 – local and global. Source: http://www.agenda21-treffpunkt. de/global/ CD 91/271/EEC EU COUNCIL DIRECTIVE of 21 May 1991 concerning urban wastewater treatment 91/271/EEC) (OJ L 135, 30.5.1991, p. 40). Source: http://eurlex.europa.eu/LexUriServ/LexUriServ.do?uri=CONSLEG:1991L0271:20081211:EN:PDF

Chris Löwer, Hblatt 2003. *Coping bei Doping – wenn der Chef die Nase voll hat – Handelsblatt* 4.3.2003 Rem. (The German title is difficult to translate– i.e. (1) When the boss is fed up; (2) When the boss is full with cocaine) Source: http://www.handelsblatt.com/unternehmen/strategie/ wenn-der-chef-die-nase-voll-hat;634011.

David, P. 2009. Drug addiction: a chronically relapsing brain disease Friedman, *N C J. Med.* **70**(1):35–37.

Dornblüth, O. Pschyrembel 2007. *Social Medicine.* p. 128. Berlin: de Gruyter;

European Monitoring Centre for Drugs and Drug Addiction (EMCDDA) 2009. Annual Report - The State of the drug problem in Europe Source: http://www.emcdda.europa.eu/publications/annual-report/2009

European School Survey Project on Alcohol and other Drugs (ESPAD) 2007. Source: http://www.espad.org/sa/site.asp?site=622

European Union (EU) Action Plan on Drugs 2009–2012. Source: http://eurlex.europa.eu/LexUriServ/LexUriServ.do?uri=OJ:C:2008:326:0007:0025:EN:PDF

Forster, B. 1984. *Upper and Middle Class Adolescent Drug Use: Patterns and Factors Advances in Alcohol & Substance Abuse*, **4**(2):27–36. Source: http://www.informaworld.com/smpp/content~db=all~content=a904726322 abstract.

Frost, N., Griffiths, P. *et al.* 2008. EMCDDA Insights No. 9 —Assessing Illicit Drugs in Wastewater, Lisbon 2008.

Hart, C. L., Krauss, R. M. *et al.* 2008. Human drug addiction is more than faulty decision-making. *Behav. Brain Sci.* **31**(4):448–449.

Koob, G. F., and Le Moal M. 2005. *Neurobiology of Addiction*. Orlando: Academic Press.

Lisbon Treaty 2007. The Treaty of Lisbon EU 2007 C306/02 Source: http://europa.eu/ lisbon_treaty/full_text/index_en.htm

Manning, P. 2007. *Drugs and Popular Culture: Drugs, Media and Identity in Contemporary Society*. Oxfordshire, UK: Willan Publ.

Quensel S. 2004. *Das Elend der Suchtprävention: Analyse, Kritik, Alternative, Wiesbaden Verlag für Sozialwissenschaften. (Translation: The Misery with Addiction Prevention: Analysis, Criticism, Alternative.)*

Tchobanoglous G, Franklin, L., Burton, H., David Stense. 2002. *Wastewater Engineering Treatment and Reuse.* McGraw-Hill Series in Civil and Environmental Engineering. New York: McGraw Hill; 2002.

Williams M. L. 2008. Whose responsibility is substance abuse treatment? *Corrections Today*; **70**(6):82–84.

World Drug Report (WDR). 2009. *United Nations Office on Drugs and Crime* —World Drug Report. Source: http://www.unodc.org/documents/wdr/WDR_2009/WDR2009_eng_web.pdf.

CHAPTER 16

ASSESSING ILLICIT DRUG CONSUMPTION BY WASTEWATER ANALYSIS: HISTORY, POTENTIAL, AND LIMITATION OF A NOVEL APPROACH

ETTORE ZUCCATO and SARA CASTIGLIONI

16.1 INTRODUCTION

Current official methods to estimate illicit drug (ID) use are based on population surveys, medical records, and crime statistics. These methods are known to be biased and to underestimate the extent of ID use, mainly because they rely on users self-reporting drug use, a socially censured behavior, about which users tend to be elusive. Moreover, surveys of the general population are too time-consuming to promptly detect changing trends. International drug agencies have thus called for more reliable methods to improve their knowledge of the phenomenon (European Monitoring Centre for Drugs and Drug Addiction, 2008).

Recently, a new approach has been developed, which is based on the chemical analysis of wastewater. This approach may provide evidence-based and almost real-time estimates of ID consumption and has been proposed as a reliable tool to detect changing trends of ID use in local communities (European Monitoring Centre for Drugs and Drug Addiction, 2008). The approach consists of the analysis of the breakdown products of the ID, collectively excreted with the urine by the consumers after ingestion, and transported with the wastewater to a treatment plant. The analysis of the residues of the ID (parent compounds and metabolites) in wastewater are carried out by highly specific and sensitive methods, such as HPLC− tandem mass spectrometry (HPLC−MS/MS) (Castiglioni et al., 2006). Residues at the treatment plant can be used to back-calculate for the amount of the parent ID consumed by the population living in the area served by the plant. This approach, first applied in 2005 using cocaine as a case study (Zuccato et al., 2005), has been subsequently extended

Illicit Drugs in the Environment: Occurrence, Analysis, and Fate Using Mass Spectrometry, Edited by Sara Castiglioni, Ettore Zuccato, and Roberto Fanelli
Copyright © 2011 John Wiley & Sons, Inc.

to the analysis of heroin, amphetamines, and cannabis (Zuccato et al., 2008a) and tested in the frame of studies carried out in several countries (Huerta-Fontela et al., 2008; Boleda et al., 2009; Kasprzyk-Hordern et al., 2009, van Nuijs et al., 2009; Banta-Green et al., 2009; Postigo et al., 2010).

The application of wastewater analysis to measure collective ID use in communities can be considered an example of a new branch of environmental epidemiology called "sewage epidemiology." When applied to public health issues, this approach has the potential to extract useful epidemiological data from qualitative and quantitative profiling of biological indicators entering the sewage system (Bohannon, 2007; Zuccato et al., 2008a).

16.2 METHOD OF CALCULATION

Wastewater of urban treatment plants contains traces of exogenous substances which after the intake by the consumers are excreted with the urine as such and/or as metabolites (Bohannon, 2007). In particular, it contains measurable levels of the urinary excretion products of the ID consumed by the population. Levels of the residues at the treatment plant can be used to back-calculate for the ID consumed by the population living in the area served by the plant (Zuccato et al., 2005).

16.2.1 Wastewater Sampling

To estimate ID consumption in a community, representative composite samples of untreated urban wastewater (representative of the amount flowing in a 24-h period) were generally prepared by sampling wastewater entering a treatment plant for 24 h, in a time-dependent or in a flow-dependent manner, and then pooling the aliquots collected by an automatic, computer-controlled device. This strategy guarantees the collection of representative and reproducible samples, but, in some studies, grab samples were also used (Hummel et al., 2006; Kasprzyk-Hordern et al., 2008), as well as polar organic integrative samplers (POCIS) (Jones-Lepp et al., 2004; Bartelt-Hunt et al., 2009).

16.2.2 Selection of Drug-Target Residues

The ID considered were generally those most widely used worldwide, including cannabinoids, amphetamines, opioids, and cocaine. Methadone, an opioid agonist used to treat heroin addiction, has been also frequently included. The general criterion to select target residues for back-calculations leading to consumption estimates of these substances was to choose the most abundant and specific urinary product(s) of each active drug, with proved stability in wastewater (the "target residues").

16.2.2.1 Cannabinoids   The principal active constituent of *Cannabis sativa* is Δ^9-tetrahydrocannabinol (THC). It is absorbed through the lung during smoking and is rapidly metabolized in the liver to the active metabolite 11-hydroxy-THC, which is further oxidized to a group of compounds, including the main metabolite 11-nor-9-carboxy-Δ^9-THC (THC-COOH), which is generally chosen as the analytical target. THC-COOH is excreted in feces and urine mostly conjugated with glucuronic acid (Huestis et al., 1996), but it is readily hydrolyzed back to the free acid by β-glucuronidases of fecal bacteria in untreated wastewater (D'Ascenzo et al., 2003). This has been confirmed by specific stability studies (Castiglioni et al., 2006).

16.2.2.2 *Cocaine* In humans, only a small percentage of a cocaine dose is excreted in urine as the parent drug, while a larger amount is excreted as benzoylecgonine (BE) (Baselt, 2004), which is the metabolite more often measured in urine to obtain evidence of cocaine use in forensic medicine. BE was shown to be sufficiently stable in wastewater to allow its detection (Castiglioni et al., 2006). Therefore, most of the published studies measured both cocaine and BE in wastewater, but then used only concentrations of BE to calculate cocaine consumption. A further advantage of using a metabolite (BE) instead of the parent compound in estimating cocaine consumption, is that of excluding from calculations the amounts of unused cocaine directly disposed in wastewater. An alternative approach used concentrations of the parent compound itself, cocaine, for the back-calculation of cocaine consumption (Bones et al., 2007).

16.2.2.3 *Amphetamines* After intake, amphetamine is normally excreted in unchanged form (30%−74% of a dose) in the urine. Methamphetamine, an N-methyl derivative of amphetamine, is also excreted mainly unchanged (43% of a dose). The same is true for methylendioxymethamphetamine (MDMA, or ecstasy), a ring-substituted derivative of methamphetamine that is also excreted unchanged (65% of a dose) in the urine and partially as methylendioxyamphetamine (MDA) (7% of a dose). In turn, MDA is also a psychotropic amphetamine derivative with hallucinogenic properties at high doses, which seems to be excreted mainly unchanged (Baselt, 2004). Amphetamines are, therefore, mainly excreted in the urine in unchanged forms and generally chosen as the analytical targets.

16.2.2.4 *Opioids* In human, heroin is rapidly deacetylated to 6-acetylmorphine and to morphine in the liver. Analysis in wastewater, therefore, focused on morphine, a metabolic residue common to heroin, codeine, and morphine, and on 6-acetylmorphine (1%−3% of a dose), a specific metabolite of heroin (Baselt, 2004). Morphine is partly excreted as glucuronide conjugates (Baselt, 2004), which are readily hydrolyzed back to morphine by β-glucuronidases of fecal bacteria, as shown for glucuronide conjugates of estrogens in untreated wastewater (D'Ascenzo et al., 2003), and confirmed by specific stability studies (Castiglioni et al., 2006).

16.2.2.5 *Methadone* Methadone was first synthesized as a morphine substitute and possesses many of that drug's pharmacological properties. It is mainly excreted in urine in unchanged form (5%−50% of a dose) or is metabolized by N-demethylation with spontaneous cyclization of the resulting metabolite to 2-ethylidene-1,5-dimethyl-3,3-diphenylpyrrolidine (EDDP) (3%−25% of a dose).

Analysis in wastewater, therefore, focuses on methadone itself and its primary metabolite EDDP.

Further details on metabolism and excretion of ID in humans are given elsewhere in this book (see Chapter 2).

16.2.3 Multiresidue Analysis

ID target residues have been measured in wastewater samples with fully validated, highly selective multiresidue assays. Briefly, water samples were generally solid-phase extracted, purified, and subsequently analyzed by liquid chromatography—tandem mass spectrometry or alternative methods. These methods are reviewed elsewhere in this book (see Chapter 3).

16.2.4 From Measured Concentrations to Collective Excretion Rates

Using the approach first described for cocaine (Zuccato et al., 2005), the concentration (nanograms per liter) of a given target residue in wastewater is multiplied by the flow rate (liters per day) of influent wastewater to calculate the daily amount of each residue (grams per day) reaching a given wastewater treatment plant (WWTP). Assuming no relevant loss of wastewater along the sewage system, and given the proved stability of the chosen residue in the wastewater under study, these figures (milligrams per day) reasonably reflect the collective excretion rates for the various target residues considered.

16.2.5 From Target Residues Excretion Rates to ID Consumption Rates

The excretion load of a given target residue is subsequently used to extrapolate the amount of the active parent drug consumed by the population under study. This is done by taking into account the known fraction of a drug dose that is excreted as the chosen target residue (Table 16.1), and the molar mass ratio between the parent drug and the residue. For example, in the case of cocaine, given that about 45% of intranasal cocaine (MW 303) is excreted as benzoylecgonine (MW 289), a measured benzoylecgonine excretion rate of 1 g/day would correspond to $1/0.45 \times 303/289 = 2.33$ g/day of consumed cocaine. Table 16.1 shows the correction factors for each ID as used by Zuccato et al. (2008a). An ideal target residue is a major and exclusive excretion product (metabolite or unchanged parent drug) of the drug under study. This is the case for cocaine, amphetamines, and cannabis. In the case of heroin, consumptions are generally estimated by measuring morphine levels, but other sources, such as the therapeutic use of morphine and codeine, may account for the presence of morphine in wastewater. However, the relative contribution of codeine to wastewater morphine levels can be considered to be limited (Zuccato et al., 2008a). Hence, when back-calculating heroin from wastewater morphine, corrections must be applied to compensate for the contribution from therapeutic morphine. In a recent study carried out in Italy, Switzerland, and United Kingdom, the daily amounts of

TABLE 16.1   Drug Residues Selected as Analytical Targets for ID Monitoring in Wastewater[a]

Drug	Selected Residue	Relation of Residue to Parent Drug	Percentage of Dose Excreted as Residue[b]	Molar Mass Ratio (Parent Drug/Residue)	Correction Factor (CF)
Cocaine	Benzoylecgonine	Major metabolite	45	1.05	2.33
	Cocaine	Parent drug (minor excretion product)			
Heroin	Morphine	Major but nonexclusive metabolite	42	1.29	3.07
	6-acetylmorphine	Minor but exclusive metabolite			
Amphetamine	Amphetamine	Parent drugs and major excretion products	30	1.0	3.3
Methamphetamine	Methamphetamine		43	1.0	2.3
Ecstasy	MDMA		65	1.0	1.5
Cannabis	THC-COOH (11-nor-9-carboxy-Δ^9-tetrahydro-cannabinol)	Major metabolite of THC (cannabis active principle)	0.6	0.91	152

[a]Levels of residues were used for back-calculating consumption of the parent drugs. The correction factor takes into account the percentage of parent drug excreted as the chosen target residue, and the parent drug-to-residue molar mass ratio (with permission from Zuccato et al., 2008a).
[b]Average for the most frequent route of intake (see Methods).

wastewater morphine expected to originate from therapeutic morphine use in these countries were subtracted from the total daily amounts of wastewater morphine and the remaining wastewater morphine was assumed to originate mostly from heroin (Zuccato et al., 2008a). Since 42.5% of a dose of intravenous heroin is excreted in the urine as morphine (Baselt, 2004) and that the heroin/morphine molar ratio is 1.29, the following formula has been used to back-calculate heroin consumption: heroine = total morphine (milligrams per day), subtracted from the predicted amount of therapeutic morphine, × 3.07. Accurate updated information about local morphine use is, therefore, mandatory to estimate for heroin consumption. Alternatively, 6-acetylmorphine, a minor, but exclusive, metabolite of heroin, can be used to estimate consumption of this ID, but because of the low and variable concentrations found in wastewater, this possibility needs further investigation.

TABLE 16.2 IDs Preferential Route of Administration and Average Dose

Drug	Route of Administration	Average Dose (mg) (Active Ingredient)
Cocaine	Intranasal	**100**
Heroin	Intravenous	**30**
Amphetamines		
Amphetamine	Oral	**30**
Methamphetamine	Oral	**30**
Ecstasy	Oral	**100**
Cannabis (THC)	Smoked	**125**

16.2.6 From ID Consumption Rates to the Number of Doses Consumed

To attempt a comparison between estimates of collective drug consumption by the sewage epidemiology approach, with official figures that mainly refer to drug use prevalence, total amounts of parent drugs should be translated into the corresponding number of doses consumed. An example of average doses (by typical intake route) as reported in the literature is summarized in Table 16.2. The number of doses consumed daily is generally calculated by dividing drug loads (milligrams per day) by the average amount of active principle in a dose. The data can be then normalized for the local population size (number of people served by the WWTP) to give, for instance, the doses per day of ID consumed per 1000 people.

16.3 RESULTS AND DISCUSSION

Official figures describing prevalence and patterns of ID use are currently estimated from population surveys integrated with crime statistics, medical records, and seizure rate. These methods are designed to estimate ID consumption prevalence data, that is, the percentage of users in a population, but give only limited information on the actual frequency of use by the consumers and the amounts of ID consumed. On the other hand, wastewater analysis can give "real-time" qualitative and quantitative information on the ID actually consumed by a given community. The application of this approach has, therefore, disclosed new information previously inaccessible, such as details of the daily consumption rates and of the weekly profiles of use, and have shown, for instance, that ID consumption in a large community is constant, with typical and predictable patterns, such as a rise in cocaine consumption during the weekend. ID consumption can, therefore, be studied by sewage epidemiology to search for the most effective interventions and for planning and monitoring local drug prevention and treatment efforts.

16.3.1 Cocaine

Analysis of BE in wastewater was used for the first time to estimate cocaine consumption rates in four medium-size Italian towns (Cagliari, Cuneo, Latina, and Varese).

In the same study, BE levels were also measured in surface water samples from the River Po, the largest in northern Italy, and used to estimate consumption rates in its heavily populated basin. Consumption estimates were in the range 0.21−0.73 g/day/1000 people in the towns investigated and 0.70 g/day/1000 people in inhabitants of the River Po basin (Table 16.3). Results of this study suggested that actual cocaine consumptions in Italy were much greater than estimated by population surveys and disclosed to the scientific community the potential of this novel, evidence-based approach (Zuccato et al., 2005). This method was subsequently applied to estimate consumptions of cocaine and other IDs in Milan, Lugano (Switzerland), and London (Zuccato et al., 2008a). Daily load profiles measured in Milan are shown in Fig. 16.1. Consumption estimates (mean± standard deviation), expressed in grams per day per 1000 people were 0.91 ± 0.15 in Milan, 0.62 ± 0.12 in Lugano, and 0.69 ± 0.04 in London. In Milan and Lugano, where weekly profiles were also monitored, results revealed a constant and reproducible consumption of the drug during the week, with a typical increase at the weekend. A further study carried out in Italy estimated consumption loads ranging from 38.02 to 72.64 g/day in the city of Florence (Mari et al., 2009). Cocaine consumption have also been estimated by wastewater analysis in Belgium, Spain, and Wales. Van Nuijs et al. (2009) carried out an extensive campaign sampling 41 WWTPs in various locations in Belgium. Samplings were repeated twice during a week, on a Wednesday, and a Sunday. Cocaine consumption estimates were 0.06 to 1.83 (mean 0.43) g/day/1000 people, with the highest values detected in large cities such as Antwerp, Charleroi, and Brussels and the lowest in small towns. Further, consumptions were higher on Sunday than on Wednesday, confirming a rise of cocaine use on weekends. Consumption rates in Spain have been reported by Huerta-Fontela et al. (2008) and Postigo et al. (2010). In the first study, grab samples were collected from 42 WWTPs of various sizes in northeast Spain and cocaine consumption rates were estimated from undetectable to 5.4 (mean 1.4) g/day/1000 people. In the second study, seven WWTPs in the Ebro River Basin were sampled and consumption rates of 0.6 to 3.1 (mean 1.8) g/day/1000 people 15−64 years old were estimated, which roughly corresponded to 0.4 to 2.2 (mean 1.3) g/day/1000 people of the general population. Another study was carried out in two cities in south Wales, where consumption rates of 0.6 and 1.2 g/day/1000 people were estimated (Kasprzyk-Hordern et al., 2009).

16.3.2 Opiates

Consumption rates were estimated in Milan, Lugano, and London starting from the total wastewater morphine, after subtracting the fraction presumably originating from local use of therapeutic morphine (Zuccato et al., 2008a). Daily load profiles measured in Milan are shown in Fig. 16.1. Wastewater morphine from heroin consumption accounted for 23, 32, and 68 mg/day/1000 inhabitants in Milan, Lugano, and London, respectively. Considering that 42.5% of a dose of intravenous heroin is excreted in the urine as morphine (Baselt, 2004) and the heroin/morphine molar ratio is 1.29, this corresponded to 70, 100, and 210 mg heroin/day per 1000 inhabitants in Milan, Lugano, and London, respectively. Boleda et al. (2009) and Postigo et al. (2010) measured morphine, respectively, in 15 and 7 WWTPs in Spain and

TABLE 16.3   Consumption Estimates (grams per day per 1000 people) of Major ID Based on Wastewater Analysis Published in the Recent Literature

	grams per day per 1000 people	References
Cocaine		
Cagliari-Cuneo-Latina-Varese (IT)	0.21−0.73	Zuccato et al., 2005
Belgium	0.43	Van Nuijs et al., 2009
UK (south Wales)	0.6−1.2	Kasprzyk-Hordern, 2009
Lugano (Switzerland)	0.62	Zuccato et al., 2008a
London	0.69	Zuccato et al., 2008a
River Po basin (IT)	0.7	Zuccato et al., 2005
Milan	0.91	Zuccato et al., 2008a
Spain (Ebro River basin)	1.3	Postigo et al., 2010
Spain (northeast)	1.4	Huerta-Fontela et al., 2008
Heroin		
Spain (Ebro River basin)	0.061	Postigo et al., 2010
Milan	0.07	Zuccato et al., 2008a
Lugano	0.1	Zuccato et al., 2008a
Spain (Catalonia)	0.138	Boleda et al., 2009
London	0.21	Zuccato et al., 2008a
THC		
Spain (Ebro River basin)	0.5	Postigo et al., 2010
Milan	3	Zuccato et al., 200a
Spain (Catalonia)	3.5	Boleda et al., 2009
Lugano	6.6	Zuccato et al., 2008a
London	7.6	Zuccato et al., 2008a
Amphetamine		
Lugano	Nd	Zuccato et al., 2008a
Milan	0.009	Zuccato et al., 2008a
Spain (northeast)	0.076	Huerta-Fontela et al., 2008
London	0.08	Zuccato et al., 2008a
Spain (Ebro River basin)	0.193	Postigo et al., 2010
South Wales	2.5	Kasprzyk-Hordern et al., 2009
Methamphetamine		
Lugano	Nd	Zuccato et al., 2008a
Spain (Ebro River basin)	0.001	Postigo et al., 2010
London	0.006	Zuccato et al., 2008a
Milan	0.01	Zuccato et al., 2008a
Spain (northeast)	0.024	Huerta-Fontela et al., 2008
MDMA (ecstasy)		
Milan	>0.01	Zuccato et al., 2008a
Lugano	>0.01	Zuccato et al., 2008a
London	>0.01	Zuccato et al., 2008a
Spain (Ebro River basin)	0.06	Postigo et al., 2010
Spain (northeast)	0.4	Huerta-Fontela et al., 2008

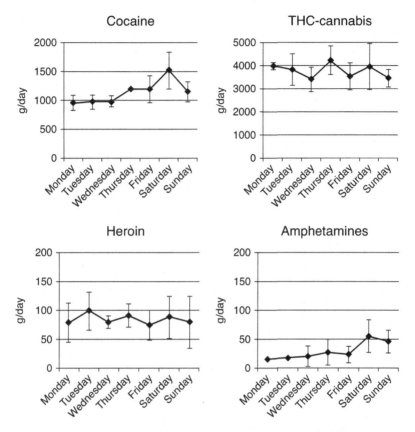

FIGURE 16.1 Consumption profile (grams per day, means ± SD) of cocaine, THC, heroin and amphetamines ($\sum$ amphetamine, methamphetamine, and MDMA) in Milan. (Modified, from Zuccato et al., 2008a.)

estimated a heroin consumption, respectively, of 138 mg/day per 1000 inhabitants and 61 mg/day per 1000 inhabitants age 15−64 years (roughly corresponding to 43 mg/day/1000 people). Consumption rates in doses/day/1000 people were also calculated, considering an average content of pure active drug in a typical dose of 30 mg for intravenous heroin (United Nations Office of Drugs and Crime, 2004). Means ± SD were 2.3 ± 0.2, 3.3 ± 0.5, and 7.0 ± 1.2 doses/day per 1000 inhabitants in Milan, Lugano, and London, respectively (Zuccato et al., 2008a) and 4.6 and 1.4 doses/day per 1000 inhabitants in Spain (Boleda et al., 2009; Postigo et al., 2010). Extrapolation to whole countries would suggest that heroin consumption was higher in the United Kingdom than in Italy, Switzerland, and Spain, which is in line with information from the United Nations Office of Drugs and Crime (2004), who reported an annual prevalence of heroin use for the age group 15−64 years in England and Wales (0.9%), which was greater than those in the other countries. However, consumption estimates by wastewater analysis in Milan were lower than in Lugano and Spain, while the corresponding prevalence of use was higher in Italy (0.8%), than in Switzerland (0.6%) (United Nations Office of Drugs and Crime, 2006) and Spain (0.4%) (European Monitoring Centre for Drugs and Drug Addiction,, 2009). This discrepancy might result from local differences in consumption, different time trends, and difficulties in extrapolating data from single cities or regions to the whole country. One year later, a further study was carried out in Milan. Results showed that heroin consumption estimated in the population served by the local WWTP (1,250,000 inhabitants) decreased from a mean of 2800 doses/day in 2006 to 1900 doses/day in 2007 (Zuccato et al., 2008b).

16.3.3 Cannabis

Surveys have been carried out in Milan, Lugano, and London in 2005−2006 (Zuccato et al., 2008a), and in Spain in 2007−2008 (Boleda et al., 2009; Postigo et al., 2010). Daily load profiles measured in Milan are shown in Fig. 16.1. THC consumption estimates were 3.0, 6.6, and 7.6 g/day per 1000 inhabitants in Milan, Lugano, and London, respectively, while in Spain, Boleda et al. (2009) and Postigo et al. (2010) estimated a consumption of 3.5 and 0.5 g/day per 1000 inhabitants, respectively. Cannabis consumption rates expressed as doses (mean $\pm$ SD) were 24 $\pm$ 2, 53 $\pm$ 12, and 61 $\pm$ 26 doses/day per 1000 inhabitants in Milan, Lugano, and London, respectively (Zuccato et al., 2008a) and 27 (Boleda et al., 2009) or 4 doses/day per 1000 inhabitants (Postigo et al., 2010) in Spain. These figures are in good agreement with data from the UNODC, which, in 2006, reported an annual prevalence for cannabis use in the age group 15−64 years of 11% in the United Kingdom, 10% in Switzerland, and 7% in Italy (United Nations Office of Drugs and Crime, 2006), while the annual prevalence estimate for Spain in 2008 was 10.1% (European Monitoring Centre for Drugs and Drug Addiction, 2009). Results of a second sampling campaign carried out in 2007, showed a decrease of cannabis consumption in Milan, where the estimated use of THC in the population served by the local WWTP (1,250,000 inhabitants) decreased from 3.80 $\pm$ 0.59 kg/day (about 30.400 doses/day) in 2006 to 2.82 $\pm$ 0.20 kg/day (22,500 doses/day) in 2007 (Zuccato et al., 2008b), in line with the "strong decrease of cannabis popularity" in Europe, as reported by European Monitoring Centre for Drugs and Drug Addiction, (2009).

16.3.4 Amphetamines

Amphetamine was not detected in wastewater samples from Lugano (Switzerland) whereas levels found in Milan (Italy) and London (UK) resulted in an amphetamine use estimate of 9 and 80 mg/day/1000 people, respectively (Zuccato et al., 2008a). Much higher estimates (2500 mg/day/1000 people) were reported in South Wales (Kasprzyk-Hordern et al., 2009), and in Spain, where Huerta-Fontela et al. (2008) and Postigo et al. (2010) calculated an average use of 76 and 193 mg/day/1000 people of the general population, respectively.

Methamphetamine was not found in wastewater samples from Switzerland, whereas it was found at levels corresponding to about 6 and 10 mg/day/1000 people in London and Milan, respectively (Zuccato et al., 2008a). Consumption estimates in Spain were 24 (Huerta-Fontela et al., 2008) and 1 mg/day/1000 people (Postigo et al., 2010).

MDMA (ecstasy) consumption rates were less than 10 mg/day/1000 people in Milan, Lugano, and London (Zuccato et al., 2008a) and 400 (Huerta-Fontela et al., 2008) and 60 mg/day/1000 people (Postigo et al., 2010) in Spain.

16.3.5 Comparison of Wastewater Analysis and Population Survey Evidence

To verify whether ID consumption rates assessed from measurement of drug residues in wastewater were in line with official epidemiological data, estimates of ID consumption obtained by wastewater analysis in Milan, Lugano, and London, were compared with national prevalence figures obtained from population surveys in Italy, Switzerland, and the United Kingdom (Fig. 16.2). Comparison of the results showed

ILLICIT DRUG USE PROFILES

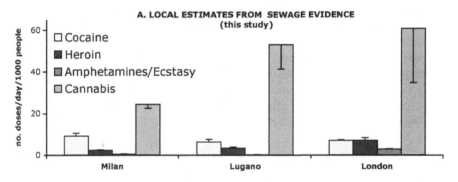

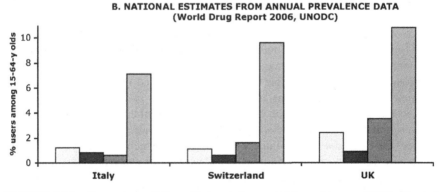

FIGURE 16.2 Comparison of ID use (doses per day per 1000 people, mean ± SD) estimated by sewage analysis (A) and national profiles of drug use (percentage of users among 15- to 64-year-olds) based on annual prevalence data in the countries under study (B) (UNODC, 2006). (With permission, from Zuccato et al., 2008a.)

that wastewater estimates reasonably matched with official figures and provided further complementary information (Zuccato et al., 2008a). In fact, results of the wastewater approach were in agreement with national prevalence figures in indicating that the most used drug by far was cannabis. The relative importance of cocaine and heroin use was also similarly represented by the two methods. However, wastewater analysis, but not prevalence estimates, suggested that amphetamines were the least used drugs in all locations. A possible reason for this discrepancy may lie in the intrinsic differences in the two approaches wWastewater analysis offers direct evidence of consumption rates, but no indication about the number of users. Population surveys provide information on the number of users, without specific reference to use patterns and amounts consumed. As the number of users is generally defined as the percentage of people who admit to having used a drug in a given interval, population surveys might overestimate the use of drugs that are occasionally used by many (e.g., the amphetamines), rather than steadily used by a few.

Overall these findings suggest that wastewater measurements provide objective, direct evidence of collective ID consumption, which are comparable with data estimated by epidemiological methods. Wastewater analysis was also used to estimate cocaine consumption at national scale level in Belgium and Spain, providing results in accordance with official statistics (van Nuijs et al., 2009; Postigo et al., 2010). Moreover, in a recent study from United States, the distribution of wastewater-derived drug index loads of cocaine, methamphetamine, and MDMA was found to correspond with expected epidemiological drug patterns and this highlighted the potential of this approach to improve the measurement of the true level and distribution of a range of drugs (Banta-Green et al., 2009).

16.3.6 Potential and Limitation of This Novel Approach

Current methods to estimate ID consumption are known to be inaccurate, relying on the individual willingness to self-report use. Moreover, surveys of the general population are too time-consuming to promptly detect changing trends. Alternatively, wastewater analysis may provide evidence-based and almost real-time estimates of ID consumption in local communities. Therefore, this approach to drug consumption monitoring has some advantages. First, wastewater analysis uses objective measures providing evidence-based pictures of the amount and type of ID collectively consumed in a community, thus avoiding possible bias associated with self-reporting. Second, these results can be obtained in near real time, because drug profiling by wastewater analysis can be completed in few days after sampling. Population surveys are expensive, which limits the frequency at which they can be conducted and require long time to be performed. Third, wastewater analysis allows comparison of consumption patterns in different communities, to highlight possible differences in ID use, and to better understand the extent of the drug problem in local populations. Fourth, results within a community, taken at different times, can be compared, to identify changing trends and to develop, target, and evaluate preventive interventions on the population.

However, although wastewater analysis is promising, it still needs to be refined before it can really become a general tool for monitoring drug abuse. For instance, this method has some sources of variability, such as pharmacokinetic and metabolism, route of ingestion, drug purity, and other criticisms that may be addressed in future studies to improve the accuracy of the calculations and assumptions. In particular, efforts could be aimed at: (1) using demographic data to cross-check population size in WWTP catchment areas, (2) improving knowledge of metabolism/kinetics of the various ID by experimental work and modeling, and (3) updating statistics on local drug use patterns for each drug (e.g., intake routes, purity, and size of typical dose). Critical work is needed to improve the current approach, and a consensus view, combining expertise from researchers, authorities, and international drug agencies, appears essential for the comparability of future studies.

REFERENCES

Banta-Green, C. J., J. A. Field, C. A. Chiaia, D. L. Sudakin, L. Power, and L. de Montigny. 2009. The spatial epidemiology of cocaine, methamphetamine and 3,4-methylenedioxymethamphetamine (MDMA) use: A demonstration using a population measure of community drug load derived from municipal wastewater. *Addiction* **104**:1874−1880.

Bartelt-Hunt, S. L., D. D. Snow, T. Damon, J. Shockley, and K. Hoagland. 2009. The occurrence of illicit and therapeutic pharmaceuticals in wastewater effluent and surface waters in Nebraska. *Environ. Pollut.* **57**:786−791.

Baselt, R. C. 2004. Disposition of Toxic Drugs and Chemicals in Man, third edition. Seal Beach, CA: Biomedical Publications.

Bohannon, J. 2007. Hard data on hard drugs, grabbed from the environment. *Science* **316**:42−44.

Boleda, R., T. Galceran, and F. Ventura F. 2009. Monitoring of opiates and cannabinoids and their metabolites in wastewater, surface water and finished water in Catalonia, Spain. *Water Res.* **43**:1126−1136.

Bones, J., K. V. Thomas, and B. Paull. 2007. Using environmental analytical data to estimate levels of community consumption of illicit drugs and abused pharmaceuticals. *J. Environ. Monit.* **9**:701−707.

Castiglioni, S., E. Zuccato, E. Crisci, C. Chiabrando, R. Fanelli, and R. Bagnati. 2006. Identification and measurement of illicit drugs and their metabolites in urban wastewaters by liquid chromatography tandem mass spectrometry (HPLC-MS-MS). *Anal. Chem.* **78**:8421−8429.

D'Ascenzo, G., A. Di Corcia, A. Gentili, R. Mancini, R. Mastropasqua, M. Nazzari, and R. Samperi. 2003. Fate of natural estrogen conjugates in municipal sewage transport and treatment facilities. *Sci. Total Environ.* **302**:199−209.

European Monitoring Centre for Drugs and Drug Addiction (EMCDDA). 2008. Insight 9: Assessing illicit drugs in wastewater; potential and limitation of a new monitoring approach, pp. 21−34. Lisbon: European Monitoring Centre for Drug and Drug Addiction.

European Monitoring Centre for Drugs and Drug Addiction (EMCDDA). The state of the drug problem in the European Union and Norway. Annual Report, 2009. Lisbon:

European Monitoring Centre for Drug and Drug Addiction. Available at: http://www. emcdda.europa.eu/publications/annual-report/2009

Huerta-Fontela, M., M. T. Galceran, J. Martin-Alonso, and F. Ventura. 2008. Occurrence of psychoactive stimulatory drugs in wastewaters in north-eastern Spain. *Sci. Total Environ.* **397**:3140.

Huestis, M. A., J. M. Mitchell, and E. J. Cone. 1996. Urinary excretion profiles of 11-nor-9-carboxy-delta 9-tetrahydrocannabinol in humans after single smoked doses of marijuana. *J. Anal. Toxicol.* **20**:441−452.

Hummel, D., D. Loffler, G. Fink, and T. A. Ternes. 2006. Simultaneous determination of psychoactive drugs and their metabolites in aqueous matrices by liquid chromatography mass spectrometry. *Environ. Sci. Technol.* **40**:7321−7328.

Jones-Lepp, T. L., D. A. Alvarez, J. D. Petty, and J. N. Huckins. 2004. Polar organic chemical integrative sampling and liquid chromatography–electrospray/ion-trap mass spectrometry for assessing selected prescription and illicit drugs in treated sewage effluents. *Arch. Environ. Contamination Toxicol.* **47**:427–439.

Kasprzyk-Hordern, B., R. M. Dinsdale, and A. J. Guwy. 2008. Multiresidue methods for the analysis of pharmaceuticals, personal care products and illicit drugs in surface water and wastewater by solid-phase extraction and ultra performance liquid chromatography-electrospray tandem mass spectrometry. *Anal. Bioanal. Chem.* **391**:1293−1308.

Kasprzyk-Hordern, B., R. M. Dinsdale, and A. J. Guwy. 2009. Illicit drugs and pharmaceuticals in the environment – forensic applications of environmental data. Part 1: estimation of the usage of drugs in local communities. *Environ. Pollut.* **157**:1773−1777.

Mari, F., L. Politi, A. Biggeri, G. Accetta, C. Trignano, M. Di Padua, and E. Bertol. 2009. Cocaine and heroin in waste water plants : A 1-year study in the city of Florence, Italy. *Forensic Sci. Int.* **198**:88−92.

Postigo, C., M. J. Lopez de Alda, and D. Barcelo. 2010. Drugs of abuse and their metabolites in the Ebro river basin: occurrence in sewage and surface water, sewage treatment plants removal efficiency, and collective drug usage estimation. *Environ Int.* **36**:75−84.

United Nations Office of Drugs and Crime (UNODC). 2004. Drug Report. Volume 2. Statistics. Available: http://www.unodc.org/pdf/WDR_2004/methodology.pdf

United Nations Office of Drugs and Crime (UNODC). 2006. World Drug Report. Volume 2. Statistics. Available at: http://www.unodc.org/pdf/WDR_2006/wdr2006_volume2.pdf.

van Nuijs, A. L. N., B. Pecceu, L. Theunis, N. Dubois, C. Charlier, P. G. Jorens, L. Bervoets, R. Blust, H. Meulemans, H. Neels, and A. Covaci. 2009. Can cocaine use be evaluated through analysis of wastewater? A nation-wide approach conducted in Belgium. *Addiction* **104**:734−741.

Zuccato, E., C. Chiabrando, S. Castiglioni, D. Calamari, R. Bagnati, S. Schiarea, and R. Fanelli. 2005. Cocaine in surface water: a new evidence-based tool to monitor community drug abuse. *Environ. Health* **4**:14 (http://www.ehjournal.net/content/4/1/14 2005).

Zuccato, E., C. Chiabrando, S. Castiglioni, R. Bagnati, and R. Fanelli. 2008a. Estimating community drug abuse by wastewater analysis. *Environ. Health Perspect.* **116**:1027−1032.

Zuccato, E., C. Chiabrando, S. Castiglioni, R. Bagnati, and R. Fanelli. 2008b. Estimating community drug use. In: EMCDDA Insight 9: Assessing illicit drugs in wastewater; potential and limitation of a new monitoring approach, pp. 21−34. Lisbon: European Monitoring Centre for Drug and Drug Addiction.

CHAPTER 17

COCAINE AND METABOLITES IN WASTEWATER AS A TOOL TO CALCULATE LOCAL AND NATIONAL COCAINE CONSUMPTION PREVALENCE IN BELGIUM

ALEXANDER L.N. VAN NUIJS, LIEVEN BERVOETS, PHILIPPE G. JORENS, RONNY BLUST, HUGO NEELS, and ADRIAN COVACI

17.1 INTRODUCTION

Prevalence of illicit drug consumption is currently estimated based on classic socioepidemiological studies using population surveys, consumer interviews, medical records, seizure rates, drug production knowledge, and crime statistics (European Monitoring Centre for Drugs and Drug Addiction, 2002, 2008a; United Nations Office of Drugs and Crime, 2009). These studies often have limitations since they include (1) specific populations (from a city, from a specific age class) or (2) the general population. While the first type of studies cannot provide nationwide estimates, because the studied population is not representative for the general population, the second type cannot offer information about local (urban versus rural areas) and age-specific trends. Moreover, classic socioepidemiological studies are, in many cases, not objective because only information originating from the consumers themselves is available. Furthermore, they are time-consuming, leading to a lack of frequently updated information in order to detect changes in drug-use patterns in a short time frame. In 2007, the United Nations Office on Drugs and Crime (UNODC) concluded in their recommendations that novel approaches are needed to estimate illicit drug consumption in a more objective way and that these approaches have to rapidly detect changes in drug abuse patterns (United Nations Office on Drugs and Crime, 2007).

As theoretical proposed by Daughton (2001), the analysis of illicit drugs and their metabolites in surface water and wastewater could have interesting implications for

Illicit Drugs in the Environment: Occurrence, Analysis, and Fate Using Mass Spectrometry, Edited by Sara Castiglioni, Ettore Zuccato, and Roberto Fanelli

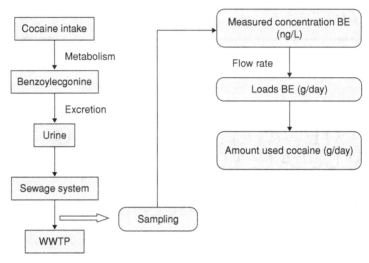

FIGURE 17.1 Schematic overview of the sewage epidemiology approach. BE, benzoylecgonine; WWTP, wastewater treatment plant.

epidemiological purposes. In his manuscript, Daughton introduced the hypothesis of estimating local and national patterns of illicit drug consumption from concentrations of (urinary) excreted illicit drugs and metabolites in wastewater. Zuccato et al. (2005) applied this theoretical idea for the first time on real waste- and surface water from four wastewater treatment plants and the River Po in Italy for the estimation of local (city or part of city) cocaine consumption and showed the great potential of this tool. There has been increased interest in the analysis of illicit drugs in waste- and surface water for the calculation of illicit drug consumption in several European countries (Bones et al., 2007; Huerta-Fontela et al., 2008; Zuccato et al., 2008; Boleda et al., 2009; Kasprzyk-Hordern et al., 2009; van Nuijs et al., 2009a–c; Mari et al., 2009), and the United States (Banta-Green et al., 2009). This approach, named "sewage epidemiology" by Zuccato and co-workers (Zuccato et al., 2008), is schematically shown for cocaine in Fig. 17.1.

The sewage epidemiology approach is objective ("a urine test for illicit drugs on a complete population") and relatively cheap. Results can be obtained in a short time frame and can provide information on local and national illicit drug consumption, which can be used in addition to information obtained with the classic socioepidemiological approaches. As already has been concluded from a multidisciplinary meeting of experts organized in 2007 by the European Monitoring Centre for Drugs and Drug Addiction (EMCDDA), it must be stressed that sewage epidemiology is not an alternative, but an additional tool that complements and extends the possibilities of existing approaches to counter the difficult problem of the monitoring of illicit drug consumption (Frost et al., 2008; European Monitoring Centre for Drugs and Drug Addiction, 2008b).

In this chapter, an overview will be given on the sewage epidemiology work that has been carried out until now in Belgium (Gheorghe et al., 2008; van Nuijs et al., 2009a–d). The focus will be on the calculation of local and national cocaine consumption based on measured concentrations of cocaine (COC) and its major metabolite, benzoylecgonine (BE), in influent wastewater. First, the analysis of these compounds with solid-phase extraction (SPE) and liquid chromatography coupled to tandem mass spectrometry (LC–MS/MS) will be reviewed. Second, the back-calculation from concentrations in wastewater to an amount of cocaine used will be explained. Third, local and national results for cocaine consumption in Belgium will be discussed. Last, general conclusions will be drawn and uncertainties of sewage epidemiology and the need for future research will be underlined.

17.2 ANALYTICAL ASPECTS OF THE DETERMINATION OF COCAINE AND METABOLITES IN WASTEWATER AND SAMPLE COLLECTION

17.2.1 Generalities

To quantify COC and metabolites in a complex matrix such as wastewater, a clean-up and concentration step have to be implemented to obtain a clean extract which contains concentrations of the three analytes in a detectable concentration without much interference. This extract will then be analyzed for the compounds of interest with an optimized and validated method based on LC–MS/MS. This section presents an overview of the method used and will be given together with a comparison and discussion with other existing and published methods.

17.2.2 Clean-up and Concentration Step

To remove solid particles from the wastewater, samples (100 mL) were first filtered over a glass filter. The filtered water was then further cleaned and concentrated based on off-line SPE. Initially, seven different SPE sorbents were tested for the extraction of COC, BE, and ecgoninemethyl ester (EME) (Gheorghe et al., 2008). Results showed the best recovery ($> 90\%$ for COC and BE with $> 70\%$ for EME) for Oasis HLB cartridges, conditioned with 3 mL methanol (MeOH) and 3 mL Milli-Q water, washed after loading (sample at pH 6) with 5% MeOH in Milli-Q water and eluted with 2×4 mL MeOH. The results are in agreement with earlier literature that reports the use of Oasis HLB cartridges for the extraction of illicit drugs from wastewater (Huerta-Fontela et al., 2007). It has to be mentioned that besides off-line SPE, other techniques have also been reported for the analysis of illicit drugs in wastewater, namely, on-line SPE proposed by Postigo et al. (2008) and large-volume injection suggested by Chiaia et al. (2008).

17.2.3 Analysis with LC–MS/MS

In most publications concerning the analysis of COC and metabolites in wastewater, reversed-phase liquid chromatography (RP–LC) coupled to MS detection was chosen for separation and quantification of illicit drugs (Castiglioni et al., 2006;

Huerta-Fontela et al., 2007; Postigo et al., 2008; Chiaia et al., 2008; Kasprzyk-Hordern et al., 2008; Bijlsma et al., 2009). For the Belgian studies, a different separation principle was used for the analyses. Gheorghe et al. (2008) compared the use of RP–LC–MS/MS with hydrophilic interaction liquid chromatography (HILIC–MS/MS) for the determination of COC, BE, and EME in wastewater. HILIC is a relatively new development in LC, which addresses the problematic retention of polar compounds in RP–LC. For these compounds, normal-phase liquid chromatography instead of chromatog-raphy (NP-LC) could be used, but, unfortunately, the mobile phase used in NP-LC contains solvents, such as hexane, which are not compatible with the necessary MS detection. HILIC is a suitable alternative, because it offers good retention for polar analytes together with a high percentage of acetonitrile (AcN) or MeOH (in a mixture with an aqueous phase) in the mobile phase and thus optimal conditions for MS detection. Especially for EME, a highly polar compound that shows very little retention in RP-LC, HILIC shows good performance (Gheorghe et al., 2008; van Nuijs et al., 2009d). HILIC separations were tested on two different columns (Zorbax Rx-SIL, 2.1 × 150 mm, 5 μm; Luna HILIC, 150 × 3 mm, 5 μm) and on two different MS detection systems (Agilent 1100 MSD ion trap with electrospray ionization and Agilent 6410 triple-quadrupole mass spectrometer with electrospray ionization) (Gheorghe et al., 2008; van Nuijs et al., 2009d). The Luna HILIC column in combination with the triple-quadrupole mass spectrometer performed the best, with a significant difference for EME (van Nuijs et al., 2009d). Quantification was done using deuterated internal standards (COC-d3, BE-d3, and EME-d3), with seven-point calibration curves. Methods were fully validated by addressing linearity, accuracy, precision, and limit of quantification (LOQ). Quality control and quality assurance during analysis of the samples was carried out by analyzing QC samples and blanks with each batch of samples and by performing an interlaboratory test (van Nuijs et al. 2009a).

17.2.4 Sampling Sites and Sample Collection

Influent 24-h composite wastewater samples were collected from 41 wastewater treatment plants (WWTPs) spread over Belgium. The chosen WWTPs cover both urban and rural regions and are all designed to serve more than 10,000 residents; the total of the 41 WWTPs cover approximately 3,700,000 residents. The following parameters were recorded for each sample: date of collection, temperature (°C), pH, and flow rate (m^3/day). Composite samples were collected in a volume-proportional way, meaning that a fixed amount of wastewater is automatically sampled when a certain volume of wastewater has passed the device. This procedure was repeated for each sample during 24 h, so that a representative sample for a complete day was acquired. To evaluate weekly variations, two samples were collected for each WWTP, one on Sunday (representative for the weekend) and one on Wednesday (representative for the week). This sampling procedure was executed during the summer–autumn of 2007 and repeated during the winter of 2007–2008 so that each WWTP was sampled on four separate days. After collection, samples were stored at pH 2 and at −20°C to prevent degradation of COC, BE, and EME during storage (Gheorghe et al., 2008; Chiaia et al., 2008).

17.3 BACK-CALCULATION OF COCAINE CONSUMPTION FROM CONCENTRATIONS OF COC, BE, AND EME IN WASTEWATER

To discriminate between direct discharge of cocaine through the sewer system and human cocaine consumption, we preferred to use only concentrations of the principal human metabolite of cocaine, benzoylecgonine, to make a back-calculation into an amount of used cocaine. The use of cocaine concentrations in the calculations would possibly lead to overestimations, because of the inclusion of (possible) discharges of cocaine powder through the drain. Moreover, it was also demonstrated by Gheorghe et al. (2008) that COC is not stable in water, with a degradation of 80% within 24 h at 20°C. In another stability study in which we tried to simulate real wastewater conditions (temperature 11°C and pH 8), a degradation of 40% for COC in 24 h was observed, while BE was stable in this time frame and under these conditions (van Nuijs et al., unpublished).

BE and EME account for 45% and 35% of the urinary recovery of a cocaine dose, respectively (Ambre et al., 1984, 1988; Cone et al., 1998). Because EME could not be quantified in any sample with the applied method, calculations for the wastewater samples from Belgium were based exclusively on BE.

The back-calculations can be summarized in three steps:

1. Concentrations of BE (nanogram per liter) are transformed into loads (grams per day) with the known flow rate (liters per day) of the wastewater stream.
2. The BE loads (grams per day) are transformed into COC equivalents with a formula proposed by Zuccato et al. (2005): COC (in grams per day) = BE loads (grams per day) × 2.33. The 2.33 factor results from the molar ratio of COC and BE (1.05) and from the fact that COC is metabolized, on average, for 45% to BE.
3. The amount of consumed COC (grams per day) is divided by the amount of residents (design capacity of the WWTP) that are served by each WWTP (the design capacity) to obtain cocaine consumption in grams per day per 1000 residents.

It has to be remarked that in a recent published HILIC–MS/MS method (van Nuijs et al., 2009d), EME could be quantified in a series of real wastewater samples. These findings will be further explored in the future.

17.4 LOCAL AND NATIONAL RESULTS OF COCAINE CONSUMPTION IN BELGIUM

17.4.1 Generalities

COC and BE were quantifiable in all influent wastewater samples with concentrations ranging 9–753 ng/L for COC and 37–2258 ng/L for BE (van Nuijs et al., 2009a). EME was not detected in any of the samples using the method described by Gheorghe et al.

TABLE 17.1 Calculation of Cocaine Consumption for Sunday Samples (Winter 2007–2008) from 41 Wastewater Treatment Plant Regions in Belgium[a,]

WWTP	Number inhabitants	Conc. BE (ng/L)	Conc. COC (ng/L)	Flow Rate (m³/day)	Loads BE (g/day)[b]	Loads COC (g/day)[c]	COC Consumption (g/day)[c]	COC Consumption (g/day per 1000 Inhabitants)
Aalst	89,847	410	158	29,768	12.21	4.71	28.438	0.317
Aartselaar	61,520	286	101	31,288	8.95	3.17	20.860	0.339
Antwerp North	69,668	248	100	34,280	8.49	3.42	19.788	0.284
Antwerp South	157,268	2130	693	57,932	123.39	40.16	287.444	1.828
Arlon	16,043	640	169	12,553	8.03	2.13	18.709	1.166
Beersel	63,531	692	273	11,614	8.04	3.17	18.733	0.295
Bruges	178,987	196	61	100,776	19.70	6.10	45.895	0.256
Brussels North	850,000	1306	453	233,303	304.69	105.69	709.785	0.835
Charleroi	138,000	1297	545	39,329	51.01	21.41	118.835	0.861
Ciney	10,000	140	70	6,000	0.84	0.42	1.955	0.196
Dendermonde	68,276	578	218	25,000	14.44	5.45	33.638	0.493
Destelbergen	57,999	392	174	17,912	7.03	3.12	16.369	0.282
Deurne	198,569	1245	394	57,088	71.07	22.49	165.568	0.834
Gembloux	37,131	160	38	8,400	1.34	0.32	3.131	0.084
Genk	68,924	863	363	45,540	39.28	16.52	91.509	1.328
Ghent	206,109	918	350	47,744	43.81	16.69	102.066	0.495
Harelbeke	111,515	803	244	33,292	26.74	8.12	62.299	0.559
Hasselt	63,333	37	9	38,712	1.43	0.35	3.337	0.053
Heist	18,123	642	284	9,770	6.27	2.77	14.611	0.806
La Louvière	29,800	549	195	12,134	6.66	2.37	15.518	0.521
Leuven	113,015	832	321	22,288	18.55	7.15	43.213	0.382
Liedekerke	92,465	648	247	27,392	17.75	6.76	41.355	0.447
Liège	26,300	1061	319	5,773	6.12	1.84	14.267	0.542
Lier	28,866	891	321	14,800	13.19	4.75	30.719	1.064
Lokeren	37,199	288	101	28,440	8.18	2.88	19.052	0.512
Mechelen North	87,452	1550	683	15,180	23.53	10.37	54.807	0.627
Menen	62,574	361	133	28,892	10.43	3.84	24.287	0.388
Mol	47,538	338	145	18,188	6.14	2.63	14.300	0.301
Mons	82,350	625	217	39,294	24.57	8.51	57.246	0.695
Morkhoven	38,211	171	67	19,358	3.31	1.29	7.711	0.202
Mouscron	27,831	228	73	11,045	2.52	0.81	5.861	0.211
Nivelles	27,000	515	133	14,400	7.42	1.92	17.279	0.640
Ostend	106,737	596	208	42,096	25.11	8.77	58.487	0.548
Roeselare	62,438	231	110	33,296	7.70	3.65	17.948	0.287
Sint-Niklaas	44,443	620	228	19,756	12.25	4.50	28.533	0.642
Sint-Truiden	44,131	393	159	16,590	6.52	2.63	15.188	0.344
Tessenderlo	41,761	350	117	16,710	5.85	1.96	13.624	0.326
Tournai	29,000	516	173	4,325	2.23	0.75	5.203	0.179
Verviers	95,000	554	161	33,355	18.48	5.36	43.046	0.453
Wavre	80,000	132	52	15,703	2.07	0.82	4.812	0.060
Wulpen	57,495	287	67	16,070	4.62	1.08	10.759	0.187

[a]WWTP, wastewater treatment plant; conc, concentration; BE, benzoylecgonine; COC, cocaine.
[b](Concentration × flow rate)/106.
[c]Loads BE × 2.33.
Reproduced from van Nuijs et al. (2009b) with the permission the publisher, Wiley-Blackwell.

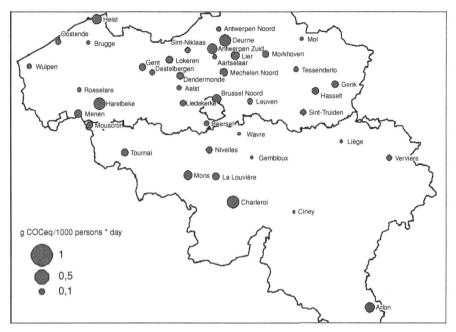

FIGURE 17.2   Geographical distribution of cocaine consumption (in grams per day per 1000 inhabitants) for Wednesday samples (sampling campaign II) from 41 wastewater treatment plant regions in Belgium.

(2008). The results were in agreement with the literature (Zuccato et al., 2005; Chiaia et al., 2008; Huerta-Fontela et al., 2008; Kasprzyk-Hordern et al., 2008; Postigo et al., 2008; Bijlsma et al., 2009).

Cocaine consumption for each WWTP region for a weekday (Wednesday) and a weekend (Sunday) and for both sampling campaigns (summer and winter), was calculated with the above mentioned strategy. Table 17.1 gives an overview of calculations from concentrations of COC and BE to amounts of used cocaine for the Sunday samples of each of the 41 WWTP regions in the winter 2007–2008. Figure 17.2 gives the graphical presentation of the results of cocaine consumption in the winter 2007–2008 for the Wednesday samples of the WWTP regions. The calculated cocaine consumption for each WWTP region for each sampling day and for each sampling campaign is shown in Table 17.2.

17.4.2   Daily Variations in Cocaine Consumption

From the statistical comparison of the cocaine consumption on a weekday and a weekend it can be concluded that there is a significant increase of cocaine use in the weekend for each WWTP region. In an additional three-week sampling campaign in the WWTP region of Brussels (daily collection of a 24-h composite sample), a

TABLE 17.2 Cocaine Consumption in grams per day per 1000 Inhabitants (Mean ± Standard Deviation) for 41 WWTP Regions and for Four Sampling Days in Belgium[a]

WWTP	Sunday 1 (summer)	Sunday 2 (winter)	Mean Sunday	Wednesday 1 (summer)	Wednesday 2 (winter)	Mean Wednesday	Mean total[b]
Aalst	0.158	0.317	0.237 ± 0.112	0.229	0.210	0.220 ± 0.013	0.228 ± 0.066
Aartselaar	0.272	0.339	0.305 ± 0.048	0.232	0.172	0.202 ± 0.042	0.254 ± 0.070
Antwerp North	0.379	0.284	0.332 ± 0.067	0.252	0.165	0.209 ± 0.061	0.270 ± 0.088
Antwerp South	1.829	1.828	1.829 ± 0.001	1.269	0.681	0.975 ± 0.416	1.402 ± 0.548
Arlon	0.763	1.166	0.965 ± 0.286	0.653	0.615	0.634 ± 0.027	0.799 ± 0.253
Beersel	0.259	0.295	0.277 ± 0.026	0.180	0.220	0.200 ± 0.029	0.239 ± 0.050
Bruges	0.298	0.248	0.273 ± 0.035	0.098	0.083	0.090 ± 0.011	0.182 ± 0.108
Brussels North	0.802	0.835	0.819 ± 0.023	0.610	0.545	0.578 ± 0.046	0.698 ± 0.142
Charleroi	0.686	0.861	0.774 ± 0.124	0.652	0.841	0.746 ± 0.134	0.760 ± 0.106
Ciney	0.050	0.196	0.123 ± 0.103	0.217	0.076	0.146 ± 0.100	0.134 ± 0.084
Dendermonde	0.663	0.493	0.578 ± 0.120	0.379	0.359	0.369 ± 0.014	0.473 ± 0.139
Destelbergen	0.376	0.282	0.329 ± 0.066	0.190	0.231	0.211 ± 0.029	0.270 ± 0.080
Deurne	1.189	0.834	1.011 ± 0.251	0.949	0.748	0.849 ± 0.142	0.930 ± 0.191
Gembloux	0.059	0.093	0.076 ± 0.024	0.104	0.086	0.095 ± 0.013	0.086 ± 0.019
Genk	0.793	1.328	1.060 ± 0.379	1.128	0.345	0.737 ± 0.554	0.899 ± 0.430
Ghent	0.470	0.495	0.483 ± 0.018	0.295	0.325	0.310 ± 0.021	0.396 ± 0.101
Harelbeke	0.553	0.559	0.556 ± 0.004	0.533	0.771	0.652 ± 0.168	0.604 ± 0.112
Hasselt	0.507	0.047	0.277 ± 0.325	0.552	0.331	0.442 ± 0.157	0.359 ± 0.229
Heist	0.314	0.864	0.589 ± 0.389	0.718	0.584	0.651 ± 0.095	0.620 ± 0.234
La Louvière	0.528	0.524	0.526 ± 0.003	0.217	0.339	0.278 ± 0.086	0.402 ± 0.151
Leuven	0.446	0.382	0.414 ± 0.045	0.225	0.250	0.238 ± 0.018	0.326 ± 0.106
Liedekerke	0.338	0.447	0.393 ± 0.077	0.220	0.250	0.235 ± 0.021	0.314 ± 0.102
Liège	0.358	0.543	0.450 ± 0.130	0.187	0.096	0.141 ± 0.064	0.296 ± 0.197
Lier	nm	1.065	1.065	Nm	0.487	0.487	0.776 ± 0.408
Lokeren	0.357	0.512	0.435 ± 0.110	0.420	0.354	0.387 ± 0.046	0.411 ± 0.074

Mechelen North	0.721	0.627	0.674 ± 0.067	0.393	0.468	0.430 ± 0.053	0.552 ± 0.149
Menen	0.277	0.450	0.363 ± 0.122	0.375	0.401	0.388 ± 0.018	0.375 ± 0.073
Mol	0.332	0.301	0.317 ± 0.022	0.132	0.100	0.116 ± 0.022	0.216 ± 0.117
Mons	0.241	0.695	0.468 ± 0.321	0.140	0.517	0.328 ± 0.266	0.398 ± 0.254
Morkhoven	0.200	0.174	0.187 ± 0.018	0.223	0.301	0.262 ± 0.055	0.224 ± 0.055
Mouscron	0.165	0.211	0.188 ± 0.032	0.171	0.287	0.229 ± 0.082	0.209 ± 0.056
Nivelles	0.435	0.640	0.537 ± 0.145	0.487	0.311	0.399 ± 0.125	0.468 ± 0.136
Ostend	0.194	0.548	0.371 ± 0.250	0.290	0.200	0.245 ± 0.064	0.308 ± 0.166
Roeselare	0.285	0.288	0.286 ± 0.002	0.244	0.136	0.190 ± 0.076	0.238 ± 0.071
Sint-Niklaas	0.270	0.642	0.456 ± 0.263	0.578	0.282	0.430 ± 0.209	0.443 ± 0.195
Sint-Truiden	0.250	0.344	0.297 ± 0.066	0.277	0.280	0.279 ± 0.002	0.288 ± 0.040
Tessenderlo	Nm	0.327	0.327	Nm	0.204	0.204	0.265 ± 0.087
Tournai	0.359	0.313	0.336 ± 0.032	0.132	0.365	0.248 ± 0.165	0.292 ± 0.110
Verviers	0.397	0.453	0.425 ± 0.040	0.269	0.268	0.268 ± 0.001	0.347 ± 0.093
Wavre	0.131	0.060	0.096 ± 0.050	0.112	0.070	0.091 ± 0.030	0.093 ± 0.034
Wulpen	0.418	0.188	0.303 ± 0.163	0.092	0.140	0.116 ± 0.034	0.210 ± 0.145

[a]WWTP: waste water treatment plant, nm: not measured,
[b]mean (weekend1, weekend2, week1, week2).

constant cocaine consumption from Monday to Friday (week) followed by a peak in the period Saturday–Sunday (weekend) was demonstrated (van Nuijs et al., 2009c). Other studies applying the sewage epidemiology approach in Italy and Spain also observed higher cocaine consumption during the weekend, confirming our results (Zuccato et al., 2008; Huerta-Fontela et al., 2008). Such observation of higher consumption of cocaine during weekends is supported by findings of the EMCDDA, which report on the recent surge of cocaine as replacement for ecstasy and other "party drugs" (European Monitoring Centre for Drugs and Drug Addiction, 2008a).

17.4.3 Cocaine Consumption and Urbanization

The local results in Belgium suggest a higher cocaine use in WWTP regions with high urbanization compared with WWTP regions that cover more rural and, thus, less urbanized areas. In WWTP regions receiving wastewater from large and strong urbanized cities in Belgium, namely Antwerp, Charleroi, Mechelen, and Brussels, higher cocaine consumption was observed in comparison with more rural regions, such as Wulpen, Mol, Menen, Ciney, and Gembloux (Tables 17.1 and 17.2 and Fig. 17.2).

If we focus on the results of cocaine consumption in Antwerp, the role of urbanization can be clearly seen. For the Antwerp region, three WWTP regions were included in the study. WWTP Deurne and Antwerp South, which receive wastewater mainly from the centre of Antwerp (high urbanicity), and WWTP Antwerp North, which receives wastewater from the northern part of Antwerp, covering a more rural (including the Antwerp harbor) area (van Nuijs et al. 2009a). The difference in cocaine consumption (Table 17.1 and 17.2 and Fig.17.2) between these areas is clear and confirms that cocaine consumption is higher in urbanized regions. This was also evidenced by a recent study from United States comparing wastewater results with socioepidemiological data (Banta-Green et al., 2009). Yet, we should be cautious in making generalizations from such results, since many other factors than urbanization can be related to cocaine consumption.

17.4.4 Extrapolation of Local Results to a Yearly Nationwide Prevalence of Cocaine

The sampling campaign in Belgium covered wastewater originating from approximately 3,700,000 residents (35% of the total Belgian population), including regions with high and low urbanization. With these features, it seemed appropriate to extrapolate local results to the general population and to estimate a yearly prevalence of cocaine consumption in Belgium. In a first step, weekly cocaine consumption was calculated for each WWTP region, assuming that a week consists of two days of a weekend and five weekdays. This assumption is supported by observations in the three-week sampling campaign carried out in the WWTP Brussels North (van Nuijs et al. 2009c). The mean value for the Sunday samples and Wednesday samples of each WWTP region was thus multiplied by 2 and 5, respectively, and summed. The weekly results were then multiplied by 52 to obtain yearly cocaine consumption for each WWTP region. In a next step, all yearly results were summed to obtain the yearly cocaine consumption by the 3,700,000 residents. In a last step, this number was multiplied by 2.86 to result in an estimate of the amount of cocaine that is consumed yearly by the total Belgian population (10,584,534 residents). Results suggested a use of 1.88 tons per year of cocaine in Belgium for 2007–2008. From this absolute amount, we estimated a prevalence of cocaine consumption in Belgium, for which several assumptions had to be made:

1. An average cocaine dose is 100 mg (Lamkaddem and Roelands, 2007)
2. An average cocaine user consumes 0.650 g/week or 33.8 g/year (Everingham and Rydell, 1994; Cohen and Sas, 1994).

Taking these assumptions into account, the calculated 1.88 tons of cocaine corresponded to 55,770 users [(1.88 × 106 g)/33.8 g], which is about 0.53% of the total Belgian population (55,770/10,584,534) or 0.80% (55,770/6,976,743) and 1.32% (55,770/4,230,029) of the Belgian population aged 15–64 or 15–44 years, respectively.

17.4.5 Comparison with Existing Estimates

An important question is if this estimate of cocaine consumption in Belgium is realistic. The EMCDDA reported an average value for the "last year cocaine prevalence" in the age group 15–64 years in 2007 in Europe of 1.2%, with a range from 0.1% (in Greece) to 3% (in Spain) (European Monitoring Centre for Drugs and Drug Addiction, 2008). The calculated prevalence for Belgium from wastewater analysis for this age group – (0.80%) seems thus realistic and fits in this European picture, where Belgium can be placed as a country with medium cocaine consumption. However, because of the lack of national studies of cocaine use in Belgium, it is very difficult to really validate this estimate. The only comparison that can be made is with an estimate of the UNODC, which, in 2004, reported a last year prevalence of cocaine consumption in Belgium of 0.9% of the population aged 15–64 years (United Nations Office of Drugs and Crime, 2004). It is important to mention that this estimate was based only on local studies, special population groups, and/or law enforcement agency assessments, since general prevalence studies have never been executed in Belgium. As a conclusion, we can state that from the little available information of cocaine use in Belgium, the estimate of the cocaine prevalence derived from wastewater data appears logical and fits in the European picture of cocaine consumption prevalence. In this way, sewage epidemiology shows its great potential to provide additional information about cocaine consumption both at local and national level.

Of course there are limitations and uncertainties about the calculations made. The extrapolation from four days to a complete year can probably be improved and the calculation of prevalence from a yearly consumption is subjected to several assumptions that have to be refined or updated in future research. The latter are based on research in the 1990s on drug consumption and is it possible that the patterns of drug use could have changed since then. However, the purpose of the study was to explore the potential of wastewater analysis as an informational tool to estimate local and, for the first time, also national cocaine use, rather than giving an exact calculation of cocaine prevalence.

17.5 CONCLUSIONS AND FUTURE TRENDS AND RESEARCH

The sewage epidemiology work carried out in Belgium shows that this approach provides objective and realistic data of illicit drug consumption both at the local and nationwide scale. Compared to the socioepidemiological approaches, wastewater analysis has a higher degree of objectivity, is relatively inexpensive, and can offer results in a short time frame. Therefore, it can be used to trace hot-spots of illicit drug use and to evaluate prevention campaigns and targeted actions in a short period of time.

However, it is important to mention that sewage epidemiology has to be seen as a tool to monitor illicit drug consumption synergistically to the classic socioepidemiological studies. Combining both approaches and tuning both methodologies will lead to a better and more detailed understanding of the illicit drug problem and use patterns. Therefore, collaborations between epidemiologists, criminologists, doctors, toxicologists, environmental scientists, and policy makers have to be established in the near future.

Sewage epidemiology is a relatively new scientific field and it is still in an early stage of development. The presented work demonstrates the great potential of the approach, but there has to be stressed that sewage epidemiology is far from optimized and validated. The presented work shows results for only cocaine. The same working strategy should be followed for the estimation of local and nationwide consumption of other drugs of abuse, such as amphetaminelike stimulants, opiates, and cannabis.

Because of various uncertainties in sewage epidemiology, extensive future research is needed. Until now, there is a high degree of uncertainty related to the degradation or adsorption of illicit drugs and metabolites in the sewer system from the toilet to the WWTP. If so, the measured concentrations have to be corrected for this loss, leading to a higher calculated illicit drug use. To optimize the back-calculation from concentration to amount of used drug, new pharmacokinetic studies have to be carried out to have more extensive knowledge about metabolic patterns of illicit drugs. In the back-calculations, more than one metabolite (e.g., EME for cocaine) should be included, if possible, to obtain more accurate calculations. Furthermore, future research should also focus on WWTP related parameters, such as the flow rate and the amount of served inhabitants. It is yet unclear what the error is on the measured flow rate and the amount of served residents by a WWTP is until now estimated through the design capacity of the WWTP, while in reality this parameter is not as stable as previously thought. To calculate prevalences, sociological studies at the average dose, the purity, and the daily amount consumed by an average user have to be established to update, confirm, or replace the values that are now assumed in the calculations.

ACKNOWLEDGMENTS

Alexander van Nuijs and Dr. Adrian Covaci acknowledge the Flanders Scientific Funds for Research (FWO) for their grants. The authors also like to thank the Belgian government and the Belgian Science Policy (Belspo) for their financial support (COWAT project DR 00-0047). A. Vandelannoote and Aquafin are acknowledged for their help with the sampling campaigns.

REFERENCES

Ambre, J., M. Fischman, and T. Ruo. 1984. Urinary excretion of ecgonine methyl ester, a major metabolite of cocaine in humans. *J. Anal. Toxicol.* **8**:23–25.

Ambre, J., T. Ruo, J. Nelson, and S. Belknap. 1988. Urinary excretion of cocaine, benzoylecgonine and ecgonine methyl ester in humans. *J. Anal. Toxicol.***12**:301–306.

Banta-Green, C. J., J. A. Field, A. C. Chiaia, D. L. Sudakin, L. Power, and L. de Montigny. 2009. The spatial epidemiology of cocaine, methamphetamine, and 3,4-methylenedioxymethamphetamine (MDMA) use: A demonstration using a population measure of community drug load derived from municipal wastewater. *Addiction* **104**:1874–1880.

Bijlsma, L., J. V. Sancho, E. Pitarch, M. Ibanez, and F. Hernandez F. 2009. Simultaneous ultra-high-pressure liquid chromatography-tandem mass spectrometry determination of amphetamine and amphetamine-like stimulants, cocaine and its metabolites, and a cannabis metabolite in surface water and urban wastewater. *J. Chromatogr. A* **1216**:3078–3089.

Boleda, M. R., M. T. Galceran, and F. Ventura. 2009. Monitoring of opiates, cannabinoids and their metabolites in wastewater, surface water and finished water in Catalonia, Spain. *Water Res.* **43**:1126–1136.

Bones, J., K. V. Thomas, and B. Paull. 2007. Using environmental analytical data to estimate levels of community consumption of illicit drugs and abused pharmaceuticals. *J. Environ. Monitor.* **9**:701–707.

Castiglioni, S., E. Zuccato, E. Crisci, C. Chiabrando, R. Fanelli, and R. Bagnati. 2006. Identification and measurement of illicit drugs and their metabolites in urban wastewater by liquid chromatography-tandem mass spectrometry. *Anal. Chem.* **78**:8421–8429.

Chiaia, A. C., C. Banta-Green, and J. Field. 2008. Eliminating solid phase extraction with large-volume injection LC/MS/MS: analysis of illicit and legal drugs and human urine indicators in US wastewaters. *Environ. Sci. Technol.* **42**:8841–8848.

Cohen, P., and A. Sas. 1994. Cocaine use in Amsterdam in non deviant subcultures. *Addict Res.* **2**:71–94.

Cone, E. J., A. Tsadik, J. Oyler, and W. D. Darwin. 1998. Cocaine metabolism and urinary excretion after different routes of administration. *Therap. Drug Monit.* **20**:556–560

Daughton CG. 2001. Illicit drugs in municipal sewage. In Pharmaceuticals and Personal Care Products in the Environment, Scientific and Regulatory Issues, edited by C. G. Daughton and T. L. Jones-Lepp, pp. 348–364. Washington, DC: American Chemical Society.

European Monitoring Centre for Drugs and Drug Addiction (EMCDDA). 2008a. The state of the drug problem in the European Union and Norway. Annual Report. Lisbon: EMCDDA. Available at http://www.emcdda.europa.eu/publications/annual-report/2008.

European Monitoring Centre for Drugs and Drug Addiction (EMCDDA). 2002. EM-CDDA project CT.99.EP.08 B. Handbook for surveys on drug use among the general population. Lisbon: EMCDDA. Available at http://www.emcdda.europa.eu/html.cfm/index58052EN.html

European Monitoring Centre for Drugs and Drug Addiction (EMCDDA). 2008b. EMCDDA Insights Series No 9. Assessing illicit drugs in wastewater. Potential and limitations of a new monitoring approach.

Everingham, S. S., and P. C. Rydell. 1994. Modelling the Demand for Cocaine. Santa Monica, CA: RAND Corporation.

Frost, N., P. Griffiths, and R. Fanelli. 2008. Peering into dirty waters: the potential and implications of a new approach to monitoring drug consumption. *Addiction* **103**:1239–1241.

Gheorghe, A., A. van Nuijs, B. Pecceu, L. Bervoets, P. G. Jorens, R. Blust, H. Neels, and A. Covaci A. 2008. Analysis of cocaine and its principal metabolites in waste and surface water using solid-phase extraction and liquid chromatography-ion trap tandem mass spectrometry. *Anal. Bioanal. Chem.* **391**:1309–1319.

Huerta-Fontela, M., M. T. Galceran, J. Martin-Alonso, F. Ventura. 2007. Ultraperformance liquid chromatography-tandem mass spectrometry analysis of stimulatory drugs of abuse in wastewater and surface waters. *Anal. Chem.* **79**:3821–3829.

Huerta-Fontela, M., M. T. Galceran, J. Martin-Alonso, and F. Ventura. 2008. Occurrence of psychoactive stimulatory drugs in wastewaters in north-eastern Spain. *Sci. Total Environ.* **397**:31–40.

Kasprzyk-Hordern, B., R. M. Dinsdale, and A. J. Guwy. 2009. Illicit drugs and pharmaceuticals in the environment – forensic applications of environmental data. Part 1: estimation of the usage of drugs in local communities. *Environ. Pollut.* **157**:1773–1777.

Kasprzyk-Hordern, B., R. M. Dinsdale, and A. J. Guwy. 2008. Multiresidue methods for the analysis of pharmaceuticals, personal care products and illicit drugs in surface water and wastewater by solid-phase extraction and ultra performance liquid chromatography-electrospray tandem mass spectrometry. *Anal. Bioanal. Chem.* **391**:1293–1308.

Lamkaddem, B., and M. Roelands 2007. Belgian National Report on Drugs 2007. Brussels: Scientific Institute of Public Health, Epidemiology Unit.

Mari, F., L. Politi, A. Biggeri, G. Accetta, C. Trignano, M. Di Padua, and E. Bertol. 2009. Cocaine and heroin in waste water plants: A 1-year study in the city of Florence, Italy. *Forensic Sci. Int.* **189**:88–92.

Postigo, C., M. J. Lopez de Alda, and D. Barcelo. 2008. Fully automated determination in the low nanogram per litre level of different classes of drugs of abuse in sewage water by on-line solid-phase extraction-liquid chromatography-electrospray-tandem mass spectrometry. *Anal. Chem.* **80**:3123–3134.

United Nations Office of Drugs and Crime (UNODC). 2004. The World Drug Report. Vienna: United Nations Office of Drugs and Crime.

United Nations Office of Drugs and Crime (UNODC). 2007. Commission on Narcotic Drugs, Draft Resolution E/CN.7/2007/L.16/Rev.1. Available at http://www.unodc.org/unodc/commissions/CND/07-reports.html

United Nations Office of Drugs and Crime (UNODC). 2009. The World Drug Report. Vienna: United Nations Office of Drugs and Crime. Available at http://www.unodc.org/unodc/en/data-and-analysis/WDR-2009.html.

van Nuijs, A. L. N., B. Pecceu, L. Theunis, N. Dubois, C. Charlier, P. G. Jorens, L. Bervoets, R. Blust, H. Neels, and A. Covaci. 2009a. Cocaine and metabolites in waste and surface water across Belgium. *Environ. Pollut.* **157**:123–129.

van Nuijs, A. L. N., B. Pecceu, L. Theunis, N. Dubois, C. Charlier, P. G. Jorens, L. Bervoets, R. Blust, H. Meulemans, H. Neels, and A. Covaci. 2009b. Can cocaine use be evaluated through analysis of wastewater? A nation-wide approach conducted in Belgium. *Addiction* **104**:734–741.

van Nuijs, A. L. N., B. Pecceu, L. Theunis, N. Dubois, C. Charlier, P. G. Jorens, L. Bervoets, R. Blust, H. Neels, and A. Covaci. 2009c. Spatial and temporal variations in the occurrence of cocaine and benzoylecgonine in waste- and surface water from Belgium and removal during wastewater treatment. *Water Res.* **43**:1341–1349.

van Nuijs, A. L. N., I. Tarcomnicu, P. G. Jorens, L. Bervoets, R. Blust, H. Neels, and A. Covaci. 2009d. Analysis of drugs of abuse in wastewater by hydrophilic interaction liquid chromatography–tandem mass spectrometry. *Anal. Bioanal. Chem.* **395**:819–828.

Zuccato, E., C. Chiabrando, S, Castiglioni, D. Calamari, R. Bagnati, S. Schiarea, and R. Fanelli R. 2005. Cocaine in surface waters: a new evidence-based tool to monitor community drug abuse. *Environ. Health* **4**:14–20.

Zuccato, E., C. Chiabrando, S. Castiglioni, R. Bagnati, and R. Fanelli. 2008. Estimating community drug abuse by wastewater analysis. *Environ. Health Perspect.* **116**:1027–1032.

CHAPTER 18

MEASUREMENT OF ILLICIT DRUG CONSUMPTION IN SMALL POPULATIONS: PROGNOSIS FOR NONINVASIVE DRUG TESTING OF STUDENT POPULATIONS

DEEPIKA PANAWENNAGE, SARA CASTIGLIONI, ETTORE ZUCCATO, ENRICO DAVOLI, and M. PAUL CHIARELLI

18.1 INTRODUCTION

The abuse of illicit drugs is a global problem from which no society is immune (Castiglioni et al., 2006; Drug Abuse, 2009). In the United States alone, approximately 40 million serious illnesses and injuries result from drug abuse annually. Drug abuse contributes to major social problems such as reckless driving, crime, and family violence, such as spousal or child abuse (Drug Abuse). According to the National Survey on Drug Use and Health (NSDUH), it is estimated that about 20.1 million Americans aged 12 or older (8.1% of the population) had consumed an illicit drug within a month of completing the survey. Approximately one-half of these people admitted to driving under the influence of drugs within the last month as well (Substance Abuse and Mental Health Services Administration, 2009).

Drug testing of individuals began in earnest in the late 1960s to discourage the use of illicit drugs. Positive drug tests were and are used to disqualify potential job applicants and, in some instances, terminate employment. One of the first groups subjected to mandatory drug testing were persons undergoing methadone treatment in federally mandated programs treating heroin and opiate addiction (Coombs et al., 1991). Methadone is a drug that eases withdrawal symptoms associated with opiate use. Individuals wanting access to methadone were required to undergo urine tests to show progress in overcoming their opiate addiction. During the Vietnam War, the US Department of Defense authorized urine testing of soldiers participating in

Illicit Drugs in the Environment: Occurrence, Analysis, and Fate Using Mass Spectrometry, Edited by Sara Castiglioni, Ettore Zuccato, and Roberto Fanelli
Copyright © 2011 John Wiley & Sons, Inc.

combat operations to discourage the use of freely available heroin originating from other parts of southeast Asia. The US Navy instituted mandatory drug testing of all personnel after a fighter plane crashed on the carrier USS Nimitz in May 1981, which resulted in the deaths of nine people and injuries to 42 others. The drug testing mandate was issued after it was determined in autopsies that nine of the crew members on deck had cannabinoids in their systems. It was also determined by autopsy that the pilot was flying under the influence of bromopheniramine, an antihistamine that causes drowsiness, which had not been prescribed by his physician (US Navy Judge Advocate General investigation and report, 1982). More recently, the drug testing of athletes participating in the Olympics or other professional sporting events has been mandated by various governing bodies to ensure that no participant has an advantage in competition because of the use of illegal or nonsanctioned drugs. Today, the most widely tested group of people work in the transportation industry. Most of the accepted analytical methods used to test individual drug use have been based on the analysis of urine extracts by GC/MS (Jones, 2001). Lately, methods for individual drug testing based on LC and tandem mass spectrometry have become more prevalent, because little or no sample pretreatment is required prior to analysis (Pizzolato et al., 2007).

Recently, many public secondary schools in the United States have considered implementing mandatory drug testing for all students in attendance (Health, kids, and parenting; Las Vegas Review-Journal online, 2010). However, in practice, implementing drug testing of all students has run into some resistance. As a result, most secondary schools that have implemented drug testing can only do so when they have the parents' permission or when students seek administrative approval to participate in extracurricular activities, such as an athletic competition. Some private secondary schools in the United States require parental consent to test students for drugs as a condition of admission. Mandatory drug screening of individuals will no doubt discourage illicit drug use. However, public resistance to individual testing is likely to persist because it is personally invasive. Here we demonstrate an alternative or complementary form of drug testing of small targeted populations that is based on the sampling and analysis of illicit drugs and their metabolites isolated from wastewater. This method can provide information regarding the overall drug use within demographically well-defined populations, such as students in secondary schools or universities. Wastewater testing can provide information on the design of anti-drug campaigns and other public health initiatives. Drug testing based on wastewater analysis is likely to be more palatable because it does not violate the personal privacy of those being tested.

To date, efforts to measure drug consumption and abuse in targeted populations, such as young adults, have relied on information gathered from questionnaires (such as Substance Abuse and Mental Health Services Administration, 2009, mentioned above). The utility of information gathered from surveys is limited because individuals have a tendency to underreport illicit drug use for fear of the breaking the law. Analytical methods based on the measurement of illicit drugs and their metabolites are an attractive alternative to survey-based data, because they are unaffected by human bias encountered in data collected by self-reporting. The potential for mass spectrometry-based analytical methods to differentiate drug use in different selected populations was first suggested by the GC/MS analysis of cocaine on paper currency (Oyler et al., 1996). In this study, dollar bills were collected from banks in 14 different geographic locations and analyzed for cocaine. Dollars taken from Baltimore, Maryland, Minneapolis, Minnesota, and Portsmouth, Ohio were found to have a factor of 10 or more cocaine than the other 11 localities tested. The data was interpreted to suggest that these three locations had the highest concentration of cocaine users. The successful analysis of drugs on money has prompted the development of analytical methods for the direct analysis of drugs on different surfaces and in different matrices using laser desorption (Lowe et al., 2009) and desorption electrospray ionization (Kauppila et al., 2007).

The discovery of pharmaceuticals, endocrine disruptors, and other personal care products accumulating in the water supply (Daughton and Ternes, 1999; Ternes et al., 2001) suggested that the LC–MS/MS analysis of illicit drugs and their metabolites may be possible as well. Cocaine was the first drug of abuse analyzed in wastewater and natural water. Zuccato et al. (2005) quantified cocaine and benzoylecgonine (the main metabolite of cocaine) in river (surface) water and untreated wastewater in 2005. The information provided by the quantification allowed the authors to estimate the cocaine consumption in the urban areas along the Po River. A year later, Castiglioni et al. (2006) demonstrated the quantification of 16 different drugs and metabolites in influent taken from wastewater treatment plants in Lugano and Milano. Some of these compounds could be found in the effluent as well, suggesting that these metabolites could accumulate in the environment if conditions at water treatment facilities were not monitored regularly. Following these initial studies of drug and drug metabolite distribution in different water sources, a number of studies have been carried out in different localities to assess the environmental impact of these micropollutants and the potential of LC and tandem mass spectrometry as a tool for monitoring drug consumption in targeted communities. To date, the size and demographic composition of the population surveyed has been dictated by the area served by a wastewater treatment plant whose water is sampled for analysis. Here we demonstrate the feasibility of testing very small populations whose composition is determined by a school campus or workplace. We have detected several illicit drugs in the wastewater run-off from a campus serving a small population (less than 1000 students).

18.2 EXPERIMENTAL

Most of the experimental procedures used in this study are described in detail elsewhere (Castiglioni et al., 2006) and are briefly summarized here. Illicit drugs and their metabolites extracted from wastewater were analyzed by LC–MS/MS (API 3000 triple-quadrupole mass spectrometer) and isotope dilution mass spectrometry. The mass spectrometer was equipped with a turbo ion-spray source (Applied Biosystems-Sciex, Thornhill, Ontario, Canada) and an HPLC system consisting of two Series 200 pumps and a Series 200 autosampler (Perkin-Elmer, Norwalk, CT). Drugs and metabolites were divided into four groups for concurrent quantification based on their chromatographic and ionization properties. A Waters Xterra C18 column, 2.1 × 100 mm was used for all separations.

Wastewater samples were collected at an educational institution in northern Illinois on different days during regular class sessions, final exams, and summer break. The wastewater pipe was accessed under a manhole cover approximately 3 m below the ground. A container was lowered below the building output pipe to collect the water as it emptied into the municipal sewer system. Samples (up to 500 mL) were collected every 20 min for 2 h and pooled together for most analyses. Composite samples were collected between 10 AM and 12 PM (AM) and between 12 PM and 2 PM (PM).

Each 2-h composite wastewater sample was acidified to pH 2 using 37% hydrochloric acid and filtered through Whatman glass microfiber filter papers (47-mm 1.6 μm GF/A). After filtering, water samples (100-mL) were spiked with 20 ng of each deuterated, internal standard, and then subjected to solid-phase extraction using mixed reversed-phase cation-exchange cartridges (Oasis-MCX, 60 mg, Waters Corp., Milford, MA, USA). Cartridges were conditioned using 6 mL of methanol, 3 ml of 18 MΩ water, and 3 mL of water acidified to pH 2 prior to the extraction. Cartridges were then eluted with 3 mL of methanol and 3 mL of a 2% ammonia solution in methanol and eluates were pooled and dried under a nitrogen stream. Eluates were concentrated to dryness and then reconstituted in 200 μL of water for LC–MS/MS analysis. Tap water samples (1 L) were collected at different times and analyzed to estimate the concentrations of drugs and metabolites entering the campus.

All standard compounds were acquired from Sigma Aldrich (St. Louis, MO USA), Cerulliant (Round Rock, TX USA), or Cambridge Isotope Laboratories (Andover, MA USA) and used without further purification.

18.3 RESULTS AND DISCUSSION

18.3.1 Identities of Drugs in School Wastewater Runoff

Our initial experiments were undertaken to demonstrate that illicit drugs and metabolites could be detected in wastewater collected from a small population. Wastewater samples were taken from a school campus on different days during the academic year. These studies of a small population were encouraged by the detection of drugs in natural and wastewater drawn from treatment plants serving large metropolitan areas (Castiglioni et al., 2008). Even though the population we sampled is much smaller than the water treatment plant studies, the volume of water in which the drugs and metabolites are dispersed is smaller as well, suggesting that the concentrations of drugs found in the water taken from the campus may be comparable to concentrations found in wastewater treatment plants. We tested the campus wastewater output for the presence of 16 different drugs and metabolites (Castiglioni et al., 2006). There are eight compounds that were detected in multiple water samples. The relative standard deviation (RSD) associated with these measurements varied from approximately 1% −10% as the concentration of the analyte in the sample decreases

TABLE 18.1 **Measured Concentrations (ng/L) of Drugs and Their Metabolites during the Regular Class Sessions (2008)**

Substances	Concentration (ng/L)								
	Wed Oct 1 AM	Wed Oct 1 PM	Fri Oct 3 AM	Mon Oct 13 PM	Tue Oct 14 AM	Wed Oct 22 PM	Fri Oct 31 AM	Fri Oct 31 PM	Sat Nov 1 AM
THC-COOH	3.6	0.0	27.9	43.8	177.1	54.2	0.0	81.5	0.0
Amphetamine	14.6	0.0	4.6	55.9	7.5	11.3	67.8	140.4	70.2
MDA	0.0	173.9	0.0	0.0	0.0	0.0	0.0	0.0	0.0
MDMA	0.0	3266.0	0.0	8.4	0.0	0.0	0.0	0.0	0.0
Benzoylecgonine	2.6	2.0	3.8	2.3	10.0	1.9	2.5	2.9	2.1
Cocaine	4.5	2.4	13.8	6.2	3.7	2.3	3.5	10.5	3.3
Codeine	16.1	62.8	71.4	21.1	0.0	12.0	19.5	0.0	0.0
Morphine	0.0	0.0	0.0	0.0	0.0	0.0	0.0	0.0	0.0
6-Acetylmorphine	0.0	0.0	0.0	0.0	0.0	0.0	0.0	0.0	0.0
Morphine-3β-glucuronide	0.0	29.1	41.9	27.1	4.6	8.8	0.0	0.0	0.0

from 1 to 0.1 ng/mL. The recovery of the analyte in the solid phase extraction process was generally $> 80\%$.

Tables 18.1, 18.2, and 18.3 list concentration measurements made at different times during regular classes, final exams, and the summer when few students were present, respectively. Δ^9-Tetrahydrocannabinol (THC) is the major psychoactive compound in marijuana. THC is rapidly oxidized to 11-hydroxy THC and then to 11-nor-9-carboxy-Δ^9-THC, which is the main metabolite excreted. The main metabolic route of cocaine involves the hydrolysis of ester linkages to produce benzoylecgonine (BE). 6-Acetylmorphine is the main specific metabolite of heroin (3,6-diacetylmorphine). Codeine is O_3-methylmorphine. Morphine-3β-glucuronide is a metabolite formed *in vivo* by the attachment of a sugar to the O_3 oxygen, which facilitates elimination of morphine from the body. All three of these morphine-based derivatives are narcotics. 3,4-Methylenedioxymethamphetamine (MDMA) and 3,4-methylenedioxyamphetamine (MDA) are amphetamine derivatives that are commonly referred to as ecstasy.

The data summarized in Tables 18.1–18.3 clearly indicates that the majority of the drug consumption was because of the students. The concentrations in Table 18.3 indicate that those drugs and metabolites sampled during the summer days at the times indicated were not present in detectable amounts. However, measurable quantities were also detected on some occasions. This because there were administrators and building maintenance staff present during the summer; the main building was also open during business hours so that anyone off the street had access to the facilities. The data in Table 18.3 shows that morphine and two of its derivatives, morphine-3β-glucuronide and 6-acetylmorphine were all present in the water system at the same time, on one occasion, and then not detected at any other time. This suggests that these concentrations may represent the narcotic usage of a single individual, probably

TABLE 18.2 Measured Concentrations (ng/L) of Drugs and Their Metabolites during Finals Week (2008)

	Monday Day 1		Tuesday Day 2		Wednesday Day 3		Thursday Day4		Friday Day 5		Monday Day 6	
	AM	PM	AM	PM	AM	PM	AM	PM	AM	PM	AM	PM
Substances												
THC-COOH	71.0	119.7	18.2	69.2	19.3	23.5	37.0	372.9	93.1	22.3	4.4	24.1
Amphetamine	9.6	153.6	7.4	0.0	91.2	27.3	125.2	62.4	29.8	12.1	43.1	4.5
Benzoylecgonine	3.6	2.1	34.7	1.7	1.5	2.0	3.5	3.8	4.5	2.5	1.2	1.3
Cocaine	4.1	4.0	5.5	7.9	2.5	3.2	3.1	4.3	6.0	9.6	4.3	4.1
Codeine	0.0	0.0	0.0	0.0	0.0	0.0	0.0	25.6	0.0	0.0	11.5	0.0
Morphine	6.5	32.5	0.0	0.0	0.0	11.1	0.0	0.0	0.0	0.0	43.1	0.0
Morphine-3β-glucuronide	33.5	10.4	0.0	565.5	8.5	11.0	7.3	91.8	0.0	0.0	0.0	0.0

Concentration (ng/L)

TABLE 18.3   Concentrations (ng/L) of Drugs and Their Metabolites during the Summer Break (2008)

Substances	Concentration (ng/L) and Sampling Date[a]		
	Mon Jun 09	Wed June 11	Thu Jul 10
THC-COOH	40.3	0.0	7.1
Amphetamine	0.0	0.0	0.0
MDA	0.0	0.0	0.0
MDMA	0.0	0.0	0.0
Benzoylecgonine	6.2	1.3	1.3
Cocaine	4.2	1.2	1.5
Codeine	6.1	0.0	0.0
Morphine	95.4	0.0	0.0
6-Acetylmorphine	7.7	0.0	0.0
Morphine-3β-glucuronide	63.6	0.0	0.0

[a]Samples were acquired in the morning only.

consuming heroin, since 6-acetylmorphine is a minor, but specific, metabolite of this drug. There are several factors that must be considered in understanding how these concentration measurements are reflective of the overall drug usage in this targeted population; they are discussed below. The data suggests that the three most frequently used drugs in this small population are cannabis, cocaine, and amphetamine. Most of the discussion that follows concerning the significance of these measurements will focus on these three drugs.

18.3.2   Calculation of Total Drug Quantities Sampled

In order to calculate the total consumption of a particular drug by the population, one needs to know both concentration and the total volume of water being sampled. There are two ways in which the volume associated with a measured concentration can be estimated in this particular application. One is dependent on how the wastewater is collected and dispersed. In most buildings that serve a large number of people such as those on this campus, wastewater coming from different sources (sinks, urinals, showers, and toilets, etc.) is first collected in a tank within the building that emits its contents once it is full. The tanks (typically one or two per building) that service these buildings have a maximum volume of 50 gallons (190 L). Therefore, one could calculate a total amount of drug by assuming that the concentration being measured represents what is contained in 190 L (or 380 L) and correlate the total amount of drug with the number of users by dividing the total amount of a drug by the average dosage. The frequency with which the tanks are discharged is proportional to the number of individuals using the facilities as well. Such an approach might be appropriate for calculating the number of doses when the instantaneous flow rate of the wastewater into the municipal sewer is small. Dosage calculations, such as these, may be most appropriate during a time when few people are present on campus.

The average daily flow rate was about 2.9×10^5 L/day during the school year and 1.5×10^5 L/day during the summer. These daily flow rates were estimated from the monthly billing by the water utility (we are assuming all the water that enters the building through the plumbing exits to the sewer). The flow rate of water going into the municipal sewer will vary depending on how often the collection tanks are discharged. In order to achieve these daily flow rates, the tanks will discharge every 1–2 min during the day.

The second way one could estimate the total quantity of a drug is to assume that the concentration is constant over the time that the wastewater is sampled. Dosages could be calculated based on the average volume of water that passes through the pipe during the time that the samples are being collected and pooled. For these particular analyses, three samples were taken and pooled over a 2-h period. If one uses the highest concentrations of the three most persistent drugs to estimate the number of doses, we find that the quantity of drug or representative metabolite that passes through the pipe in 2 h was less than an estimated single dose of all drugs, except THC. The highest 2-h concentrations for the three most persistent drugs and/or metabolites were found during finals week (Table 18.2). The highest concentration of the THC–carboxylic acid found was on Thursday afternoon during finals (372.9 ng/L). For amphetamine, the highest concentration found was 153.6 ng/L on the first Monday morning of the examination period. The highest BE concentration determined was 34.7 ng/L on Tuesday morning. Based on the flow rates stated above, this means that 153.6 ng/L $\times$ 2/24 h $\times$ 2.9×10^5 L/day = 3.7 mg of amphetamine was collected in a 2-h period. Based on the maximum concentrations stated for the other two drugs, the total amounts of BE and THC carboxylic acid were 0.8 and 9.0 mg, respectively. We extracted and analyzed 1 L of tap water on five consecutive days to determine the background of levels of the various drugs in the water entering the campus. Only BE and cocaine were detected; their average concentrations were determined to be 0.67 and 0.43 ng/L, respectively. These results suggest that background levels of these compounds do not contribute significantly to our measurements.

18.3.3 Estimating Differences/Changes in Drug Use

The concentrations of the different drugs found (Tables 18.1.1–18.3) and the average daily flow rates suggest that less than one whole dose of any of these three drugs passes through the wastewater system over the course of day, except for THC. This result is significant because, to date, comparisons between different populations (in different urban centers) have been made by comparing the total number of doses of a particular drug per 1000 people (Zuccato et al., 2008; Postigo et al., 2008). Correlations between amounts of drugs consumed and drug/metabolites in wastewater have been made taking into account two factors (Zuccato et al., 2008). One is the percentage of drug itself that is excreted or as a target analyte (parent or metabolite). The second factor is the molar mass ratio of a particular metabolite structure to that of the parent compound consumed by the user. The quantity of drugs consumed for the purpose of making comparisons is calculated assuming that the concentration of the drug in question or metabolite is the same throughout the treatment plant where the water is sampled. Such an assumption may cause an underestimation if a significant percentage of the compound being analyzed is present in the sediment or dissolved solids. The amounts of the three drugs present in the wastewater system we are sampling are calculated below.

An average dose of cocaine is 100 mg, which results in a maximum of 54 mg of BE upon excretion (Baselt and Cravey, 1989; Inaba et al., 1978) over the course of a day. The amount of BE that passes through a wastewater pipe over the course of a day even when calculated using the largest concentrations of metabolite found corresponds to less than one dose. Amphetamines are most often acquired as prescription drugs, so the fraction of the determined concentration that represents illicit drug use is not immediately obvious. It is possible that a student with a prescription for an ADHD (attention deficit hyperactivity disorder) drug, such as Adderall, could sell some tablets to their fellow students to help them study for exams. Adderall capsules are manufactured with total amphetamine doses ranging from 5 to 30 mg. Typically, 30% of the dose is excreted as the parent compound (Zuccato et al., 2008). For a 30-mg dose, this corresponds to 9 mg in wastewater. The total amount of amphetamine flowing through the pipe for the highest amphetamine concentration measured was 3.7 mg, as discussed above.

The amount of THC in average dose is harder to estimate. An accepted dosage for the sake of calculating total drug is 125 mg (Zuccato et al., 2008). The amount of THC absorbed is more much harder to estimate than drugs that are ingested in a pill or powder form, because much of the THC can be exhaled and thus not metabolized. The amount of THC in different strains of marijuana can vary a great deal as well. Approximately 0.6% of a dose is excreted as the carboxylic acid. As a result, the total measured quantity of the carboxylic acid that represents one dose is only 800 ng, suggesting that the cannabis is used far more frequently than cocaine or amphetamine in this population. This is roughly consistent with the results obtained by the National Survey on Drug Use and Health (NSDUH) that suggests in the United States among individuals aged 12 and over marijuana is the most frequently abused drug (Substance Abuse and Mental Health Services Administration). The survey states that over a 1-month period, 15.2 million individuals used marijuana compared to 1.9 million that used cocaine. The survey also suggests that marijuana is used far more frequently than cocaine. Fifteen percent of individuals admitting to using marijuana within a month of taking the survey, indicated that they used marijuana on 300 or more days within the past year. Unusually, large concentrations of MDMA (3266 ng/L —Table 18.1) and morphine-3β-glucuronide (565.5 ng/L—Table 18.2) were detected in two different wastewater samples. The observation of the high glucuronide concentration at a time when no 6-acetylmorphine is detected suggests the glucuronide resulted from a large therapeutic dose of morphine, rather than heroin (6-acetylmorphine is specific for heroin). The concentration of MDMA detected at one single time suggests that an individual may have disposed of a single dose at once, although without a simultaneous flow rate measurement, one cannot be sure. The flow rate of water into the sewer varies during the day and it is reasonable to assume that the peak flow rates would occur during a time of day when students are on campus. If the average flow rate over the 3-h period approached 1.9 million liters per day (above the average of 2.9×10^5 L/day), then a single dose may have passed through the water supply over that time. Observations of instantaneously high concentrations underscore the need for flow rate measurements conducted at the time of sampling to help distinguish drug concentrations that result from a single act of disposal from the concentrations that result from the metabolism of many individuals.

Our data suggests that measuring drug consumption in small, demographically well-defined populations (1000 individuals) will have to be approached differently than for populations whose size is defined by the area serviced by a wastewater treatment plant. Since the concentrations of drug and/or metabolite being measured are so small, the best way to describe changes in drug consumption is to compare the absolute quantities of drugs in wastewater rather than the number of doses. Estimating the number of users (implied by the number of doses) in a population such as this is not possible, because individuals will excrete a whole dose over a period of several days. The amount contributed by a particular user will vary depending upon when (and how much) a drug was consumed relative to the time of excretion. Therefore, estimating the total drug consumption in a small population is easier than estimating the total number of users. As a result, the success of a public health initiative targeted at a particular school may be defined by the changes in the absolute quantity of drugs measured at different times.

18.4 CONCLUSION

We have demonstrated the measurement of several illicit drugs and their metabolites in a wastewater stream coming from the campus of an educational institution using an analytical method based on liquid chromatography and tandem mass spectrometry. The concentrations of these drugs varied depending upon the time of the year samples were taken (final exam period > regular class session >> summer). The results suggest that drug measurements targeted at small populations may be used to monitor changes in drug consumption over time.

REFERENCES

Baselt, R. C., and R. H. Cravey. 1989. *Disposition of Toxic Drugs and Chemicals in Man*, third edition. Chicago, IL: Yearbook Medical Publishers.

Castiglioni, S., E. Zuccato, C. Chiabrando, R. Fanelli, and R. Bagnati. 2008. Mass spectrometric analysis of illicit drugs in wastewater and surface water. *Mass Spectrom. Rev.* **27**(4):378–394.

Castiglioni, S., E. Zuccato, E. Crisci, C. Chiabrando, R. Fanelli, and R. Bagnati. 2006. Identification and measurement of illicit drugs and their metabolites in urban wastewater by liquid chromatography-tandem mass spectrometry. *Anal.Chem.* **78**:8421–8429.

Coombs, R. H., W. J. West, editors. 1991. *Drug Testing: Issues and Options.* New York: Oxford.

Daughton, C. G., and Ternes, T. A. 1999. Pharmaceuticals and personal care products in the environment: agents of subtle change? *Environ. Health Perspect.* **107**:907−938.

Drug Abuse. http://www.nlm.nih.gov/medlineplus/drugabuse.html. Accessed 2009 July 19.

Health, kids and parenting. http://www.msnbc.msn.com/id/20631668/ns/health-kids_and_parenting/ Accessed 2010 January 21.

Las Vegas Rev.-Journal online. http://www.lvrj.com/news/7183316.html Accessed 2010 January 21.

Inaba, T., D. J. Stewart, and W. Kalow. 1978. Metabolism of cocaine in man. *Clin. Pharmacol. Therap.* **23**:547−552.

Jones, A.W. 2001. Heroin use by motorists in Sweden confirmed by analysis of 6-acetylmorphine in urine. *J. Anal. Toxicol..* **25**(5):353−355.

Kauppila, T. J., N. Talaty, T. Kuuranne, T. Kotiaho, R. Kostiainen, and R. G. Cooks. 2007. Rapid analysis of metabolites and drugs of abuse from urine samples by desorption electrospray ionization-mass spectrometry. *Analyst* **132**:868−875.

Lowe, R. D., G. E. Guild, P. Harpas, P. Kirkbride, P. Hoffmann, N. H. Voelcker, and H. Kobus. 2009. Rapid drug detection in oral samples by porous silicon assisted laser desorption/ionization mass spectrometry. *Rapid Commun. Mass Spectrom.* **23**:3543−3548.

Oyler, J., W. D. Darwin, and E. J. Cone. 1996. Cocaine contamination of United States paper currency. J. Anal. *Toxicol.* **20**:213−216.

Pizzolato, T. M., de Alda, M. J. L., and D. Barcelo. 2007. LC-based analysis of drugs of abuse and their metabolites in urine. *Trends Anal. Chem..* **26**(6):609−624.

Postigo, C., M. J. Lopez de Alda, and D. Barcelo. 2008. Analysis of drugs of abuse and their human metabolites in water by LC-MS2: A non-intrusive tool for drug abuse estimation at the community level. *Trends Anal. Chem.* **27**(11):1053−1069.

Substance Abuse and Mental Health Services Administration. 2009. Results from the 2008 National Survey on Drug Use and Health: National Findings (Office of Applied Studies,NSDUH Series H-36, HHS Publication No. SMA 09-4434). Rockville, MD. Available online at http://www.oas.samhsa.gov/nsduh/2k8nsduh/2k8Results.pdf Accessed 2009 Sep 02.

Ternes, T., M. Bonerz, and T. Schmidt. 2001. Determination of neutral pharmaceuticals in wastewater and rivers by liquid chromatography-electrospray tandem mass spectrometry. *J. Chromatogr.* **938**:175−185.

U.S. Navy Judge Advocate General investigation and report. 1982. Final endorsement (11th) from the Secretary of the Navy. June 29.

Zuccato, E., C. Chiabrando, S. Castiglioni, D. Calamari, R. Bagnati, S. Schiarea, and R. Fanelli. 2005. Cocaine in surface waters: a new evidence-based tool to monitor community drug abuse. *Environ Health* **4**:14−21.

Zuccato, E., C. Chiabrando, S. Castiglioni, R. Bagnati, and R. Fanelli. 2008. Estimating community drug abuse by wastewater analysis. *Environ. Health Perspect.* **116**(8):1027−1032.

SECTION VI

CONCLUSIONS

CHAPTER 19

CONCLUSIONS AND FUTURE PERSPECTIVES

ROBERTO FANELLI

The two key words with which I can summarize the contents of this book are mass spectrometry and drug abuse. I had the unique opportunity to follow the application and the explosive trend of the use of mass spectrometry in the field of life sciences from the beginning and was lucky enough to be part of the recent exciting application of this technique in the field of drug abuse monitoring. Therefore, I am glad to offer some impressions about the specific contribution the reader can find in this book and, to make more general comments about the state of the art of the problems addressed.

This book collects all the main research that has emerged in about the last 5 years since the first publication on the successful attempt to estimate community drug abuse by wastewater analysis. The matter is of outmost importance because modern society, with its rapid changes in social behaviors, needs to have tools able to describe and monitor real-time trends in illicit behaviors, which can have a negative impact on community life. The idea of monitoring social behaviors and diseases through the observation of urban wastewater is not new ("Sewer Sociology. The Days of our Lives." 2006, ADA Corporation) spanning from the simple measure of sewer flows during weekdays and holidays back to the origins of modern epidemiology when Doctor John Snow discovered that sewage water contaminating tap water was the cause of cholera outbreaks in London. (http://www.ph.ucla.edu/epi/snow/fatherofepidemiology.html#ONE). However, only in recent years has the availability of powerful mass spectrometry-based analytical methodologies made possible a breakthrough in the possibilities to investigate, in detail, the content of sewer waters and to identify behavioral markers.

Mass spectrometry is a unique technique with the capacity of unambiguously identifying the chemical structure of compounds present in complex matrices in

Illicit Drugs in the Environment: Occurrence, Analysis, and Fate Using Mass Spectrometry, Edited by Sara Castiglioni, Ettore Zuccato, and Roberto Fanelli
Copyright © 2011 John Wiley & Sons, Inc.

parts per billion or even parts per trillion concentrations and to obtain, when needed, quantitative information about their presence. Sewage waters represent the sum of what we ingest, use, metabolize, and excrete from our body, in an extreme diluted sample containing thousands of different compounds, each one carrying information about its nature, origin, and, therefore, potentially useful data on their significance at the population level. It is not surprising, therefore, that the application of such a powerful analytical technique to the description of a wide diffuse and hidden social behavior like the consumption of illicit drugs has emerged, creating a striking interest among those working in the field of drug abuse, the analytical chemistry community, and the press, which always craves new data describing social behaviors and trends.

The state of the art of analytical chemistry of illicit drugs in the environment as described in this book and in several publications shows that is possible to identify and measure, with a high degree of confidence, drugs, metabolites, and degradation products in environmental matrices. However, information arising from the availability of such data is still difficult to obtain and to gather in a significant way. There are still open issues that need to be addressed, which, when solved, will completely fulfill the potential of the "sewage epidemiology" approach. One of these is the availability of a direct quantitative marker of human presence that will allow the estimation of how many people are present at a certain time in the territory served by a specific sewage treatment plant, acting as contributors to the content of sewage waters; this is not an easy task, since we require a substance that is sufficiently stable during the trip from the origin to the sample collection and that is excreted in a rather constant rate by humans. Currently, we are still using indirect information, such as the amount of water or electricity consumed on a certain day. Another point in need of a better quantitative assessment is what we know about the metabolism and excretion of compounds that we ingest or to which we are exposed. Since this is key information to back-calculate the exposure from sewage water concentrations, the reliability of the exercise is bound to the availability of robust data about the behavior of xenobiotic in the human body and to the diversity of people, in a case by case basis.

To really fulfill the potential of this new methodology, one should try to intensify the cross-relationships with classical epidemiological techniques, which is not an easy task, since epidemiology is based on what happened in the past, but is slow in obtaining new data and analyzing it. Sewage epidemiology, on the other hand, can be used for real-time data, but lacks detailed knowledge about the people contributing to the composite samples that arrive at wastewater treatment plants. It is easy to see how a comparison and a melding of the two techniques would end up in a powerful tool to study social problems at the population level.

Can we try to figure out what could be the future of this story? As we physically reached the moon 100 hundred years after Julius Verne in his book (From the Earth to the Moon -1865) was dreaming of it, which prompted the human race to dream and create programs of more challenging trips to outer planets, so we are now in the position of planning new, challenging applications of the "sewage epidemiology" approach, since we have demonstrated that we have the necessary technology and knowledge on how to draw useful information from wastewater. There are thousands of wastewater treatment plants serving small and large communities all over the developed world, which allows one to think of them as unique diffuse sources of important information easily available on a daily basis,. This permits real-time epidemiological considerations that can be used in both small- and large-scale applications, as needed. How to extend our observations beyond illicit drugs monitoring is just a matter of using our imagination and putting our technology to work on what we think is more interesting and/or needed. A few examples that I find more interesting will follow.

Pharmaceuticals and their metabolites are classes of compounds that are well described as arriving in substantial amounts to wastewater treatment plants where, in many cases, they are not completely degraded and therefore enter surface waters. Hence, they become a new problem of water contamination both for ecotoxicological concerns and for human exposure through tap water. However, the amount of pharmaceuticals arriving through sewage waters brings some precious information not yet mined. One consideration is that people who are sick or who think they are sick are consuming pharmaceuticals. Thus, if we could do for pharmaceuticals what has been done for illicit drugs, we could monitor single active principles consumed daily and have a real-time measure of the health state of the population or the perception they have of their health. One could say that this could be done by collecting sales data, but these data are just sales data not real consumer data since the compliance to prescriptions is not 100% foolproof and there are huge amounts of pharmaceuticals sitting in our drawers waiting for their expiration date.

A second never-explored field is the theoretical possibility of monitoring, at the population level, the exposure to food and air contaminants. This is generally done by measuring, at a national or local level, food and air concentrations of selected contaminants and through simple statistical values of food consumed. For instance, it is possible to calculate population exposure to contaminants. This type of monitoring is very expensive and does not take into consideration the enormous variability in food and air contaminations in time and for different types of products. The possibility of measuring exposure biomarkers, such as metabolites of food contaminants and air pollutants, known to be present in urine, at the level of sewage wastewaters would allow the monitoring of real-time population exposure and open the possibility of comparison between exposure in different locations and to make correlations between exposure and epidemiological data using morbidity and mortality data. Besides exposure to pollutants, our health is also bound to our lifestyles, which are based on genetic traits. Therefore, one could measure using the sewage epidemiology approach some of the excreted biomarkers proposed to be indicative of an increased risk for cancer or cardiovascular diseases. If one could find a relationship between disease incidence and the increased presence of a risk biomarkers one could rapidly go to the individual level to identify the population subgroups at increased risk for disease.

Last, but not least, our responsibility toward the sustainability of our presence on the planet must be based, among other things, on the reasoned use of consumer items chosen among those that do not increase the environmental chemical burden. A reasoned classification of the type and amount of consumer chemicals arriving at wastewater plants and surviving the treatment could provide useful information to the industry and to consumers on how to orient production and use toward a more environmentally sustainable life.

WILEY-INTERSCIENCE SERIES IN MASS SPECTROMETRY

Series Editors

Dominic M. Desiderio
Departments of Neurology and Biochemistry
University of Tennessee Health Science Center

Nico M. M. Nibbering
Vrije Universiteit Amsterdam, The Netherlands

John R. de Laeter • *Applications of Inorganic Mass Spectrometry*

Michael Kinter and Nicholas E. Sherman • *Protein Sequencing and Identification Using Tandem Mass Spectrometry*

Chhabil Dass • *Principles and Practice of Biological Mass Spectrometry*

Mike S. Lee • *LC/MS Applications in Drug Development*

Jerzy Silberring and Rolf Eckman • *Mass Spectrometry and Hyphenated Techniques in Neuropeptide Research*

J. Wayne Rabalais • *Principles and Applications of Ion Scattering Spectrometry: Surface Chemical and Structural Analysis*

Mahmoud Hamdan and Pier Giorgio Righetti • *Proteomics Today: Protein Assessment and Biomarkers Using Mass Spectrometry, 2D Electrophoresis, and Microarray Technology*

Igor A. Kaltashov and Stephen J. Eyles • *Mass Spectrometry in Biophysics: Confirmation and Dynamics of Biomolecules*

Isabella Dalle-Donne, Andrea Scaloni, and D. Allan Butterfield • *Redox Proteomics: From Protein Modifications to Cellular Dysfunction and Diseases*

Silas G. Villas-Boas, Ute Roessner, Michael A.E. Hansen, Jorn Smedsgaard, and Jens Nielsen • *Metabolome Analysis: An Introduction*

Mahmoud H. Hamdan • *Cancer Biomarkers: Analytical Techniques for Discovery*

Chabbil Dass • *Fundamentals of Contemporary Mass Spectrometry*

Kevin M. Downard (Editor) • *Mass Spectrometry of Protein Interactions*

Nobuhiro Takahashi and Toshiaki Isobe • *Proteomic Biology Using LC-MS: Large Scale Analysis of Cellular Dynamics and Function*

Agnieszka Kraj and Jerzy Silberring (Editors) • *Proteomics: Introduction to Methods and Applications*

Ganesh Kumar Agrawal and Randeep Rakwal (Editors) • *Plant Proteomics: Technologies, Strategies, and Applications*

Rolf Ekman, Jerzy Silberring, Ann M. Westman-Brinkmalm, and Agnieszka Kraj (Editors) • *Mass Spectrometry: Instrumentation, Interpretation, and Applications*

Christoph A. Schalley and Andreas Springer • *Mass Spectrometry and Gas-Phase Chemistry of Non-Covalent Complexes*

Riccardo Flamini and Pietro Traldi • *Mass Spectrometry in Grape and Wine Chemistry*

Mario Thevis • *Mass Spectrometry in Sports Drug Testing: Characterization of Prohibited Substances and Doping Control Analytical Assays*

Sara Castiglioni, Ettore Zuccato, and Roberto Fanelli • *Illicit Drugs in the Environment: Occurrence, Analysis, and Fate Using Mass Spectrometry*

INDEX

Printed in the United States
By Bookmasters